Boltzmann $\qquad w = N! \prod_{j=1}^{n} \dfrac{g_j^{N_j}}{N_j!} \qquad\qquad \dfrac{\quad}{g_j} = \dfrac{\quad}{Z} e^{\quad}, \qquad_{j=1}$

Maxwell-
Boltzmann $\qquad w = \prod_{j=1}^{n} \dfrac{g_j^{N_j}}{N_j!} \qquad\qquad \dfrac{N_j}{g_j} = \dfrac{N}{Z} e^{-\varepsilon_j/kT}, \quad Z = \sum_{j=1}^{n} g_j e^{-\varepsilon_j/kT}$

Fermi-Dirac $\qquad w = \prod_{j=1}^{n} \dfrac{g_j!}{N_j!(g_j - N_j)!} \qquad \dfrac{N_j}{g_j} = \dfrac{1}{e^{(\varepsilon_j - \mu)/kT} + 1}$

Bose-Einstein $\qquad w = \prod_{j=1}^{n} \dfrac{(N_j + g_j - 1)!}{N_j!(g_j - 1)!} \qquad \dfrac{N_j}{g_j} = \dfrac{1}{e^{(\varepsilon_j - \mu)/kT} - 1}$

ENTROPY

$$S = k \ln w$$

MAXWELL DISTRIBUTION OF MOLECULAR SPEEDS

$$N(v)\,dv = 4\pi N \left(\frac{m}{2\pi kT}\right)^{3/2} v^2 e^{-mv^2/2kT}\,dv$$

$$\bar{v} = \left(\frac{8kT}{\pi m}\right)^{1/2} \qquad v_{rms} = \left(\frac{3kT}{m}\right)^{1/2} \qquad v_m = \left(\frac{2kT}{m}\right)^{1/2}$$

STIRLING'S APPROXIMATION

$$\ln N! \approx N \ln N - N, \quad N \text{ large}$$

Classical and Statistical Thermodynamics

Ashley H. Carter
Drew University

Prentice Hall
Upper Saddle River, New Jersey 07458

Library of Congress Cataloging-in-Publication Data

Carter, Ashley H.
　　Classical and statistical thermodynamics / Ashley H. Carter.
　　　　p.　cm.
　　Includes bibliographical references and index.
　　ISBN 0-13-779208-5
　　1. Thermodynamics. 2. Statistical thermodynamics. I. Title.

QC311.C39 2001　　　　　　　　　　　　　　　　　00-027105
536'.7—dc21　　　　　　　　　　　　　　　　　　　　　CIP

Executive Editor: Alison Reeves
Art Director: Jayne Conte
Cover Designer: Bruce Kensalaar
Manufacturing Manager: Trudy Pisciotti
Buyer: Michael Bell
Senior Marketing Manager: Erik Fahlgren
Editorial Assistant: Christian Botting
Production Supervision/Composition: WestWords, Inc.

 © 2001 by Prentice-Hall, Inc.
Upper Saddle River, New Jersey 07458

Printed in the United States of America

10　9

ISBN 0-13-779208-5

Prentice-Hall International (UK) Limited, *London*
Prentice-Hall of Australia Pty. Limited, *Sydney*
Prentice-Hall Canada Inc., *Toronto*
Prentice-Hall Hispanoamericana, S.A., *Mexico*
Prentice-Hall of India Private Limited, *New Delhi*
Prentice-Hall of Japan, Inc., *Tokyo*
Pearson Education Asia Pte. Ltd.
Editora Prentice-Hall do Brasil, Ltda., *Rio de Janeiro*

To my wife Eva

Contents

14. The Classical Statistical Treatment of an Ideal Gas   261

15. The Heat Capacity of a Diatomic Gas   277

20.   Information Theory   373

Appendices

A.   Review of Partial Differentiation   391

B.   Stirling's Approximation   401

C.   Alternative Approach to Finding the Boltzmann Distribution   403

D.   Various Integrals   407

Preface

This book is intended as a text for a one-semester undergraduate course in thermal physics. Its objective is to provide third- or fourth-year physics students with a solid introduction to the classical and statistical theories of thermodynamics. No preparation is assumed beyond college-level general physics and advanced calculus. An acquaintance with probability and statistics is helpful but is by no means necessary.

The current practice in many colleges is to offer a course in classical thermodynamics with little or no mention of the statistical theory—or vice versa. The argument is that it is impossible to do justice to both in a one-semester course. On the basis of my own teaching experience, I strongly disagree. The standard treatment of temperature, work, heat, entropy, etc. often seems to the student like an endless collection of partial derivatives that shed only limited light on the underlying physics and can be abbreviated. The fundamental concepts of classical thermodynamics can easily be grasped in little more than half a semester, leaving ample time to gain a reasonably thorough understanding of the statistical method.

Since statistical thermodynamics subsumes the classical results, why not structure the entire course around the statistical approach? There are good reasons not to do so. The classical theory is general, simple, and direct, providing a kind of visceral, intuitive comprehension of thermal processes. The physics student not confronted with this remarkable phenomenological conception is definitely deprived. To be sure, the inadequacies of classical thermodynamics become apparent upon close scrutiny and invite inquiry about a more fundamental description. This, of course, exactly reflects the historical development of the subject. If only the statistical picture is presented, however, it is my observation that the student fails to appreciate fully its more abstract concepts, given no exposure to the related classical ideas first. Not only do classical and statistical thermodynamics in this sense complement each other, they also beautifully illustrate the physicist's perpetual striving for descriptions of greater power, elegance, universality, and freedom from ambiguity.

Chapters 1 through 10 represent a fairly traditional introduction to the classical theory. Early on emphasis is placed on the advantages of expressing the fundamental laws in terms of state variables, quantities whose differentials are exact. Accordingly, the search for integrating factors for the differentials of work and heat is discussed. The elaboration of the first law is followed by chapters on applications and consequences. Entropy is presented both as a useful mathematical variable and as a phenomenological construct necessary to explain why there are processes permitted by the first law that do not occur in nature. Calculations are then given of the change in entropy for various reversible and irreversible processes. The thermodynamic potentials are broached via the Legendre transformation following elucidation of the rationale for having precisely four such quantities. The conditions for stable equilibrium are examined in a section that rarely appears in undergraduate texts. Modifications of fundamental relations to deal with open systems are treated in Chapter 9 and the third law is given its due in Chapter 10.

The kinetic theory of gases, treated in Chapter 11, is concerned with the molecular basis of such thermodynamic properties of gases as the temperature, pressure, and thermal energy. It represents, both logically and historically, the transition between classical thermodynamics and the statistical theory.

The underlying principles of equilibrium statistical thermodynamics are introduced in Chapter 12 through consideration of a simple coin-tossing experiment. The basic concepts are then defined. The statistical interpretation of a system containing many molecules is observed to require a knowledge of the properties of the individual molecules making up the system. This information is furnished by the quantum mechanical notions of energy levels, quantum states, and intermolecular forces. In Chapter 13, the explication of classical and quantum statistics and the derivation of the particle distribution functions is based on the method of Lagrange multipliers. A discussion of the connection between classical and statistical thermodynamics completes the development of the mathematical formulation of the statistical theory. Chapter 14 is devoted to the statistics of an ideal gas. Chapters 15 through 19 present important examples of the application of the statistical method. The last chapter introduces the student to the basic ideas of information theory and offers the intriguing thought that statistical thermodynamics is but a special case of some deeper, more far-reaching set of physical principles.

Throughout the book a serious attempt has been made to keep the level of the chapters as uniform as possible. On the other hand, the problems are intended to vary somewhat more widely in difficulty.

In preparing the text, my greatest debt is to my students, whose response has provided a practical filter for the refinement of the material presented herein.

A.H.C.
Drew University

ACKNOWLEDGMENTS

In addition to my students at Drew University, I owe thanks to two colleagues and friends, Professors Robert Fenstermacher and John Ollom, who have encouraged me at every turn during the writing of this book. I am indebted to Professor Mark Raizen of the University of Texas at Austin, who reviewed the manuscript and used it as the text in his thermal physics course; his comments were invaluable.

I am especially grateful to Professor Roy S. Rubins of the University of Texas at Arlington for his thoughtful and thorough critique. I also received useful feedback from other reviewers, whose suggestions contributed substantially to an improved text. They are Anjum Ansari, University of Illinois at Chicago; John Jaszczak, Michigan Technological University; David Monts, Mississippi State University; Hugh Scott, Oklahoma State University; Harold Spector, Illinois Institute of Technology at Chicago; Zlatko Tesanovic, Johns Hopkins University.

I thank my editor Alison Reeves and her assistants, Gillian Buonanno and Christian Botting, for their support, guidance, and patience. Production editors Richard Saunders and Patrick Burt of WestWords Inc. were particularly helpful. Finally, I am extremely grateful to Heather Ferguson, who turned my lecture notes into a first draft, and to Lori Carucci and her daughters Amanda and Brigette, who prepared the final manuscript.

Without all of these people, the book would never have seen the light of day.

Chapter *1*

The Nature of Thermodynamics

1.1 WHAT IS THERMODYNAMICS?

Thermodynamics is the study of heat in the field of physics. The central concept of thermodynamics is temperature. Since temperature is not expressible in terms of basic mechanical quantities such as mass, length, and time, it is evidently a fundamental notion that sets thermodynamics apart from other branches of physics.

The development of thermodynamics provides some of the most fascinating chapters in the history of science. It began at the start of the Industrial Revolution when it became important to understand the conversion of heat into mechanical work. The experiments of Joule, Hirn, and others, and the theoretical studies of Helmholtz resulted in the principle of the conservation of energy when applied to thermal phenomena. The principle became the first law of thermodynamics. Mayer postulated the equivalence of heat and work and made an estimate of the mechanical equivalence of heat.

The subsequent progress of thermodynamics owes much to the Frenchman, Sadi Carnot, whose treatise of 1824 led to one of the most far-reaching principles of physical science: Carnot's theorem, which is, in effect, the second law of thermodynamics. Actually, the principle emerged before the first law. It was the outgrowth of Carnot's interest in the practical question of the efficiency of steam engines. His work, put in simple mathematical form by Clapeyron, attacked the more fundamental problem of the efficiency of heat engines in general.

The concept of entropy began to appear quite early in papers by Clausius and William Thomson (Lord Kelvin), but it was not until 1865 that Clausius saw fit to give it a name and a full definition. Later, Nernst and Planck added a third law, which is a statement about the behavior of thermodynamic quantities, including entropy, at the temperature of absolute zero.

In this period classical thermodynamics was worked out in essentially its present form. It is a phenomenological theory, describing the macroscopic properties of matter, most of which are amenable to direct measurement. No assumptions are made about the fine structure of material substances. No attempt is made to explain underlying causes or to provide a mechanistic description. As a

Figure 1.1 Ludwig Boltzmann, 1844–1906, whose work led to an understanding of the macroscopic world on the basis of molecular dynamics. (Courtesy of American Institute of Physics/Emilio Segrè Visual Archives.)

consequence, the theory has the advantages of great simplicity, broad generality, and a close connection between experimental results and familiar concepts. The noted experimentalist P.W. Bridgman said, "The laws of thermodynamics have a different feel from most of the other laws of physics. There is something palpably verbal about them—they smell of their human origin."

Toward the end of the nineteenth century, when the atomic nature of matter began to be understood, ways were found to express the pressure, temperature, and other macroscopic properties of a gas in terms of average values of the properties of molecules, such as their kinetic energy. The results, with which Maxwell's name is closely associated, came to be called the kinetic theory of gases.

The kinetic theory eventually expanded into the far more comprehensive statistical mechanics (or statistical thermodynamics) of Boltzmann and Gibbs, and ultimately encompassed the ideas of quantum mechanics. The statistical approach takes account of the molecular constitution of matter and reveals a deeper foundation on which the laws of thermodynamics exist. The connection between the classical and statistical theories is associated with the fact that macroscopic measurements are averages of the behavior of astronomical numbers of particles.

Thermodynamics is by no means a closed subject. In recent years it has been possible to extend the statistical theory to include nonequilibrium processes and even nonlinear effects. At the present time, the relationship between thermodynamics and information theory, and the study of chaotic behavior in thermodynamic systems, attract a great deal of attention. Modern ideas, furthermore, have been shown to be applicable to a wider variety of phenomena than hitherto suspected.

Thermodynamics has captured the imagination of many of the greatest minds of science. Einstein, who was captivated by the subject, wrote: "A theory is the more impressive the greater the simplicity of its premises, and the more extended its area of applicability. Therefore, the deep impression which classical thermodynamics made upon me. It is the only physical theory of universal content concerning which I am convinced that, within the applicability of its basic concepts, it will never be overthrown." That is quite a statement, indeed!

1.2 DEFINITIONS

In developing the basic ideas of thermodynamics the originators of the theory were careful not to be so concise as to render definitions sterile. The following brief definitions will be illustrated by examples in future sections.

Thermodynamics can be described as the study of *equilibrium properties* in which *temperature* is an important variable. All of the words in italics need to be defined.

In thermodynamics we are concerned with a *system,* some portion of the physical world. The system could be a container of gas, a piece of metal, a magnet. The system must not interact chemically with the vessel that contains it. (The behavior of a liquid must not be influenced by the test tube that holds it.) A system may exchange energy with other systems, which constitute the *surroundings* of the given system. The system, together with its surroundings, comprise a *universe*.

We classify systems as to whether they are open, closed, or isolated. An *open* system can exchange mass and energy with its surroundings. A *closed* system cannot exchange mass with its surroundings, but can exchange energy in other forms. An *isolated* system cannot exchange mass or energy in any form with its surroundings; it is completely cut off from other systems.

The quantities we use to describe the macroscopic behavior of a system are called *properties,* observable characteristics of a system. Other names are thermodynamic variables or thermodynamic coordinates. An extremely important concept is that of a *state variable,* a property whose differential is exact.

Properties are extensive or intensive. An *extensive* property is proportional to the mass. An example is the volume V; if the mass is doubled, the volume is doubled (assuming that the density remains constant). An *intensive* property is independent of the mass. Temperature T is an intensive property; its value is not affected by a change of mass. Pressure P and density ρ are further examples of intensive properties.

Inherently extensive properties are given by capital letters. Inherently intensive properties are denoted by lowercase letters. There are two important exceptions: The temperature T is always capitalized to avoid confusion with the time t; and the pressure P is capitalized to distinguish it from the probability p.

An extensive property can be converted to an intensive property by dividing by the mass. This is called a *specific value:*

$$\text{Specific value} \equiv \frac{\text{value of the extensive property}}{\text{mass of the system}}.$$

In this text, we will go back and forth between extensive properties and their corresponding specific values.

1.3 THE KILOMOLE

The kilomole is a unit of mass defined as follows:

1 kilomole $\equiv$ mass m in kilograms equal to the molecular weight.

Thus one kilomole of oxygen gas (O_2) is equal to 32 kg. (The mole is a unit of mass more familiar to chemists; one mole is equal to the mass in grams. Thus a mole of oxygen gas is 32 g.)

1.4 LIMITS OF THE CONTINUUM

We tacitly assume that classical thermodynamics is a continuum theory, that properties vary smoothly from point to point in the system. But if all systems are made up of atoms and molecules (as our definition of the kilomole implies), it is reasonable to ask: how small a volume can we be concerned with and still have confidence that our continuum theory is valid?

We can obtain an approximate answer to this question by invoking Avogadro's law, which states that at standard temperature and pressure (0°C and atmospheric pressure), one kilomole of gas occupies 22.4 m^3 and contains 6.02×10^{26} molecules. (The latter is called Avogadro's number, N_A.) Then

$$\frac{6.02 \times 10^{26} \text{ molecules kilomole}^{-1}}{22.4 \text{ m}^3 \text{ kilomole}^{-1}} = 2.69 \times 10^{25} \frac{\text{molecules}}{\text{m}^3}.$$

This molecular density is sometimes called Loschmidt's number. Using it we can easily show that a cube one millimeter on each side contains roughly 10^{16} molecules, whereas a cube one nanometer (10^{-9} m) on a side has a very small probability of containing even one molecule. We can therefore be reasonably certain that classical thermodynamics is applicable down to very small macroscopic (and even microscopic) volumes, but ultimately a limit is reached where our theory will break down.

1.5 MORE DEFINITIONS

The *state* of a system is defined as a condition uniquely specified by a set of properties. Examples of such properties are pressure, volume, and temperature. The question arises: how many properties are required to specify the state of a thermodynamic system? By "required number" we mean that every time a system with the given properties is subjected to a particular environment, every feature of its subsequent behavior is identical. In classical mechanics, if the displacement and velocity of a system are known at some instant of time, as well as the forces acting on it, the behavior of the system is predicted for all times. Most frequently, the thermodynamic state of a single component system is also specified by two independent variables.

An *equilibrium state* is one in which the properties of the system are uniform throughout and do not change with time unless the system is acted upon by external influences. A *non-equilibrium state* characterizes a system in which gradients exist and whose properties vary with time (the atmosphere and the oceans are examples). *State variables* are properties that describe equilibrium states. An *equation of state* is a functional relationship among the state variables for a system in equilibrium. A *path* is a series of states through which a system passes.

In introductory physics, a "change of state" is frequently used to denote a transition from a liquid to a gas, or from a solid to a liquid, etc. In thermodynamics, such a change is referred to as a *change of phase* or *phase transformation*.

If the pressure P, the volume V, and the temperature T are the state variables of the system, the equation of state takes the form

$$f(P, V, T) = 0. \tag{1.1}$$

This relationship reduces the number of independent variables of the system from three to two. The function f is assumed to be given as part of the specification of the system. It is customary to represent the state of such a system by a point in three-dimensional P-V-T space. The equation of state then defines a surface in this space (Figure 1.2). Any point lying on this surface represents a state in equilibrium. In thermodynamics a state automatically means a state in equilibrium unless otherwise specified.

The boundary between a system and its surroundings through which changes may be imposed is called a *system wall*. An *adiabatic wall* is a boundary that permits no heat interaction with the surroundings. The word comes from

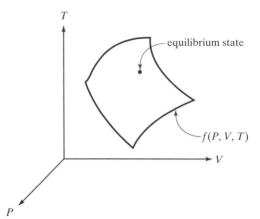

Figure 1.2 Surface in P-V-T space. Points on the surface represent equilibrium states of the system whose equation of state is $f(P, V, T) = 0$.

the Greek *adiabatos*, meaning "not going through." An isolated system is adiabatically contained. However, a system with an adiabatic wall is not necessarily isolated. Mechanical interactions can take place through adiabatic walls. For example, material can be removed or added, the volume can change, a magnetic field can be applied, etc. A *diathermal wall* is a boundary that freely allows heat to be exchanged. The Greek word *diathermos* means "heat through."

If a system strongly interacts with its walls, the problem becomes complicated. Think of pancake batter sticking to the griddle, or sulfuric acid in an iron vessel!

A *process* is a change of state expressed in terms of a path along the equation of state surface in Figure 1.2. In a *cyclical process*, the initial and final states are the same. A *quasi-static process* is a process in which, at each instant, the system departs only infinitesimally from an equilibrium state. That is, changes of state are described in terms of differentials. An example is the gradual compression of a gas.

A *reversible process* is a process whose direction can be reversed by an infinitesimal change in some property. It is a quasi-static process in which no dissipative forces such as friction are present. All reversible processes are quasi-static, but a quasi-static process is not necessarily reversible. For example, a slow leak in a tire is quasi-static but not reversible. A reversible process is an idealization; friction is always present. An *irreversible process* involves a finite change in a property in a given step and includes dissipation (energy loss). All natural processes are irreversible.

In many processes, some property of the system remains constant. An *isobaric process* is a process in which the pressure is constant. If the volume is constant, the process is *isochoric*. And if the temperature doesn't change, the process is called *isothermal*.

Suppose a piston enclosing a gas is immersed in a heat bath, so that the gas is kept at constant temperature. The gas is slowly compressed (Figure 1.3).

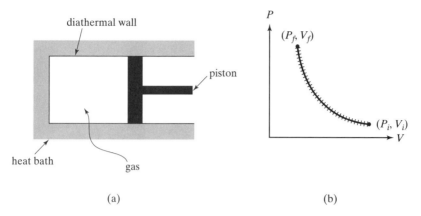

Figure 1.3 Example of a quasi-static, reversible and isothermal process.
(a) A piston *slowly* compresses a gas held at constant temperature. (b) A *P-V*
diagram representing the process; (P_i, V_i) and (P_f, V_f) denote the initial and
final states, respectively.

The process is quasi-static, reversible, and isothermal and can be described by
a path in a *P-V* diagram connecting the initial and final states through a series
of infinitesimal changes.

  If, instead of a gradual compression, the piston is given a violent push,
sound waves or shock waves and turbulence are generated, accompanied by
strong temperature and pressure gradients. The process is an irreversible, non-
equilibrium process that cannot be represented by a path in the *P-V* plane.
Only the end points can be plotted, representing the initial state and the final
state reached after equilibrium has been eventually reestablished.

1.6 UNITS

The International System of Units (SI) will be used, almost exclusively. The
system is based on the fundamental units of length, mass, and time—the meter
(m), the kilogram (kg) and the second (s), respectively. In this system, the unit
of pressure is the pascal (Pa), equal to 1 Nm^{-2}.

  Other units of pressure in common use are

$$1 \text{ bar} \equiv 10^5 \text{ Pa},$$
$$1 \text{ atmosphere (atm)} \equiv 1.01 \times 10^5 \text{ Pa},$$
$$1 \text{ torr} \equiv 133.3 \text{ Pa}.$$

The atmosphere and the torr (named after Torricelli) are derived units. The
atmosphere is based on the use of a *manometer* to measure pressure (Figure 1.4).

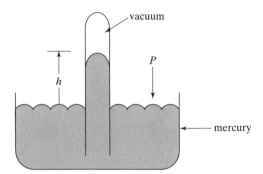

Figure 1.4 A simple mercury
manometer used to measure
the atmospheric pressure.

The height of a column of mercury is balanced by the force exerted by
the gas on the liquid surface. A pressure of 1 atmosphere causes the mercury
to rise to a height of 76 cm. That is, $P = h\rho g$, where P is the pressure of the
atmosphere, h is the height of the mercury column, ρ is the density of mercury,
and g is the acceleration of gravity. Thus

$$1 \text{ atm} = (0.76 \text{ m})\left(1.36 \times 10^4 \frac{\text{kg}}{\text{m}^3}\right)\left(9.8 \frac{\text{m}}{\text{s}^2}\right) = 1.01 \times 10^5 \text{ Pa.}$$

The torr is then defined as the pressure of 1 mm of mercury:

$$1 \text{ torr} = \frac{1}{760} \text{ atm} = 133.3 \text{ Pa.}$$

1.7 TEMPERATURE AND THE ZEROTH LAW OF THERMODYNAMICS

Temperature is a more subtle property than pressure. Its origin is the so-called
zeroth law of thermodynamics.* The zeroth law is based on experiments (as
are all physical laws) and is concerned with properties of systems in thermal
equilibrium, that is, systems in equilibrium connected by a diathermal wall.
The law states:

> *If two systems are separately in thermal equilibrium with a third system,*
> *they are in equilibrium with each other.*

Let systems A, B, and C each consist of a mass of fluid in an insulated container. We shall use C as a reference. We choose to describe the state of each

---

*There are some who believe that this should not be dignified by calling it a "law."

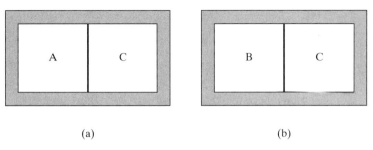

(a)   (b)

Figure 1.5 Systems in thermal contact: (a) systems A and C in thermal equilibrium; (b) systems B and C in thermal equilibrium.

system in terms of the state variables P and V. We place A in contact with C through a diathermal wall as in Figure 1.5(a).

The system C can be thought of as a thermometer. Its state is given by the pair of variables (P_c, V_c). From observations, if we choose a particular value for P_A, then V_A will be uniquely determined (only one variable is independent). The condition under which A and C are in *equilibrium* may be expressed by the equation

$$F_1(P_A, V_A, P_C, V_C) = 0, \tag{1.2}$$

where F_1 is some function of the four variables. We assume that this equation can be solved for P_C:

$$P_C = f_1(P_A, V_A, V_C). \tag{1.3}$$

Next we place system B and C in thermal contact (Figure 1.5(b)). For equilibrium,

$$F_2(P_B, V_B, P_C, V_C) = 0 \tag{1.4}$$

or

$$P_C = f_2(P_B, V_B, V_C). \tag{1.5}$$

Equating Equations (1.3) and (1.5), we obtain the condition under which A and B are separately in equilibrium with C:

$$f_1(P_A, V_A, V_C) = f_2(P_B, V_B, V_C). \tag{1.6}$$

But, according to the zeroth law, A and B are then in equilibrium with each other, so that

$$F_3(P_A, V_A, P_B, V_B) = 0. \tag{1.7}$$

Solving for P_A, we obtain

$$P_A = f_3(V_A, P_B, V_B). \tag{1.8}$$

Now, Equation (1.6) can also be solved for P_A, in principle:

$$P_A = g(V_A, P_B, V_B, V_C). \tag{1.9}$$

Equation (1.9) states that P_A is determined by four variables, whereas Equation (1.8) says that it is a function of only three. This can only mean that the functions f_1 and f_2 in Equation (1.6) contain V_C in such a form that it cancels out on the two sides of the equation. For example,

$$f_1 = \phi_1(P_A, V_A)\zeta(V_C) + \eta(V_C),$$
$$f_2 = \phi_1(P_B, V_B)\zeta(V_C) + \eta(V_C).$$

When the cancellation is performed, we have

$$\phi_1(P_A, V_A) = \phi_2(P_B, V_B). \tag{1.10}$$

Extending the argument to additional systems, we get

$$\phi_1(P_A, V_A) = \phi_2(P_B, V_B) = \phi_3(P_C, V_C) = \dots . \tag{1.11}$$

For any system in thermal equilibrium with a given system, we can choose to write

$$\phi(P, V) = T, \tag{1.12}$$

where we define T as the empirical *temperature* and Equation (1.12) is the equation of state of the system. Equation (1.11) then says that systems in thermal equilibrium with one another have the same temperature. Thus *temperature* is a property of a system that determines if thermal equilibrium exists with some other system.

The next section contains a scheme for measuring the temperature. Let C be a "thermometer," an arbitrarily selected standard body. Then the relative temperature of the systems A and B can be compared without bringing them into contact with each other.

1.8 TEMPERATURE SCALES

To assign a numerical value to the temperature of a system, we choose as a thermometer a substance that has a so-called *thermometric property* that changes with temperature and is easily measured. An example is the volume of a fluid that expands on heating (think of the familiar liquid-in-glass thermometer).

We choose a thermometric property X that is linearly related to the temperature over as wide a range as possible:

$$X = aT + b. \tag{1.13}$$

Here a and b are constants. We choose reference points to define the scale. Prior to 1954, two fixed points were used, the ice and steam points of water. The scale was determined by assigning the numerical values 0°C (C for Celsius) to the ice point and 100°C to the steam point.

An excellent choice for the thermometric property X is the pressure of a gas. The constant-volume gas thermometer, shown in Figure 1.6, is a practical method for measuring the temperature of an object.

Several observations were made when this method was used (Figure 1.7):

a. The P-T curve is very nearly linear over a wide range of temperature.

b. The curve is increasingly linear as the pressure decreases.

c. A linear extrapolation of the plot gives $P = 0$ at $T = -273.15°C$. This turns out to be true for all gases although the slope of the curve is different for different gases.

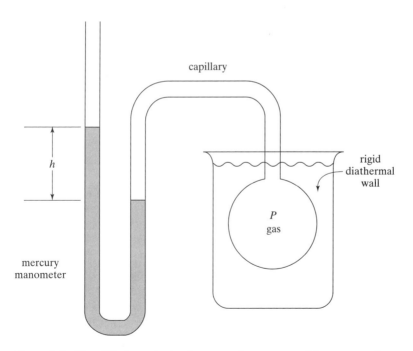

Figure 1.6   Simplified constant-volume gas thermometer. The pressure is given by $P = h\rho g$, as in Figure 1.4.

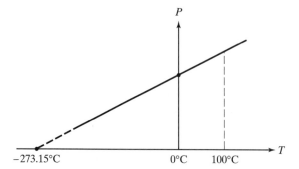

Figure 1.7 Pressure versus temperature for a gas thermometer. The ice and steam points of water are fixed points. The temperature at zero pressure is obtained by linear extrapolation.

For a gas thermometer, then,

$$P = aT + b.$$

The zero pressure point gives

$$0 = -273.15a + b.$$

Thus

$$P = a[T(\,^\circ\text{C}) + 273.15]. \tag{1.14}$$

We can now choose an absolute scale such that there are 100°C between the ice and steam points. We want

$$P = aT(\text{K}), \tag{1.15}$$

where K is the abbreviation for kelvin, the unit of temperature on the absolute scale (no degree sign). It follows that

$$T(\text{K}) - T(\,^\circ\text{C}) = 273.15. \tag{1.16}$$

Now we need only one reference point to define the slope a in Equation (1.15). In 1954 the reference point was taken to be the triple point of water, the pressure and temperature at which ice and liquid water coexist in equilibrium with saturated vapor. The triple point temperature of water is 0.01°C, that is, 0.01 degrees above the ice point at atmospheric pressure. Equations (1.15) and (1.16) give

$$\frac{T(K)}{273.16} = \frac{P}{P_{TP}}, \qquad (1.17)$$

where P_{TP} is the pressure at the triple point. A better definition, taking into account the convergence of the P-T curves for gases at low temperature (Figure 1.8), is

$$T(K) = 273.16 \lim_{P_{TP} \to 0} \frac{P}{P_{TP}}. \qquad (1.18)$$

The curves of the figure are constructed in the following way. An amount of gas in the bulb of a constant-volume gas thermometer is exposed to a cell containing water at the triple point. The pressure is recorded. With the volume held constant, the bulb is surrounded with boiling water at atmospheric pressure, the gas pressure is measured, and the temperature at the steam point, T_S, is calculated using Equation (1.17). Then some of the gas is removed from the bulb, the lower triple point pressure is measured, and the new values of the pressure and temperature at the steam point are found. The process is continued and the resulting curve is extrapolated to the ordinate. Although the curves for different gases have different slopes, they are discovered to intersect in the limit as P_{TP} approaches zero. The triple point reference is precisely reproducible and absolute zero is precisely defined.

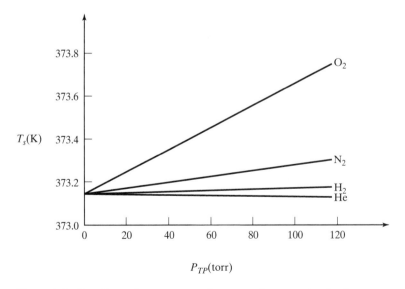

Figure 1.8   Readings of a constant-volume gas thermometer for the temperature of condensing steam as the density of the gas (and hence P_{TP}) is lowered. The extrapolated value of T_S is the same for all gases.

Two other temperature scales are in use. The familiar Fahrenheit scale is related to the Celsius scale by the equation

$$T(°F) = \frac{9}{5}T(°C) + 32. \tag{1.19}$$

On this scale the ice point is 32°F and the steam point is 212°F. The Rankine scale is derived from the Fahrenheit scale:

$$T(R) = T(°F) + 459.67. \tag{1.20}$$

The Rankine scale is widely used in engineering; it does not use the "degree" symbol.

We shall see later that it is possible to define the absolute scale of temperature independently of the thermometric properties of gases.

PROBLEMS

1-1 Classify the following systems as open, closed, or isolated:
   (a) A mass of gas in a container with rigid, impermeable, diathermal walls.
   (b) A mass of gas in a container with rigid, impermeable, adiabatic walls.
   (c) A sugar solution enclosed by a membrane permeable only to water that is immersed in a large container of water.

1-2 Using the terms defined in the chapter, characterize the following processes as completely as possible:
   (a) The temperature of a gas, enclosed in a cylinder provided with a frictionless piston, is slowly increased. The pressure remains constant.
   (b) A gas, enclosed in a cylinder provided with a piston, is slowly expanded. The temperature remains constant. There is a force of friction between the cylinder wall and the piston.
   (c) A gas enclosed in a cylinder provided with a frictionless piston is quickly compressed.
   (d) A piece of hot metal is thrown into cold water. (Assume that the system is the metal, which neither contracts nor expands.)
   (e) A pendulum with a frictionless support swings back and forth.
   (f) A bullet is stopped by a target.

1-3 On a plot of volume versus temperature, draw and label lines indicating the following processes, each proceeding from the same initial state V_0, T_0.
   (a) An isothermal expansion.
   (b) An isothermal compression.
   (c) An isochoric increase in temperature.

1-4 Estimate the pressure you exert when standing on the floor. Express your answer in atmospheres, in pascals, and in torr. Repeat the calculation for spiked heels.

1-5 Let the resistance R of a piece of wire be a thermometric property for measuring the temperature T. Assume that

$$R = aT + b,$$

where a and b are constants. The resistance of the wire is found to be 5 ohms when it is at the temperature of melting ice and 6 ohms when it is at the temperature at which water boils at atmospheric pressure. If the ice point is taken as $100°$ and the boiling point as $500°$ on a particular scale, what is the temperature on this scale when $R = 5.4$ ohms?

1-6 The following table gives the observed values of the pressure P of a gas in a constant-volume gas thermometer at an unknown temperature T and at the triple point of water as the mass of the gas used is reduced.

P_{TP} (torr)	100	200	300	400
P (torr)	127.9	256.5	385.8	516.0

Determine T in kelvins to two decimal places by considering the limit $P_{TP \to 0}(P/P_{TP})$. What is this temperature in $°C$?

1-7 The resistance R of a doped germanium crystal is related to the temperature T through the equation

$$\log R = 4.70 - 3.92 \log T,$$

when R is in ohms, T is in kelvins, and the logarithm is taken to the base 10. In a liquid helium cryostat, the resistance is measured to be 218 ohms. What is the temperature?

1-8 A thermocouple consists of two wires made of dissimilar metals that are joined together to form an electrical circuit. A thermal electromotive force (emf) $\mathcal{E}$ is generated when the two junctions are at different temperatures. When one junction is held at the ice point and the other is at a Celsius temperature of T, the thermometric function is given by

$$\mathcal{E} = aT + bT^2,$$

where $\mathcal{E}$ is in millivolts. Calculate the constants a and b for a thermocouple reading 60 mV at $200°C$ and 40 mV at $400°C$. What temperature corresponds to a reading of 30 mV?

1-9 **(a)** Consider the linear relationship between the thermometric property X and the temperature T given by

$$X = aT + b.$$

Suppose that the ice and steam points are used as fixed points with temperatures of $0°$ and $100°$ respectively. Show that

$$T = 100\left[\frac{X - X_i}{X_s - X_i}\right].$$

(b) If, instead, the thermometric function is chosen as

$$T = a\ln X + b,$$

show that on the new scale

$$T = 100\left[\frac{\ln{(X/X_i)}}{\ln{(X_s/X_i)}}\right].$$

1-10 At what temperature do the Fahrenheit and Kelvin scales give the same reading?

1-11 The temperature of the normal boiling point of nitrogen is 77.35 K. Calculate the corresponding value of the temperature on the (a) Celsius, (b) Fahrenheit, and (c) Rankine scales.

Chapter 2

Equations of State

2.1 INTRODUCTION

In thermodynamics, we add to the basic dimensions of mass, length and time the "fourth dimension" of temperature. We saw in Chapter 1 that the concept of temperature is intimately associated with the notion of systems in thermal equilibrium. Perhaps the simplest example of a thermodynamic system is a homogeneous fluid—say, a gas or vapor.* Its state, as we have observed, can be described by an equation of the form

$$f(P, V, T) = 0. \tag{2.1}$$

The equation of state connects the three fundamental state variables, only two of which are independent. It is an expression of the results of experiments. Every system has its own equation of state.

We note that the equation of state does not involve time. Classical thermodynamics deals mainly with systems in thermal, mechanical, and chemical equilibrium—that is, *thermodynamic equilibrium*—and does not concern itself with the rate at which a process takes place.

The mechanical variables occur in "canonically conjugate" pairs, one an extensive variable, the other intensive. Thus the volume V (extensive) is "conjugate" to the pressure P (intensive). The situation is analogous to that in mechanics, where the generalized position coordinates q_k and the generalized momenta p_k form canonically conjugate pairs.

In thermodynamics, the quantity PdV is the differential of work (with units of energy). We are led to ask if there is an extensive state variable canonically conjugate with the temperature whose product has the unit of energy. This very important question will be addressed in Chapter 6.

---

*The laws of thermodynamics apply to all forms of matter, but they are most easily formulated and understood when applied to gases.

2.2 EQUATION OF STATE OF AN IDEAL GAS

The equation of state of a system composed of m kg of a gas whose molecular weight is M is given approximately by

$$PV = \frac{m}{M} RT, \tag{2.2}$$

where R is a universal constant, having the same value for all gases:

$$R = 8.314 \times 10^3 \frac{J}{\text{kilomole} \cdot K}. \tag{2.3}$$

Since $n \equiv m/M$ is the number of kilomoles of the gas, we can write

$$PV = nRT. \tag{2.4}$$

This equation is called the equation of state of an ideal gas or perfect gas. It includes the laws of Boyle, Gay-Lussac, Charles, and Avogadro, which were discovered over a period of 200 years.

In 1811 Avogadro postulated that at a given temperature and pressure equal volumes of all gases contain equal numbers of molecules. The reason for this is that the molecules in a given sample of gas have negligible volumes compared with the volume of the sample itself.

In Equation (2.4) we note that the extensive variable V divided by n, the number of kilomoles of the gas, is the specific volume v. Thus the equation of state can be written $Pv = RT$. The projections of the surface $f(P, v, T) = 0$ on the P-v plane, the P-T plane, and the v-T plane are shown in Figure 2.1. We use whatever diagram is most appropriate for the process we are interested in.

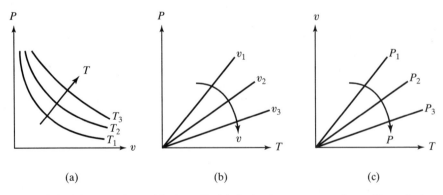

Figure 2.1 Diagrams for an ideal gas. In (a) the isotherms are equilateral hyperbolae; in (b) the isochores are straight lines; and in (c) the isobars are also straight lines.

2.3 VAN DER WAALS' EQUATION FOR A REAL GAS

The characteristic equation of an ideal gas represents the behavior of real gases fairly well for high temperatures and low pressures. However, when the temperature and pressure are such that the gas is near *condensation,* important deviations from the ideal gas law are observed.

Among the numerous equations of state that have been introduced to represent the behavior of real gases, that of van der Waals is especially interesting because of its simplicity and because it satisfactorily describes the behavior of many substances over a wide range of temperature and pressure.

Van der Waals derived his equation from considerations based on kinetic theory, taking into account to a first approximation the size of a molecule and the cohesive forces between molecules. His equation of state is

$$\left(P + \frac{a}{v^2}\right)(v - b) = RT, \tag{2.5}$$

where a and b are characteristic constants for a given substance. For $a = b = 0$, Equation (2.5) reduces to the equation of state for an ideal gas.

The term a/v^2 arises from the intermolecular forces due to the overlap of electron clouds. The constant b takes into account the finite volume occupied by the molecules; its effect is to subtract from the volume term.

Multiplication of Equation (2.5) by v^2 yields the equation

$$Pv^3 - (Pb + RT)v^2 + av - ab = 0. \tag{2.6}$$

The result is a cubic equation in v with three roots, only one of which needs to be real. In Figure 2.2 some isotherms calculated from the van der Waals equation have been drawn.

As T increases, the curves approach $Pv =$ constant. The correction terms in the van der Waals equation become less important. For $T < T_C$, there is a local maximum and minimum value of P. The transition between the two types of curves is a curve having an inflection point CP, the so-called *critical curve,* $T = T_C$. Imagine that the volume is decreased by loading a piston confining the gas in a cylinder. Suppose one moves along an isotherm for which $T < T_C$. After we reach a maximum of the curve, the pressure begins to fall. This is an unstable region since the pressure no longer increases as the volume is diminished.

Actually, this portion of the isotherm is not really traversed at all because the gas undergoes a *change of phase.** Assume that the compression

---

*The transition from a gas to a liquid, from a liquid to a solid, etc. is termed a "phase change" or a "phase transformation."

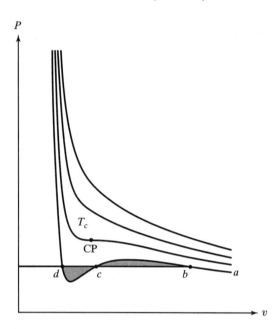

Figure 2.2 Isotherms for a
van der Waals gas.

begins at point a in the diagram. Part of the gas begins to liquefy at b and the
pressure remains *constant* as the volume is further decreased as long as the
temperature is held constant. Between b and d, liquid and vapor are in equilibrium. Finally, at d liquefaction is complete. Thereafter the curve rises steeply
because it takes a large increase of pressure to compress a liquid.

The curves with no extrema $(T \geq T_C)$ can have no regions of this kind.
Above T_C it is impossible to liquefy a gas, no matter how large the pressure is.
The interface between a liquid and a vapor is not discernible.

The critical values v_C, T_C, and P_C of a substance can be expressed in terms
of the constants a and b that appear in van der Waals' equation. For $T = T_C$,

$$P_C = \frac{RT_C}{v_C - b} - \frac{a}{v_C^2}. \tag{2.7}$$

The point at which the tangent to the curve is horizontal is given by

$$\left(\frac{\partial P}{\partial v}\right)_C = 0 = -\frac{RT_C}{(v_C - b)^2} + \frac{2a}{v_C^3}, \tag{2.8}$$

and the point of inflection is the point at which the rate of change of the slope
is zero:

$$\left(\frac{\partial^2 P}{\partial v^2}\right)_C = 0 = \frac{2RT_C}{(v_C - b)^3} - \frac{6a}{v_C^4}. \tag{2.9}$$

Solving these equations yields

$$v_C = 3b,$$

$$T_C = \frac{8a}{27Rb},$$

$$P_C = \frac{a}{27b^2}. \tag{2.10}$$

Evidently, $P_C v_C = \frac{3}{8} R T_C$ for a van der Waals gas at the critical point. The relations of Equation (2.10) suggest that the equation of state of a van der Waals gas can be written in terms of the critical variables, eliminating the constants a and b. This is indeed the case (see Problem 2-5).

2.4 P-v-T SURFACES FOR REAL SUBSTANCES

Real substances can exist in the gas phase only at sufficiently high temperatures and pressures. At low temperatures and high pressures transitions occur to the liquid phase and the solid phase. The P-v-T surface for a pure substance includes these phases as well as the gas phase.

Figure 2.3 is a schematic diagram of the P-v-T surface for a substance that contracts on freezing.

Notice the regions (solid, liquid, gas or vapor) in which the substance can exist in a single phase only. Elsewhere two phases can exist simultaneously in equilibrium, and along the so-called *triple line,* all three phases can coexist.

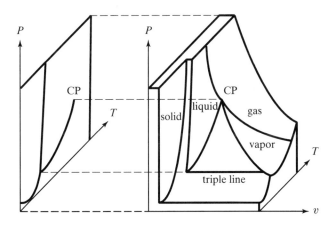

Figure 2.3 P-v-T surface for a substance that contracts on freezing.

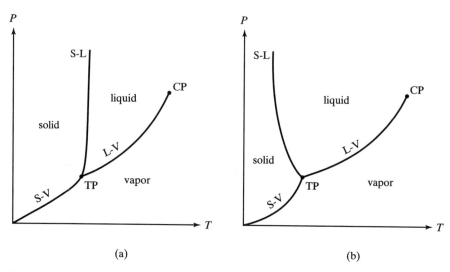

(a)   (b)

Figure 2.4 *P-T* diagrams for (a) a substance that contracts on freezing; and (b) a substance that expands on freezing.

The physical distinction between the various phases is straightforward. A solid has a definite volume and shape. A liquid has a fixed volume but not a fixed shape. A gas has neither a fixed volume nor shape. A vapor is a gas at any temperature less than the critical temperature.

The projections of the surfaces on the *P-T* plane are of special interest (Figure 2.4). The L-V curve is the vapor pressure curve, for which the liquid and vapor phases coexist in equilibrium. It is also known as the "saturated vapor" or boiling-point curve. The S-L curve is the freezing point curve, and the S-V curve is the sublimation curve.

At $T = T_C$ the specific volumes of the saturated liquid and vapor become equal. For $T > T_C$ there is no separation into two phases; the interface between the liquid and the vapor is indiscernible. By raising the temperature above T_C, then increasing the pressure, and finally reducing the temperature, one can make an "end run" around the critical point. In this way, it is possible to move from a vapor to a liquid without crossing the vapor pressure curve.

No critical point can exist for solid-liquid equilibrium (the S-L curve). That is because solids and liquids possess different symmetry properties. A normal liquid is isotropic, whereas a solid has a crystalline structure whose orientation defines a particular set of directions. The transition from one symmetry to another is strictly a discontinuous process.

For CO_2 the critical temperature is 31.0°C and the critical pressure is 72 atmospheres. The triple point occurs at −56.6°C and about 5 atmospheres. When heat is supplied to solid CO_2 (dry ice) at atmospheric pressure, it sublimes and changes directly to the vapor phase.

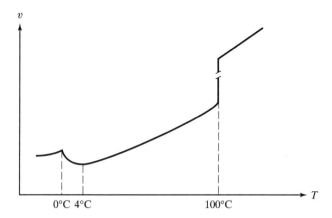

Figure 2.5   Sketch of the specific volume of water as a function of temperature.

For H_2O the critical point is 374.2°C and 207 atmospheres. The triple point is 0.01°C and 4.6 torr $(6.0 \times 10^{-3}$ atm). Thus the pressure must be reduced to a low value to reach the triple point.

Water and a few other substances *expand* on solidifying. Water expands about 10 percent on freezing. A block of ice floats with nine-tenths of its volume below the surface of the water and one-tenth above. (An iceberg floats with one-seventh of its volume above the surface; the difference is because sea water is denser than fresh water and common ice is full of air bubbles.) If water contracted on freezing, the results would be disastrous, but ice floats. During a change of state, there is always a change in volume but the temperature and pressure are constant. For example, at the freezing point of water, the change in volume is 10 percent, as noted. But at the steam point the expansion is enormous: 1 cm^3 of liquid water at 100°C becomes 1600 cm^3 of steam at this temperature. Changes of state are accompanied by changes in molecular forces. In Figure 2.5 the specific volume is sketched as a function of temperature for water. Since v is the reciprocal of the density, it follows that the maximum density of water occurs at 4°C.

2.5 EXPANSIVITY AND COMPRESSIBILITY

Suppose that the equation of state of a given substance is written in the form

$$v = v(T, P). \tag{2.11}$$

Taking the differential, we obtain

$$dv = \left(\frac{\partial v}{\partial T}\right)_P dT + \left(\frac{\partial v}{\partial P}\right)_T dP. \qquad (2.12)$$

(The subscript denotes the quantity held constant.) Two important measurable quantities are defined in terms of the partial derivatives. The expansivity, or coefficient of volume expansion, is given by

$$\beta \equiv \frac{1}{v}\left(\frac{\partial v}{\partial T}\right)_P. \qquad (2.13)$$

This is the fractional change of volume resulting from a change in temperature, at constant pressure. Similarly, the isothermal compressibility is defined as

$$\kappa \equiv -\frac{1}{v}\left(\frac{\partial v}{\partial P}\right)_T, \qquad (2.14)$$

the fractional change in volume as the pressure changes, with the temperature held constant. The negative sign is used since the volume always decreases with increasing pressure (at constant temperature); that is, the partial derivative is inherently negative, and the isothermal compressibility is a positive quantity.

For an ideal gas, $v = RT/P$, and

$$\beta = \frac{1}{v}\left(\frac{R}{P}\right) = \frac{1}{T}. \qquad (2.15)$$

The higher the temperature, the lower the expansivity. Also,

$$\kappa = -\frac{1}{v}\left(-\frac{RT}{P^2}\right) = \frac{1}{P}. \qquad (2.16)$$

The isothermal compressibility of an ideal gas is small at high pressures.

For a liquid or a solid, in contrast to a gas, β and κ are nearly constant over a fairly wide range of temperature and pressure. This experimental fact allows us to develop an approximate equation of state that is useful in many applications. Substituting Equation (2.13) and (2.14) in Equation (2.12), we obtain

$$dv = \beta v\, dT - \kappa v\, dP. \qquad (2.17)$$

We assume that the volume change is small when the temperature and pressure are changed so that, to a first approximation, $v \approx v_0$ (a constant), and β and κ are constants. Then

$$dv \approx \beta v_0\, dT - \kappa v_0\, dP.$$

Integrating, we have

$$\int_{v_0}^{v} dv = \beta v_0 \int_{T_0}^{T} dT - \kappa v_0 \int_{P_0}^{P} dP,$$

so that

$$v = v_0[1 + \beta(T - T_0) - \kappa(P - P_0)]. \tag{2.18}$$

This is an approximate equation of state for a liquid or a solid. The volume increases linearly with an increase in temperature and decreases linearly with an increase in pressure.

Suppose that the temperature of a block of copper is increased from $127°C$ to $137°C$. We wish to know what change in pressure would be necessary to keep the volume constant. For copper in this temperature range,

$$\beta \approx 5.2 \times 10^{-5} \, \text{K}^{-1} \text{ and } \kappa \approx 7.6 \times 10^{-12} \, \text{Pa}^{-1}.$$

Then, Equation (2.18) gives

$$\Delta P = \frac{\beta}{\kappa} \Delta T = \frac{5.2 \times 10^{-5}}{7.6 \times 10^{-12}} (10) = 6.8 \times 10^7 \, \text{Pa} \approx 680 \, \text{atm!}$$

2.6 AN APPLICATION

Suppose that we wish to calculate the decrease in pressure of a fluid when it is cooled from T_1 to T_2. We know that the equilibrium states of the fluid are fixed by specifying two of the state variables that are related by some equation of state

$$f(P, v, T) = 0. \tag{2.19}$$

We assume that the process in which the fluid is changed from an equilibrium state (P_1, T_1) to another equilibrium state (P_2, T_2) is isochoric—that is, the volume is unchanged. We suppose for a moment that the fluid is cooled *reversibly.* This could be achieved by placing in contact with the fluid a series of large bodies (reservoirs) ranging in temperature from T_1 to T_2 to effect a quasistatic cooling through a sequence of equilibrium states. For any one of these states we may write Equation (2.19), or, solving for P,

$$P = P(v, T). \tag{2.20}$$

Taking the differential, we obtain

$$dP = \left(\frac{\partial P}{\partial v}\right)_T dv + \left(\frac{\partial P}{\partial T}\right)_v dT. \tag{2.21}$$

Since we are assuming that the cooling takes place at constant volume, the first term on the right-hand side is zero. Thus

$$P_2 - P_1 = \int_{T_1}^{T_2}\left(\frac{\partial P}{\partial T}\right)_v dT. \tag{2.22}$$

Unfortunately, the integrand is unknown. However, using the cyclical relation given in Appendix A, we have

$$\left(\frac{\partial P}{\partial T}\right)_v \left(\frac{\partial T}{\partial v}\right)_P \left(\frac{\partial v}{\partial P}\right)_T = -1.$$

Thus

$$\left(\frac{\partial P}{\partial T}\right)_v = -\frac{1}{\left(\frac{\partial T}{\partial v}\right)_P \left(\frac{\partial v}{\partial P}\right)_T} = -\frac{\left(\frac{\partial v}{\partial T}\right)_P}{\left(\frac{\partial v}{\partial P}\right)_T} = \frac{\beta}{\kappa}.$$

Substituting this result in Equation (2.22), we obtain

$$P_2 - P_1 = \frac{\beta}{\kappa}(T_2 - T_1), \tag{2.23}$$

if β and κ are independent of T. This is negative if $T_2 < T_1$, indicating that the pressure is reduced on cooling. Note that this result could have been obtained immediately from Equation (2.18).

In practice, the cooling would not be reversible because one simply heats the fluid and lets it cool. The result would be large temperature gradients within the fluid itself and between the fluid and its surroundings. Thus the intermediate states are not equilibrium states and the equation of state cannot be applied.

However, the initial and final states *are* equilibrium states, and it doesn't matter how we go from state 1 to state 2 to determine the change in the state variable P. This is because $\Delta P = P_2 - P_1$ is *independent of the path*, being determined only by the end points. We can choose any convenient path to calculate changes in state functions for processes between a pair of equilibrium states. The most convenient path is a *reversible* path.

Our fundamental state variables P, v, and T are all exact differentials whose integrals are independent of the path. Formulating all thermodynamic relations in terms of state variables will be a goal of the theoretical development; this is called the *condition of integrability*. When we encounter a thermodynamic quantity whose differential is imperfect, we will seek an integrating factor to give us a quantity that we prefer to work with. Appendix A includes a comprehensive discussion of exact and inexact differentials.

PROBLEMS

2-1 How many kilograms of helium gas are contained in a vessel of 1 liter volume at 50°C if the pressure is one atmosphere? (The atomic weight of helium is 4.)

2-2 A tank of volume 0.5 m³ contains oxygen at a pressure of 1.5×10^6 Pa and a temperature of 20°C. Assume that oxygen behaves like an ideal gas.
(a) How many kilomoles of oxygen are in the tank?
(b) How many kilograms?
(c) Find the pressure if the temperature is increased to 500°C.
(d) At a temperature of 20°C, how many kilomoles can be withdrawn from the tank before the pressure falls to 10 percent of the original pressure?

2-3 A cylinder provided with a movable piston contains an ideal gas at a pressure P_1, specific volume v_1, and temperature T_1. The pressure and volume are simultaneously increased so that at every instant P and v are related by the equation

$$P = Av, \qquad A = \text{constant}$$

(a) Express A in terms of the pressure P_1, the temperature T_1, and the gas constant R.
(b) Make a sketch representing the process in the P-v plane.
(c) Find the temperature when the specific volume has doubled, if $T_1 = 200$ K.

2-4 For an ideal gas the slope of an isotherm is given by

$$\left(\frac{\partial P}{\partial v} \right)_T = -\frac{P}{v},$$

and that of an isochore is

$$\left(\frac{\partial P}{\partial T} \right)_v = \frac{P}{T}.$$

Show that these relations give $Pv = RT$, the equation of state.

2-5 Noting that at the critical point, the three roots of the van der Waals equation are equal (i.e., $(v - v_C)^3 = 0$), show that the critical values of the specific volume, temperature, and pressure are given by Equation (2.10). Show that, in terms of the reduced quantities $v' \equiv v/v_c$, $T' \equiv T/T_C$ and $P' \equiv P/P_C$, the van der Waals equation becomes

$$\left(P' + \frac{3}{v'^2}\right)\left(v' - \frac{1}{3}\right) = \frac{8}{3}T'.$$

2-6 Using the Berthelot equation of state,

$$P = \frac{RT}{v - b} - \frac{a}{Tv^2},$$

show that

$$v_C = 3b, \; T_C = \sqrt{\frac{8a}{27bR}}, \; P_C = \frac{1}{12b}\sqrt{\frac{2aR}{3b}}$$

(Hint: As noted in the text, $\partial P/\partial v = 0$ and $\partial^2 P/\partial v^2 = 0$ at the critical point.) Compare the numerical value of $RT_C/P_C v_C$ with the experimental values given in the following table:

Substance	$RT_C/P_C v_C$
He	3.06
H_2	3.27
O_2	3.42
CO_2	3.61
H_2O	4.29

2-7 Using the Dieterici equation of state,

$$P = \frac{RT}{v - b}e^{-a/RTv},$$

show that

$$v_C = 2b, \; T_C = \frac{a}{4Rb}, \; P_C = \frac{a}{4e^2b^2},$$

and find the numerical value of $RT_C/P_C v_C$. How does this compare with the tabulated experimental values?

2-8 (a) Making use of the cyclical relation (Equation (A.7) in Appendix A), find the expansivity β of a substance obeying the Dieterici equation of state in Problem 2-7.

(b) At higher temperatures and large specific volumes (low densities) all gases approximate an ideal gas. Show that for large values of T and v, the expression for β obtained in (a) goes over to the corresponding equation for an ideal gas.

2-9 Show that in general

$$\left(\frac{\partial \beta}{\partial P}\right)_T + \left(\frac{\partial \kappa}{\partial T}\right)_P = 0.$$

2-10 A hypothetical substance has an expansivity $\beta = 2bT/v$ and an isothermal compressibility $\kappa = a/v$, where a and b are constants. Show that the equation of state is

$$v - bT^2 + aP = \text{constant}.$$

2-11 Suppose that

$$\beta = \frac{v - a}{Tv}, \quad \kappa = \frac{3(v - a)}{4Pv}.$$

Show that the equation of state is

$$P^{3/4}(v - a) = AT,$$

where a and A are constants.

2-12 Show that β and κ are infinite at the critical point.

2-13 A glass bottle of nominal capacity 250 cm^3 is filled brim full of water at 20°C. If the bottle and contents are heated to 50°C, how much water spills over? (For water, $\beta = 0.21 \times 10^{-3}$ K^{-1}. Assume that the expansion of the glass is negligible.)

Chapter 3

The First Law of Thermodynamics

3.1 CONFIGURATION WORK

Configuration work is the work in a *reversible* process given by the product of some intensive variable (P, for example) and the *change* in some extensive variable (V, for example). Let y represent such an intensive variable and X the corresponding extensive variable. In the most general case, where more than one pair of variables may be involved,

$$\bar{d}W = \sum_i y_i dX_i, \quad i = 1, 2, \ldots n. \tag{3.1}$$

Here $\bar{d}W$ represents the work done by virtue of infinitesimal changes in the extensive variables X_i; $\bar{d}W$ is not an exact differential, as indicated by the notation (see Appendix A). The variables X_1, X_2, etc., are said to determine the *configuration* of the system, and

$$\sum_i y_i dX_i$$

is called *configuration work.*

Consider the configuration work done by the system consisting of a gas enclosed in a cylindrical container by a piston (Figure 3.1). The gas expands against the pressure associated with the external force F (possibly due to the weight of the movable part of the piston). The work done by the gas is

$$\bar{d}W = F\,dx = PA\,dx = P\,dV.$$

Here dx is the displacement of the piston, A is its cross-sectional area, P is the pressure (the intensive variable), and V is the volume (the extensive variable). Since P and dV are positive, $\bar{d}W > 0$, denoting work done *by* the system. This sign convention will be used throughout the text. We assume that the process is reversible and that the system is in equilibrium at all times. No friction (dissipation) is involved. The integration is nontrivial since $P = P(V)$ in general.

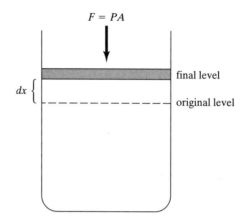

Figure 3.1 Expansion of an enclosed gas against the pressure produced by an external force.

TABLE 3.1 Various examples of configuration work.

System	Intensive Variable	Extensive Variable	$đW$
gas, liquid, or solid	P (pressure)	V (volume)	PdV
film	Γ (surface tension)	A (area)	ΓdA
electrolytic cell	$\mathcal{E}$ (electromotive force)	q (charge)	$\mathcal{E}dq$
magnetic material	B (magnetic field)	$\mathcal{M}$ (magnetization)	$Bd\mathcal{M}$
dielectric material	E (electric field)	$\mathcal{P}$ (polarization)	$Ed\mathcal{P}$

In thermodynamics configuration work is often called "PdV" work for obvious reasons. However, there are many examples of configuration work, some of which are listed in Table 3.1. For a surface film of liquid, the reversible work required to increase isothermally the surface area of the film by an amount dA is $đW = \Gamma dA$, where Γ is the surface tension. In an electrolytic cell such as a lead storage battery, the configuration work is the product of the electromotive force $\mathcal{E}$ of the cell and the amount of charge transferred in the chemical reaction. Magnetic work is done on a magnetizable material by a magnetic field. If, for example, the material is placed in a solenoid that generates a field B, the incremental work done is $Bd\mathcal{M}$, where $d\mathcal{M}$ is the change in total magnetization of the material. Analogously, an electrically susceptible material inserted between the plates of a parallel-plate capacitor experiences a change of total electrical polarization $d\mathcal{P}$; the work is $Ed\mathcal{P}$, where E is the electric field strength in the region between the plates.

The configuration work done in the change of volume of an ideal gas can be easily calculated in some special cases. For an isochoric process, $dV = 0$, so $đW = 0$. For an isobaric process,

$$W = \int PdV = P\int_{V_a}^{V_b} dV = P(V_b - V_a). \tag{3.2}$$

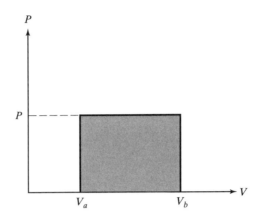

Figure 3.2   Work done
in an isobaric process.

Here V_a is the initial volume and V_b the final volume. The work done is evidently the area under the curve in a P-V diagram. This is the shaded rectangle of Figure 3.2 for the case of an isobaric process.

For an isothermal process, the temperature is constant, the P-V curve for an ideal gas is a hyperbola, and

$$W = nRT \int_{V_a}^{V_b} \frac{dV}{V} = nRT \ln\left(\frac{V_b}{V_a}\right), \tag{3.3}$$

(Figure 3.3).

In both the isobaric and the isothermal process, if $V_b > V_a$, the process is an *expansion,* and work is done *by* the system ($W > 0$). If $V_b < V_a$, the process is a *compression,* and work is done *on* the system ($W < 0$). This important sign convention will be followed throughout the text.

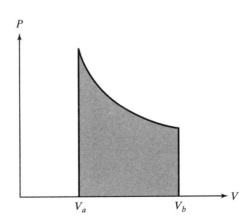

Figure 3.3   Work done
in an isothermal
process involving an
ideal gas.

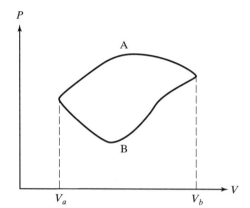

Figure 3.4 Work is path-dependent; the area under curve A is different from the area under curve B.

It must be emphasized that work is not a property of the system; W is not a state variable. This is easily seen in Figure 3.4. Since $\int P\,dV$ is the area under the curve, different results are obtained for paths A and B. Thus

   a. $\bar{d}W$ is not exact.
   b. W depends on the path of integration.
   c. The integral around a closed path is not zero; that is, $\oint \bar{d}W \neq 0$.

3.2 DISSIPATIVE WORK

Dissipative work is work done in an *irreversible* process. Such work is always done *on* the system. In the general case, both configuration work and dissipative work may be done in a process. The total work is the algebraic sum of the two kinds of work. If a process is to be *reversible,* the dissipative work must be zero. Since a reversible process is necessarily quasistatic, then to specify that a process is reversible implies that the process is quasistatic.

There are two well-known examples of dissipative work. The first is stirring work. Consider a stirrer immersed in a fluid. The stirrer and the fluid together constitute the system. The stirrer is attached to a shaft projecting through the wall of the container and an external torque is applied at the outer end of the shaft (Figure 3.5).

Regardless of the sense of rotation of the shaft, the external torque is always in the same direction as the angular displacement of the shaft, and the work done by the external torque is always *negative* (work is always done *on* the system). The work is given by

$$\bar{d}W = -\tau d\theta. \tag{3.4}$$

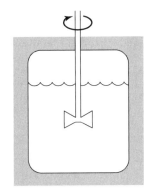

Figure 3.5   Stirring work. A stirrer is immersed in a fluid and an external torque is applied.

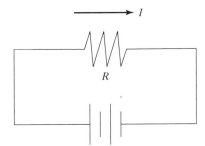

Figure 3.6   Electrical work. A current is passed through a resistor.

Clearly, reversing the sense of the torque τ reverses the direction of the angular displacement $d\theta$.

A second example of dissipative work is the work needed to maintain an electric current I in a resistor of resistance R (Figure 3.6). Quantitatively,

$$\overline{d}W = -I^2R\,dt. \tag{3.5}$$

Here $\overline{d}W$ is the work done in time dt on the system comprising the current and the resistor. Reversing the direction of current flow does not affect the sign of the work.

3.3 ADIABATIC WORK AND INTERNAL ENERGY

A fundamental result of the greatest importance is arrived at by considering adiabatic processes between two equilibrium states of a system. Having defined configuration work and dissipative work, we can think of different adiabatic paths between two states a and b of the system. In the P-V diagram of Figure 3.7, the path labeled 1, denoted by the solid line, could represent stirring work (at constant volume), followed by an adiabatic compression of the fluid. (The system is assumed to be thermally insulated from its surroundings.)

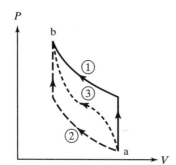

Figure 3.7 Three different adiabatic paths representing processes in which an isolated system is carried from state a to state b.

Correspondingly, path 2, portrayed by the dashed line, could depict an adiabatic compression followed by electrical work at constant volume. The dotted line of path 3 could delineate some combination of mechanical and electrical work performed simultaneously.

Now, it is an experimental fact that the work done in the three processes is the same. To be sure, experiments have not been performed for all possible adiabatic processes between all equilibrium states of all possible systems. However, the whole edifice of thermodynamics is consistent with the conclusion that

The total work done in all adiabatic processes between any two equilibrium states is independent of the path.

That is,

$$W_{ad} = \int_a^b \bar{d}W_{ad}$$

is path-independent and $\bar{d}W_{ad}$ is an exact differential. (We can remove the bar from the differential, but only if we restrict ourselves to adiabatic work.) It follows that we can define a property of the system, a state variable U, such that

$$\int_a^b dU = U_b - U_a = -\int_a^b \bar{d}W_{ad} = -W_{ad}.$$

In differential form,

$$dU = -\bar{d}W_{ad}. \tag{3.6}$$

We note that dU is the negative of the adiabatic work done *by* the system, so it is the adiabatic work done *on* the system. We define U as the *internal energy* of the system. Thus Equation (3.6) states that the increase in internal energy is equal to the work done on the system in any adiabatic process. Alternatively, the work done on the system (with no heat flow) results in an increase in its internal energy.

To summarize, experiments show that an energy function of state can be defined by the amount of work done in an adiabatically enclosed system. This is not to say, however, that *any* two states of a system can be coupled by an adiabatic path.

3.4 HEAT

We wish now to extend our ideas to processes that are not adiabatic. If the system is not contained within adiabatic walls, it is nevertheless possible to bring about a given change of state in quite different ways. A beaker of water may be brought from 20°C to 100°C by electrical work performed on a resistor immersed in it, or, alternatively, by lighting a Bunsen burner under it. The latter process involves *no work at all* (Figure 3.8).

For any change between given states, dU can always be uniquely determined by carrying out an experiment under adiabatic conditions, for which $dU = -\dbar W_{ad}$. If the conditions are not so specialized, then $dU \neq -\dbar W_{ad}$, in general. Instead we may write the equation

$$dU = \dbar Q - \dbar W,$$

or

$$\dbar Q = \dbar W - \dbar W_{ad}.$$

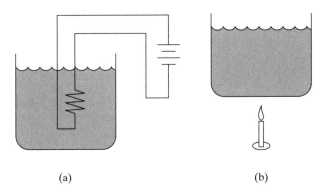

(a)   (b)

Figure 3.8   Two ways of raising the temperature of water:
(a) by electrical work; (b) by applying heat.

In other words, we can define heat flow as follows:

> *Heat flow into the system is equal to the total work done by the system minus the adiabatic work done.*

Then

$$đQ = dU + đW, \tag{3.7}$$

which is the first law of thermodynamics in differential form. In words, the first law states that

> *The heat supplied is equal to the increase in internal energy of the system plus the work done by the system. Energy is conserved if heat is taken into account.*

Note that heat is not a property (state variable) of the system; only the internal energy is.

It can be shown that the quantity $đQ$ exhibits the properties that are commonly associated with heat. These properties are summarized as follows:

1. The addition of heat to a body changes its state.
2. Heat may be conveyed from one body to another by conduction, convection, or radiation.
3. In a calorimetric experiment by the method of mixtures, heat is conserved if the experimental bodies are adiabatically enclosed.

We note that since U is a state variable,

$$\oint dU = 0.$$

That is, in a *cyclic* process, the change in internal energy is zero so that the work performed is equal to the heat absorbed by the system.

One useful way of looking at the first law follows from transposing the terms in Equation (3.7):

$$dU = đQ - đW.$$

The increase in internal energy of a system equals the heat flow into the system minus the work done by the system. Think of a bank: $đQ$ is a deposit, $đW$ a withdrawal. What's left in the bank is dU. The sign convention for $đQ$ is $đQ > 0$ for heat flow *into* a system, and $đQ < 0$ for heat flow *out of* the system.

The temperature of a body alone is what determines whether heat will be transferred from it to another body with which it is in contact, or vice versa. A large block of ice at 0°C has far more internal energy than a cup of hot water; yet when the water is poured on the ice some of the ice melts and the water becomes cooler, which signifies that energy has passed from the water to the ice.

When the temperature of a body increases, it is customary to say that *heat* has been added to it; when the temperature decreases, it is customary to say that heat has been removed form it. When no work is done, $\Delta U = Q$, which says that the internal energy change of the body is equal to the heat transferred to it from the surroundings. One definition of heat is:

> *Heat is energy transferred across the boundary of a system as a result of a temperature difference only.*

This definition is not entirely satisfactory, however. Changes of state (e.g., from ice to water or from water to steam) involve the transfer of heat to or from a body *without* any change in temperature. It should also be kept in mind that heat transfer is not the *only* way to change the temperature of matter: a body that has work done on it may become hotter as a result, and a body that *does* work on something else may become cooler. A more precise definition of heat is:

> *Heat is the change in internal energy of a system when no work is done on or by the system.*

3.5 UNITS OF HEAT

Since heat is a form of energy, the correct SI unit of heat is the joule (J). However, the kilocalorie is widely used. It is defined as follows:

> 1 kilocalorie (kcal) ≡ the heat required to raise the temperature of 1 kg of water from 14.5°C to 15.5°C.

The relationship between the joule and the kilocalorie is:

$$1 \text{ J} = 2.39 \times 10^{-4} \text{ kcal},$$

$$1 \text{ kcal} = 4184 \text{ J}.$$

The *calorie* that dieticians use is the same as the kilocalorie and is sometimes written "Calorie" to distinguish it from the ordinary, smaller calorie associated with the cgs system of units.

3.6 THE MECHANICAL EQUIVALENT OF HEAT

We have noted that if the configuration work *and* the dissipative work are both zero,

$$\Delta U = Q,$$

or the heat flow *in* increases the internal energy. Suppose, on the other hand, that the configuration work is zero and that work is done with a stirrer immersed in a fluid kept at constant volume and thermally insulated. Then

$$\Delta U = |W_d|,$$

where W_d is the dissipative work done *on* the system and is inherently negative. The right-hand sides of the equations in the two examples give rise to the same change in internal energy. Thus work and heat are "equivalent" in this sense.

However, heat and work can be differentiated from a microscopic viewpoint. When energy is added to a system in the form of heat, the *random* motion of the constituent molecules is increased. Consider gas in a cylinder. If heat is added through diathermal walls, this increases the random kinetic energy of the molecules which means a rise in temperature. Now let the gas be confined in a cylindrical piston with adiabatic walls. When the piston is pushed in, the molecules striking the piston are accelerated in the direction of its travel. However, any organized motion initially imparted to these molecules is rapidly randomized by collisions, either with the walls or with other gas molecules. Again, the increase in random kinetic energy appears as an increase in temperature.

3.7 SUMMARY OF THE FIRST LAW

1. Energy is conserved. Heat is energy transferred to a system causing a change in its internal energy minus any work done in the process.
2. The quantity U is a generalized store of energy possessed by a thermodynamic system which can be changed by adding or subtracting energy in any form.
3. The internal energy U is a state variable; it is extensive.
4. The first law can be expressed in differential form as

$$đQ = dU + đW.$$

5. For a reversible process, $đW$ is solely configuration ("PdV") work, so that

$$đQ = dU + PdV.$$

3.8 SOME CALCULATIONS OF WORK

The first law can be written

$$đW = đQ - dU.$$

For a reversible process involving a fluid or solid, $đW = PdV$. Let the equation of state of the substance be $V = V(T, P)$. Then

$$dV = \left(\frac{\partial V}{\partial T}\right)_P dT + \left(\frac{\partial V}{\partial P}\right)_T dP.$$

Recall that the expansivity and isothermal compressibility are given, respectively, by

$$\beta = \frac{1}{V}\left(\frac{\partial V}{\partial T}\right)_P,$$

$$\kappa = -\frac{1}{V}\left(\frac{\partial V}{\partial P}\right)_T.$$

Thus

$$dV = \beta V dT - \kappa V dP,$$

and

$$đW = \beta PV dT - \kappa PV dP \tag{3.8}$$

for a reversible process. For the special case of an ideal gas, $\beta = 1/T$ and $\kappa = 1/P$, and

$$đW = nRdT - nRT\frac{dP}{P}. \tag{3.9}$$

In general, Equation (3.9) can be integrated, but that does not mean that the result will be independent of the path.

For an isothermal expansion, $dT = 0$ and

$$W = -nRT \int_{P_a}^{P_b} \frac{dP}{P} = nRT \ln\left(\frac{P_a}{P_b}\right),$$

where P_a is the initial value of the pressure and P_b is its final value. For an expansion, $P_a > P_b$, so $W > 0$ and work is done by the system.

If the expansion is isobaric, $dP = 0$ and

$$W = nR(T_b - T_a) = P(V_b - V_a).$$

Again, work is done by the system.

Equation (3.8) is especially useful for determining the work done in compressing a solid. Suppose that a 10 g block of copper is isothermally (and quasistatically) compressed at a temperature of 0°C. If the initial pressure is 1 atm and the final pressure is 1000 atm, how much work is done on the block of copper? We assume that whereas the pressure change is large, the volume change is small, $V_b \approx V_a = V$. Then

$$đW = -\kappa PV\,dP,$$

and

$$W = -\kappa V \int_{P_a}^{P_b} P\,dP = -\frac{\kappa V}{2}(P_b^2 - P_a^2). \tag{3.10}$$

For copper, $\kappa = 7.5 \times 10^{-12}\,\mathrm{Pa}^{-1}$ at 0°C and the density ρ is 8.93 g(cm)$^{-3}$. Thus

$$V = \frac{m}{\rho} = \frac{10\,\mathrm{g}}{8.93\,\mathrm{g/cm^3}} \times \frac{1\,\mathrm{m^3}}{10^6\,\mathrm{cm^3}} = 1.12 \times 10^{-6}\,\mathrm{m^3}.$$

Since $P_a^2 \ll P_b^2$, and $P_b = 1.01 \times 10^8\,\mathrm{Pa}$ we find that

$$W = -\frac{1}{2}(7.5 \times 10^{-12})(1.12 \times 10^{-6})(1.01 \times 10^8)^2 = -4.3 \times 10^{-2}\,\mathrm{J},$$

a comparatively small amount of work.

At this stage of our study, we cannot say much about what happens to the heat and the internal energy in these processes. We need some more basic concepts, which we shall introduce in subsequent chapters.

PROBLEMS

3-1 Ten kilomoles of an ideal gas are compressed isothermally and reversibly from a pressure of 1 atmosphere to 10 atmospheres at 300 K. How much work is done?

3-2 Steam at a constant pressure of 30 atmospheres is admitted to the cylinder of a steam engine. The length of the stroke is 0.5 m and the diameter of the cylinder is 0.4 m. How much work is done by the steam per stroke?

3-3 An ideal gas originally at a temperature T_1 and pressure P_1 is compressed reversibly against a piston to a volume equal to one-half its original volume. The temperature of the gas is varied during the compression so that at each instant the relation $P = AV$ is satisfied, A = constant.
(a) Draw a diagram of the process in the P-V plane.
(b) Find the final temperature T_2 in terms of T_1.
(c) Find the work done on the gas in terms of n, R, and T_1.

3-4 Ice at $0°C$ and at a pressure of 1 atmosphere has a density of 916.23 kg m^{-3}, while water under these conditions has a density of 999.84 kg m^{-3}. How much work is done against the atmosphere when 10 kg of ice melts into water?

3-5 (a) Derive the general expression for the work per kilomole of a van der Waals gas in expanding reversibly and at a constant temperature T from a specific volume v_1 to a specific volume v_2.
(b) Find the work done when 2 kilomoles of steam expand from a volume of 30 m^3 to a volume of 60 m^3 at a temperature of $100°C$. Take $a = 5.80 \times 10^5$ Jm^3(kilomole)$^{-2}$ and $b = 3.19 \times 10^{-2}$ m^3(kilomole)$^{-1}$.
(c) Find the work of an ideal gas in the same expansion.

3-6 The temperature of an ideal gas at an initial pressure P_1 and volume V_1 is increased at constant volume until the pressure is doubled. The gas is then expanded isothermally until the pressure drops to its original value, where it is compressed at constant pressure until the volume returns to its initial value.
(a) Sketch these processes in the P-V plane and in the P-T plane.
(b) Compute the work in each process and the net work done in the cycle if $n = 2$ kilomoles, $P_1 = 2$ atm, and $V_1 = 4$ m^3.

3-7 Two kilomoles of a monatomic ideal gas are at a temperature of 300 K. The gas expands reversibly and isothermally to twice its original volume. Calculate (a) the work done by the gas; (b) the heat supplied. (Note that for an ideal gas $U \propto T$, so $\Delta U = 0$ in an isothermal process.)

3-8 A gas is contained in a cylinder fitted with a frictionless piston and is taken from the state a to the state b along the path acb shown in Figure 3.9. 80 J of heat flow into the system and the system does 30 J of work.

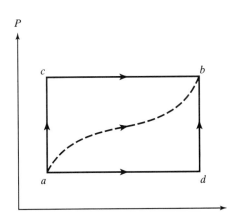

Figure 3.9 P-V diagram for Problem 3-8.

(a) How much heat flows into the system along the path *adb* if the work done by the gas is 10 J?

(b) When the system is returned from *b* to *a* along the curved path, the work done on the system is 20 J. What is the heat transfer?

(c) If $U_a = 0$ and $U_d = 40$ J, find the heat absorbed in the processes *ad* and *db*.

3-9 A volume of 10 m³ contains 8 kg of oxygen at a temperature of 300 K. Find the work necessary to decrease the volume to 5 m³

(a) At constant pressure.

(b) At constant temperature.

(c) What is the temperature at the end of the process in (a)?

(d) What is the pressure at the end of the process in (b)?

(e) Sketch both processes in the *P-V* plane.

3-10 The pressure on 100 g of nickel is increased quasistatically and isothermally from zero pressure to 500 atm. Calculate the work done on the material, assuming that the density and isothermal compressibility remain constant at the values of $8.90 \times 10^3 \, \mathrm{kg \, m^{-3}}$ and $6.75 \times 10^{-12} \, \mathrm{Pa^{-1}}$, respectively.

3-11 During the ascent of a meteorological balloon its volume increases from 1 m³ to 1.8 m³, and the pressure of the ideal gas inside the balloon decreases from 1 bar to $\frac{1}{2}$ bar (1 bar = 10^5 Pa). The internal energy of the gas is $U = 800T$ joules, where *T* is in kelvins.

(a) Assume that in this process (the ascent) $V = AP + B$. Find *A* and *B*.

(b) If the initial temperature is 300 K, what is the final temperature?

(c) How much work is done by the gas in the balloon?

(d) How much heat does it absorb?

Chapter 4

Applications of the First Law

4.1 HEAT CAPACITY

Consider a process in which heat Q is added to a system, causing it to change from one equilibrium state to another with an accompanying rise in temperature ΔT. The *heat capacity* C of a system is defined as the limiting ratio of the heat absorbed divided by the temperature increase:

$$C \equiv \lim_{\Delta T \to 0} \left(\frac{Q}{\Delta T} \right) = \frac{đQ}{dT}. \tag{4.1}$$

The heat capacity is not truly a derivative because Q is not a state variable and $đQ$ is not an exact differential. We interpret $đQ$ as an infinitesimally small flow of heat and dT as the resulting change in temperature. Furthermore, "heat capacity" is a misnomer because it implies that a system is capable of holding heat, which is not; a term connoting the storage of internal energy would be more appropriate.

The heat required to produce a given temperature change is proportional to the mass of the system. We define the *specific heat capacity*, often abbreviated to "specific heat," as the heat capacity per unit mass:

$$c \equiv \frac{1}{n} \left(\frac{đQ}{dT} \right) = \frac{đq}{dT}. \tag{4.2}$$

The SI units are J kilomole^{-1} K^{-1}.

The specific heat capacity depends on the conditions under which the heating takes place — that is, on what parameters are held constant during the heating process. There are two important cases: (1) the specific heat c_v, where the heat is supplied at constant volume; and (2) the specific heat c_P, in which the heat is added at constant pressure. Thus

$$c_v \equiv \left(\frac{đq}{dT} \right)_v \quad \text{and} \quad c_P \equiv \left(\frac{đq}{dT} \right)_P. \tag{4.3}$$

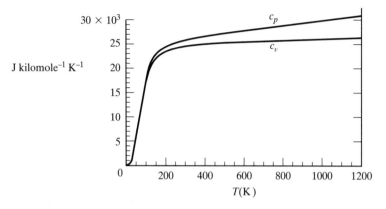

Figure 4.1 Plots of c_v and c_P for copper versus temperature at a pressure of one atmosphere. (Adapted from *Thermodynamics, Kinetic Theory, and Statistical Thermodynamics*, 3rd edition, by F.W. Sears and G.L. Salinger.)

The distinction between the two heat capacities is important chiefly for gases because of the large coefficients of thermal expansion.

In general, the specific heats are functions of temperature and can be regarded as constants only over limited ranges of temperature. As an example, consider the specific heats of a solid such as copper, shown in Figure 4.1. At low temperatures, $c_P \approx c_v$ and both tend toward zero as the temperature approaches zero. For high temperatures, c_v approaches the constant value of $25 \times 10^3 \, \text{J kilomole}^{-1} \, \text{K}^{-1}$. Many solids exhibit this behavior, an empirically established fact known as the law of Dulong and Petit, to be discussed in Chapter 16.

4.2 MAYER'S EQUATION

We wish to find the relationship between c_v and c_P for an ideal gas. For a *reversible* process, the first law is

$$dU = \dbar Q - PdV,$$

or, in terms of intensive variables,

$$du = \dbar q - Pdv. \tag{4.4}$$

Since u is a function of the state variables, we can write, say,

$$u = u(v, T). \tag{4.5}$$

Here we assume that the equation of state is

$$Pv = RT. \tag{4.6}$$

From Equation (4.5),

$$du = \left(\frac{\partial u}{\partial v}\right)_T dv + \left(\frac{\partial u}{\partial T}\right)_v dT. \tag{4.7}$$

Combining Equations (4.4) and (4.7), we have

$$đq = \left(\frac{\partial u}{\partial T}\right)_v dT + \left[\left(\frac{\partial u}{\partial v}\right)_T + P\right]dv. \tag{4.8}$$

To obtain c_v, we divide this equation by dT and hold the volume constant so that $dv = 0$. The result, which holds for any reversible process, is

$$c_v \equiv \left(\frac{đq}{dT}\right)_v = \left(\frac{\partial u}{\partial T}\right)_v. \tag{4.9}$$

We shall see that for an ideal gas, $u = u(T)$ only. This follows from the Gay-Lussac–Joule experiment (see Chapter 5). Thus

$$\left(\frac{\partial u}{\partial v}\right)_T = 0 \tag{4.10}$$

for an ideal gas. The substitution of Equations (4.9) and (4.10) in Equation (4.8) gives

$$đq = c_v dT + Pdv. \tag{4.11}$$

Differentiating Equation (4.6), we have

$$Pdv + vdP = RdT. \tag{4.12}$$

Using this in Equation (4.11) leads to

$$đq = (c_v + R)dT - vdP. \tag{4.13}$$

Now

$$c_P \equiv \left(\frac{đq}{dT}\right)_P,$$

by definition. We let the pressure P be constant in Equation (4.13) so that $dP = 0$. Then

$$\left(\frac{dq}{dT}\right)_P = c_v + R = c_P,$$

or

$$c_P = c_v + R. \tag{4.14}$$

This relation is known as Mayer's equation.* It states that over the range of variables for which the ideal gas law holds, the two specific heat capacities differ by the constant R.

We introduce the *ratio* of specific heat capacities

$$\gamma \equiv \frac{c_P}{c_v}. \tag{4.15}$$

For a monatomic gas at room temperature, $c_P = \frac{5}{2}R$, $c_v = \frac{3}{2}R$, and $\gamma = 1.67$. For a diatomic gas at room temperature, $c_P = \frac{7}{2}R$, $c_v = \frac{5}{2}R$, and $\gamma = 1.40$. These values will be derived in Chapters 14 and 15. The difference $c_P - c_v$ is equal to $R = 8.31 \times 10^3$ J kilomole^{-1} K^{-1} for a wide range of temperatures. For solids the difference is about 5×10^3 J kilomole K^{-1} at 1200 K, by way of comparison.

Finally, from the foregoing analysis we can calculate the internal energy of an ideal gas. From Equation (4.9) we can write

$$c_v = \left(\frac{\partial u}{\partial T}\right)_v = \frac{du}{dT},$$

since u depends solely on T in the case of an ideal gas. Then

$$du = c_v dT$$

or

$$u - u_0 = \int_{T_0}^{T} c_v dT.$$

If c_v is independent of the temperature,

$$u = u_0 + c_v(T - T_0). \tag{4.16}$$

---

*Julius Robert von Mayer, a German physician, discovered the equivalence of heat and work and first enunciated the principle of conservation of energy (1842).

The specific heat capacity at constant volume is nearly constant over a region of temperature extending from approximately 200 K to 1200 K for most gases. Thus Equation (4.16) is applicable for many practical problems.

4.3 ENTHALPY AND HEATS OF TRANSFORMATION

The heat of transformation is the heat transfer accompanying a phase change. A change of phase is an isothermal and isobaric process and entails a change of volume, so work is always done on or by a system in a phase change. Thus

$$w = P(v_2 - v_1)$$

is the specific work involved in the process. Now

$$du = \bar{d}q - Pdv,$$

or, for a finite change,

$$(u_2 - u_1) = l - P(v_2 - v_1),$$

so that

$$l = (u_2 + Pv_2) - (u_1 + Pv_1). \tag{4.17}$$

Here l is the latent heat of transformation per kilomole associated with a given phase change. Let

$$h \equiv u + Pv.$$

The quantity h is called the specific *enthalpy*. The origin of the word is the Greek verb *thalpein* meaning "to heat." Since u, P, and v are all state variables, h is also a state variable. (If two variables have exact differentials, their product and sum also have exact differentials.)

With this definition, Equation (4.17) becomes

$$l = h_2 - h_1,$$

which is to say that the latent heat of transformation is equal to the difference in enthalpies of the two phases.

Consider now the three possible phase changes. Let 1 denote a solid, 2 a liquid, and 3 a vapor. Then

$l_{12} = h'' - h' \to$ solid to liquid (fusion),

$l_{23} = h''' - h'' \to$ liquid to vapor (vaporization),

$l_{13} = h''' - h' \to$ solid to vapor (sublimation).

Here the *first* superscript is associated with the *second* subscript; that is, double prime is associated with the subscript 2 denoting a liquid, etc. This is merely a notational convention.

Since enthalpy is a state function,

$$\oint dh = 0;$$

the change of enthalpy in a cyclical process is zero. This can be depicted in a P-T diagram as in Figure 4.2.

Consider a cyclical process around the triple point and close enough to it so that only changes in the enthalpy occur during phase transitions. Then

solid $\to$ vapor (heat flows in), $\Delta h_1 = l_{13}$

vapor $\to$ liquid (heat flows out), $\Delta h_2 = l_{32} = -l_{23}$

liquid $\to$ solid (heat flows out). $\Delta h_3 = l_{21} = -l_{12}$

Here

$$\oint dh = 0 = \Delta h_1 + \Delta h_2 + \Delta h_3,$$

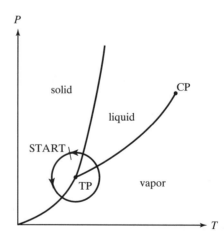

Figure 4.2 Pressure-temperature plot for a substance that contracts on freezing. The circle denotes a cyclical process around the triple point for which the total change of enthalpy is zero.

TABLE 4.1   Latent heats of vaporization and fusion for various substances.

Latent Heat of Vaporization at Steam Point, 1 atm		Latent Heat of Fusion at Melting Point, 1 atm	
Water	538 kcal/kg	Water	80 kcal/kg
Mercury	63	Mercury	3
Alcohol	207	Lead	5
Gasoline	95	Aluminum	77

so

$$l_{13} - l_{23} - l_{12} = 0$$

or

$$l_{13} = l_{23} + l_{12}. \tag{4.18}$$

(All these latent heats are intrinsically positive.) This is an important result. Some values are given in Table 4.1. The latent heat of sublimation for water is $538 + 80 = 618$ kcal kg^{-1}.

The SI unit for latent heat is J kg^{-1}. Thus, for example,

$$538 \frac{\text{kcal}}{\text{kg}} \times 4184 \frac{\text{J}}{\text{kcal}} = 2.25 \times 10^6 \frac{\text{J}}{\text{kg}}.$$

4.4 RELATIONSHIPS INVOLVING ENTHALPY

In Section 4.2 we assumed that $u = u(v, T)$. Whereas the selection of v, T from among the three fundamental state variables P, v, and T was arbitrary, we will find that this choice is a natural one both from an experimental and mathematical viewpoint. Similarly, the natural choice in the case of the thermodynamic variable h is

$$h = h(T, P). \tag{4.19}$$

Then the analysis proceeds in a manner exactly parallel to what was done in the case of the internal energy. Here

$$đq = du + P dv, \tag{4.20}$$

$$h = u + Pv, \tag{4.21}$$

and

$$dh = \left(\frac{\partial h}{\partial T}\right)_P dT + \left(\frac{\partial h}{\partial P}\right)_T dP. \tag{4.22}$$

Also, from Equation (4.21), we have

$$dh = du + Pdv + vdP,$$

or

$$Pdv = dh - du - vdP. \tag{4.23}$$

Substituting this result in Equation (4.20), we obtain

$$đq = dh - vdP. \tag{4.24}$$

Next, substitution of Equation (4.22) in Equation (4.24) yields

$$đq = \left(\frac{\partial h}{\partial T}\right)_P dT + \left[\left(\frac{\partial h}{\partial P}\right)_T - v\right]dP. \tag{4.25}$$

Since

$$c_P \equiv \left(\frac{đq}{\partial T}\right)_P,$$

we obtain, after setting $dP = 0$ in Equation (4.25), the result

$$c_P = \left(\frac{\partial h}{\partial T}\right)_P. \tag{4.26}$$

From the result of the Joule–Thomson experiment (see Chapter 5), it will be shown that

$$\left(\frac{\partial h}{\partial P}\right)_T = 0 \tag{4.27}$$

for an ideal gas. Then

$$đq = c_P dT - vdP. \tag{4.28}$$

Thus, for an ideal gas,

$$c_P = \left(\frac{\partial h}{\partial T}\right)_P = \frac{dh}{dT},$$

since, from Equation (4.27), h depends on T only. Hence we can write

$$h - h_0 = \int_{T_0}^{T} c_P dT = c_P(T - T_0)$$

if c_P is a constant. For an ideal gas, then, the specific enthalpy is given by

$$h = h_0 + c_P(T - T_0). \tag{4.29}$$

4.5 COMPARISON OF u AND h

Table 4.2 summarizes the parallel expressions involving the internal energy and the enthalpy.

TABLE 4.2   Analogous relations involving the internal energy and the enthalpy.

	Internal energy u	Enthalpy h
Reversible process:	$du = đq - Pdv$	$dh = đq + vdP$
	$c_v = \left(\dfrac{\partial u}{\partial T}\right)_v$	$c_P = \left(\dfrac{\partial h}{\partial T}\right)_P$
Ideal gas:	$đq = c_v dT + Pdv$	$đq = c_P dT - vdP$
	$\left(\dfrac{\partial u}{\partial v}\right)_T = 0$	$\left(\dfrac{\partial h}{\partial P}\right)_T = 0$
	$u - u_0 = \displaystyle\int_{T_0}^{T} c_v dT$	$h - h_0 = \displaystyle\int_{T_0}^{T} c_P dT$

4.6 WORK DONE IN AN ADIABATIC PROCESS

We have seen that the specific work w done in an isobaric process is $P(v_2 - v_1)$, where v_1 is the initial specific volume and v_2 is the final volume. Similarly, for an isothermal process in which the substance is an ideal gas,

$$w = RT \ln\left(\frac{v_2}{v_1}\right) = P_1 v_1 \ln\left(\frac{v_2}{v_1}\right). \tag{4.30}$$

We now wish to find the specific work done in an adiabatic process involving an ideal gas. Setting $đq = 0$ in Equation (4.28), we obtain

$$vdP = c_P dT. \tag{4.31}$$

We also have

$$dq = c_v dT + P dv,$$

which for $dq = 0$ yields

$$P dv = -c_v dT. \tag{4.32}$$

Dividing Equation (4.31) by Equation (4.32) gives

$$\frac{v dP}{P dv} = -\frac{c_P}{c_v} = -\gamma,$$

or

$$\frac{dP}{P} = -\gamma \frac{dv}{v}. \tag{4.33}$$

This equation can be easily integrated to give

$$P v^\gamma = K, \tag{4.34}$$

where K is a constant of integration. This is the relationship between the pressure and volume for an adiabatic process involving an ideal gas. Since $\gamma > 1$, it follows that P falls off more rapidly with v for an adiabatic process than it does for an isothermal process (for which Pv = constant).

The work done in the adiabatic process is

$$w = \int P dv = K \int_{v_1}^{v_2} v^{-\gamma} dv = \frac{1}{1-\gamma} (K v^{1-\gamma}) \Big|_{v_1}^{v_2}.$$

Now, $K = P v^\gamma$ at both limits; if we use $K = P_2 v_2^\gamma$ at the upper limit and $K = P_1 v_1^\gamma$ at the lower limit, we obtain

$$w = \frac{1}{1-\gamma} [P_2 v_2 - P_1 v_1]. \tag{4.35}$$

For an expansion, $v_2 > v_1$, $w > 0$, and the work is done *by* the gas; for a compression the work is done by the surroundings *on* the gas. Note that for a reversible adiabatic process, $w = u_1 - u_2 = c_v(T_1 - T_2)$, which is another useful expression for an ideal gas.

As an example of an adiabatic expansion, consider an ideal mona-tomic gas enclosed in an insulated chamber with a movable piston. The ini-tial values of the state variables are $P_1 = 8$ atm, $V_1 = 4$ m^3 and $T_1 = 400$ K. The final value of the pressure after the expansion is $P_2 = 1$ atm. We wish to find V_2, T_2, W, and ΔU. For an ideal monatomic gas, the ratio of specific heats γ is 5/3.

For an adiabatic process,

$$P_1 V_1{}^\gamma = P_2 V_2{}^\gamma, \quad \text{or } V_2 = V_1 \left(\frac{P_1}{P_2}\right)^{1/\gamma} = 4\left(\frac{8}{1}\right)^{3/5} = 13.9 \text{ m}^3.$$

Then T_2 is found from the ideal gas law:

$$T_2 = \left(\frac{P_2}{P_1}\right)\left(\frac{V_2}{V_1}\right)T_1 = \left(\frac{1}{8}\right)\left(\frac{13.9}{4}\right)(400) = 174 \text{ K}.$$

As the gas expands, the temperature and the pressure drop (no heat flows into or out of the system).

The work done by the system in the expansion is

$$W = \frac{1}{1 - \gamma}(P_2 V_2 - P_1 V_1)$$

$$= \frac{1}{1 - 1.67}[(1.013)(13.9) - (8)(1.013)(4)] \times 10^5$$

$$= 2.74 \times 10^6 \text{ J}.$$

Finally, since $Q = 0$, $\Delta U = -W = -2.74 \times 10^6$ J. The internal energy decreases. This result could also be found by using the expression $\Delta U = C_V(T_2 - T_1)$:

$$\Delta U = n\left(\frac{3R}{2}\right)(T_2 - T_1) = \frac{P_1 V_1}{RT_1}\left(\frac{3R}{2}\right)(T_2 - T_1)$$

$$= \frac{(8 \times 1.01 \times 10^5 \times 4)}{400}\left(\frac{3}{2}\right)(174 - 400)$$

$$= -2.74 \times 10^6 \text{ J}.$$

Work is done by the system at the expense of internal energy.

PROBLEMS

4-1 The specific heat capacity c_v of solids at low temperature is given by the Debye T^3 law:

$$c_v = A\left(\frac{T}{\theta}\right)^3.$$

The quantity A is a constant equal to 19.4×10^5 J kilomole^{-1} K^{-1} and θ is the Debye temperature, equal to 320 K for NaCl.

(a) What is the molar specific heat capacity at constant volume of NaCl at 10 K and at 50 K?

(b) How much heat is required to raise the temperature of 2 kilomoles of NaCl from 10 K to 50 K at constant volume?

(c) What is the mean specific heat capacity at constant volume over this temperature range?

4-2 The equation of state of a certain gas is $(P + b)v = RT$ and its specific internal energy is given by $u = aT + bv + u_0$.

(a) Find c_v.

(b) Show that $c_P - c_v = R$.

4-3 Show that

$$\left(\frac{\partial u}{\partial T}\right)_P = c_P - P\beta v.$$

(Refer to Equation (2.13).)

4-4 Show that

$$\left(\frac{\partial u}{\partial P}\right)_T = P\kappa v - (c_P - c_v)\frac{\kappa}{\beta}.$$

(Refer to Equations (2.13) and (2.14).)

4-5 When a material is heated, it experiences a very slight increase in mass since $E = mc^2$. Estimate the fractional change in mass when the temperature of a block of copper is raised from 300 K to 400 K. Take c_P for copper to be 2.6×10^4 J kilomole^{-1} K^{-1}. Its atomic weight is 29.

4-6 Consider oxygen as an ideal gas with $c_v = (5/2)R$. Suppose that the temperature of 2 kilomoles of O_2 is raised from 27°C to 227°C.

(a) What is the increase in internal energy?

(b) What is the increase in enthalpy?

4-7 The specific heat capacity at constant pressure of a gas varies with the temperature according to the expression

$$c_P = a + bT - \frac{c}{T^2},$$

where a, b, and c are constants. How much heat is transferred in an isobaric process in which a kilomole of gas experiences a temperature increase from T to $2T$?

4-8 Show that the following relations hold for a reversible adiabatic expansion of an ideal gas:
(a) $TV^{\gamma-1} = $ constant.
(b) $\dfrac{T}{P^{1-1/\gamma}} = $ constant.

4-9 An ideal gas undergoes an adiabatic reversible expansion from an initial state (T_1, v_1) to a final state (T_2, v_2).
(a) Show that

$$\ln\left(\frac{T_2}{T_1}\right) = (\gamma - 1)\ln\left(\frac{v_1}{v_2}\right),$$

where γ is the ratio of specific heats.
(b) If $T_2/T_1 = 2/5$ and $v_2/v_1 = 2$, show that the final state is not accessible from the initial state via any adiabatic reversible process involving any known ideal gas.

4-10 The equation of state for radiant energy in equilibrium with the temperature of the walls of a cavity of volume V is $P = aT^4/3$, where a is a constant. The energy equation is $U = aT^4V$.
(a) Show that the heat supplied in an isothermal doubling of the volume of the cavity is $(4/3)aT^4V$.
(b) Show that in an adiabatic process, VT^3 is constant.

4-11 An ideal diatomic gas, for which $c_v = 5R/2$, occupies a volume of 2 m³ at a pressure of 4 atm and a temperature of 20°C. The gas is compressed to a final pressure of 8 atm. Compute the final volume, the final temperature, the work done, the heat released, and the change in internal energy for:
(a) A reversible isothermal compression.
(b) A reversible adiabatic compression.

4-12 An ideal monatomic gas having an initial pressure P_0 and an initial volume V_0 undergoes an adiabatic expansion in which the volume is doubled.
(a) Show that the final pressure is $0.314\,P_0$.
(b) Show that the change in enthalpy in the process is $-0.93\,P_0V_0$.

4-13 One kilogram of ice at $-20°C$ is heated to $0°C$ and melted. The resulting water is heated to $100°C$, vaporizes, and the steam is further heated to $400°C$. Assume that all the processes are isobaric. Calculate the change in enthalpy in kilocalories and in joules. (The specific heats at constant pressure of ice, liquid water, and steam are $0.55, 1.00,$ and 0.48 kcal kg^{-1}, respectively.)

4-14 An automobile tire is inflated to a pressure of 270 kPa at the beginning of a trip. After three hours of high-speed driving the pressure is 300 kPa. What is the internal energy change of the air in the tire between pressure measurements? Assume that air is an ideal gas with a constant specific heat capacity $c_v = 5R/2$ and that the internal volume of the tire remains constant at 0.057 m³.

4-15 Shortly after detonation the fireball of a uranium fission bomb consists of a sphere of gas of radius 15 m and temperature 3×10^5 K. Assuming that the expansion is adiabatic and that the fireball remains spherical, estimate the radius of the ball when the temperature is 3000 K. (Take $\gamma = 1.4$.)

Chapter 5

Consequences of the First Law

5.1 THE GAY-LUSSAC–JOULE EXPERIMENT

We have postulated that in general $u = u(T, v)$. The first efforts undertaken to measure the dependence of the internal energy of a gas on its volume were made by Gay-Lussac and later by Joule in the middle of the nineteenth century. The results of the experiments showed that the internal energy is a function of T only and *does not* depend on the volume v. This, like many other properties of gases, is only approximately true for real gases and is assumed to hold exactly for ideal gases.

Before describing the experiments, we consider the following arguments. Using the cyclical and reciprocal relations for partial derivatives given in Appendix A, we can write

$$\left(\frac{\partial T}{\partial v}\right)_u = -\frac{\left(\frac{\partial u}{\partial v}\right)_T}{\left(\frac{\partial u}{\partial T}\right)_v}. \tag{5.1}$$

For a reversible process, we have (Equation (4.9))

$$c_v = \left(\frac{\partial u}{\partial T}\right)_v.$$

Thus Equation (5.1) gives

$$\left(\frac{\partial u}{\partial v}\right)_T = -c_v\left(\frac{\partial T}{\partial v}\right)_u. \tag{5.2}$$

Equation (5.2) implies that if we can measure the change in temperature of a gas as the volume increases while the internal energy is held constant, we can

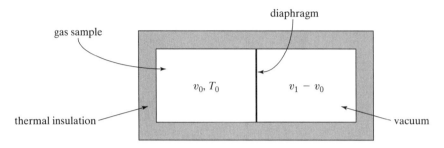

Figure 5.1   Setup for the Joule experiment prior to rupture of the diaphragm.

determine the dependence of the internal energy on the volume. The question is: how can we keep u constant during the expansion? Since

$$du = đq - đw,$$

evidently we want an adiabatic process in which no work is done.

The principle used by Joule is shown schematically in Figure 5.1. A sample of the gas under investigation is enclosed in one portion of a thermally insulated vessel, separated by a diaphragm from an initially evacuated chamber. The temperature of the gas is measured with a thermometer. When the diaphragm is ruptured, the gas undergoes a *free expansion*. Here v_0 is the initial specific volume of the gas, v_1 the final specific volume; T_0 and T_1 are the initial and final temperatures, respectively. In the expansion no work is done and the internal energy is unchanged since no energy is lost or gained.* The final temperature, which can be measured, is

$$T_1 = T_0 + \int_{v_0}^{v_1} \left(\frac{\partial T}{\partial v}\right)_u dv,$$

from basic mathematics. The subscript on the derivative is appropriate since the internal energy is held constant. The integrand is known as η, the Joule coefficient.

Joule was unable to measure any temperature change for air. Subsequent experiments using various gases indicate that the temperature decreases by at most a very small amount. Summarizing these results,

$$|\eta| = \left|\left(\frac{\partial T}{\partial v}\right)_u\right| < 0.001 \text{ K kilomole m}^{-3}.$$

---

* Although no external work is done, Pdv is finite. However, the rupture of the diaphragm results in a rapid expansion during which the gas is so highly nonuniform that we cannot assign a distinct value to the pressure. See Section 7.7 for a further discussion of free expansion.

It can be shown (Problem 5-3) that for a van der Waals gas,

$$\eta = -\frac{a}{v^2 c_v}.$$

Since the constant a is zero for an ideal gas, we conclude that η is exactly zero in that case:

$$\eta = \left(\frac{\partial u}{\partial v}\right)_u = 0 \quad \text{and} \quad u = (T) \tag{5.3}$$

for an ideal gas.

This very important experimental result can be proved theoretically. Consider Equation (4.8), which is valid for any reversible process. If we divide the equation by the temperature T, we obtain

$$\frac{dq}{T} = \frac{1}{T}\left(\frac{\partial u}{\partial T}\right)_v dT + \frac{1}{T}\left[\left(\frac{\partial u}{\partial v}\right)_T + P\right]dv. \tag{5.4}$$

In Appendix A we note that an inexact differential can be made exact if an integrating factor can be found. We shall see that the integrating factor $1/T$ makes the left-hand side of Equation (5.4) an exact differential, a function of the state variables T and v.* If $dz = M\,dx + N\,dy$, then the condition that dz be exact is

$$\frac{\partial M}{\partial y} = \frac{\partial N}{\partial x}.$$

In this case, the condition is

$$\frac{\partial}{\partial v}\left(\frac{1}{T}\frac{\partial u}{\partial T}\right) = \frac{\partial}{\partial T}\left[\frac{1}{T}\left(\frac{\partial u}{\partial v} + P\right)\right],$$

where the subscripts on the partial derivatives have been omitted for convenience. Carrying out the differentiation, we have

$$\frac{1}{T}\frac{\partial^2 u}{\partial v \partial T} = -\frac{1}{T^2}\left(\frac{\partial u}{\partial v} + P\right) + \frac{1}{T}\frac{\partial^2 u}{\partial T \partial v} + \frac{1}{T}\frac{\partial P}{\partial T},$$

or

$$\left(\frac{\partial u}{\partial v}\right)_T = T\left(\frac{\partial P}{\partial T}\right)_v - P. \tag{5.5}$$

* This fact is intimately associated with the introduction of the concept of entropy, as will be shown in Chapter 6.

For an ideal gas,

$$P = \frac{RT}{v},$$

so that

$$\left(\frac{\partial u}{\partial v}\right)_T = T\left[\frac{\partial}{\partial T}\left(\frac{RT}{v}\right)\right]_v - P$$

$$= \frac{RT}{v} - P = 0. \tag{5.6}$$

This proves that u does not depend on v. The proof depends on the validity of the assumption that $đq/T$ is an exact differential and the applicability of the analysis to an irreversible process. These matters will be discussed in Chapter 6.

5.2 THE JOULE–THOMSON EXPERIMENT

The Joule–Thomson experiment, also known as the Joule–Kelvin experiment,* gets the same result as the Joule experiment in a different way. It is important because the temperature drop in Joule's experiment is difficult to measure.

The experiment involves a *throttling process,* performed by forcing gas through a porous plug to a region of lower pressure in an insulated cylinder (Figure 5.2). A continuous throttling process can be performed by a pump that maintains a constant high pressure on one side of a porous wall and a constant low pressure on the other side. The objective is to measure the temperature difference $T_2 - T_1$. Since the process takes place in an insulated cylinder, $đq = 0$ and $u_2 - u_1 + w = 0$. The specific work done in forcing the gas through the plug is

$$w_1 = \int_{v_1}^0 P_1 \, dv = -P_1 v_1,$$

while the work done by the gas in the expansion is

$$w_2 = \int_0^{v_2} P_2 \, dv = P_2 v_2.$$

The total work is therefore

$$w = w_1 + w_2 = P_2 v_2 - P_1 v_1 = u_1 - u_2. \tag{5.7}$$

* William Thomson became Lord Kelvin at the end of the nineteenth century.

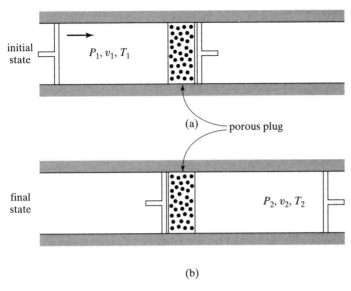

Figure 5.2   Principle of the Joule–Thomson porous plug
experiment. Here (a) corresponds to the initial state and (b) to the
final state, with the piston in the fully advanced position.

It follows that

$$u_1 + P_1 v_1 = u_2 + P_2 v_2,$$

or

$$h_1 = h_2. \tag{5.8}$$

Thus a throttling process occurs at constant enthalpy.

In the Joule–Thomson experiment, the pressure and temperature of the gas before passing through the plug are kept constant and the pressure on the other side is varied while the corresponding temperature is measured. The results give a locus of points on a $T\text{-}P$ diagram representing the isenthalpic state of the gas (Figure 5.3). The numerical value of the slope of the curve at any point is called the Joule–Thomson coefficient μ, where

$$\mu \equiv \left(\frac{\partial T}{\partial P} \right)_h. \tag{5.9}$$

The gas is cooling when μ is positive and heating when μ is negative. The point where $\mu = 0$ is called the inversion point.

The experimental result is that for most gases over a reasonably wide range of temperatures and pressures, the $T\text{-}P$ curve is approximately flat and

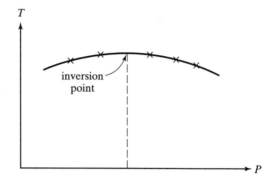

Figure 5.3 A curve of constant enthalpy for the throttling process.

$\mu \approx 0$. It can be shown in a manner completely parallel to the analysis of the Joule experiment that

$$\left(\frac{\partial h}{\partial P}\right)_T = -c_P\left(\frac{\partial T}{\partial P}\right)_h.$$

Thus, for an ideal gas,

$$\left(\frac{\partial h}{\partial P}\right)_T = 0 \quad \text{and} \quad h = h(T). \tag{5.10}$$

The results of the two experiments are equivalent; that is,

$$\left(\frac{\partial u}{\partial v}\right)_T = \left(\frac{\partial h}{\partial P}\right)_T = 0.$$

Again, the result can be proved theoretically.

5.3 HEAT ENGINES AND THE CARNOT CYCLE

The heat engine is a highly useful concept in thermodynamics and reminds one of the subject's origins. A heat engine is a system that receives an input of heat at a high temperature, does mechanical work, and gives off heat at a lower temperature.

Before discussing a practical heat engine, we consider two extreme cases. In Figure 5.4(a), work is done *on* the system and is converted to heat. The heat in turn is used to raise the internal energy of a heat reservoir. In this case, the heat out (of the system M) is equal to the work applied because the state of the system is unaltered. Examples are stirring heat or the heat generated by the flow of current through a resistor.

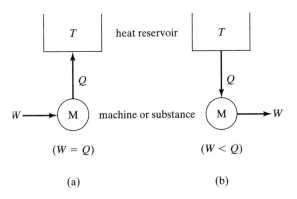

Figure 5.4   The concept of a heat engine. In (a), work
is done on the system and is converted to heat. In (b),
heat is extracted from a reservoir and is converted to
mechanical work. This configuration is not possible.

Case (b) is the reverse of case (a). Heat is extracted from a reservoir and
is converted to mechanical work by the machine. The question is: can the work
done by the system be equal to the heat in? Can the conversion be 100 percent
efficient? The answer is *no*. The second law of thermodynamics states *unequiv-
ocally* that it is impossible to construct a perfect heat engine. (This is sometimes referred to as the Kelvin statement of the second law, which will be
discussed in the next chapter.)

Thus case (b) must be modified as shown in Figure 5.5. Heat is absorbed
from the higher temperature reservoir, work is done, and heat is given up by the
engine to a lower temperature reservoir. The engine works in a cycle, so that the
state of the system is the same at the end of the cycle as at the beginning.

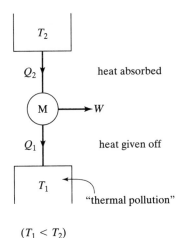

Figure 5.5   Modification of
Figure 5.4b to include a cold
temperature reservoir.

The efficiency of the engine η is equal to the work done by the system (the machine or working substance) divided by the heat absorbed Q_2:*

$$\eta = \frac{W}{Q_2} = \frac{|W|}{|Q_2|} = \frac{\text{output}}{\text{input}}. \tag{5.11}$$

In other words, the efficiency is the ratio of "what you get" to "what you pay." Applying the first law to the system, we have

$$\Delta U = Q_1 + Q_2 - W = |Q_2| - |Q_1| - |W|.$$

In writing this, we are mindful of our sign convention in which the heat flow *into* the system and the work done *by* the system are positive quantities, whereas heat flow *out* of the system and work done *on* the system are inherently negative. We assume that the engine operates in a cycle. Since the state of the system is unaltered in a cyclical process, there is no change in its internal energy (recall that U is a state variable and $\Delta U = 0$ in a cycle). The first law becomes[†]

$$W = Q_1 + Q_2, \quad \text{or} \quad |W| = |Q_2| - |Q_1|. \tag{5.12}$$

Substituting this in Equation (5.11), we obtain

$$\eta = \frac{Q_1 + Q_2}{Q_2} = 1 + \frac{Q_1}{Q_2} = 1 - \frac{|Q_1|}{|Q_2|}. \tag{5.13}$$

We now consider by far the most important case of an engine operating in a cycle—namely, the Carnot engine. We assume a four-step reversible process in which the "working substance" is an ideal gas. The four steps are an isothermal expansion (a to b in a P-V diagram), an adiabatic expansion (b to c), an isothermal compression (c to d), and an adiabatic compression (d to a), returning the gas to its original state (Figures 5.6 and 5.7).

Referring to Table 5.1, we see that the total work done in the cycle is

$$W = W_2 + W' + W_1 + W'' = \oint P \, dV.$$

Note that $\oint P \, dV \neq 0$; the work must be greater than zero if the engine is to be useful. We should remind ourselves that because P and V are state variables, $\oint dP = 0$ and $\oint dV = 0$, but this does not imply that $\oint P \, dV = 0$.

---

* The efficiency η is not to be mistaken for the Joule coefficient η.

[†] Some texts write $W = Q_2 - Q_1$, meaning $W = |Q_2| - |Q_1|$. The omission of the absolute value signs often leads to confusion.

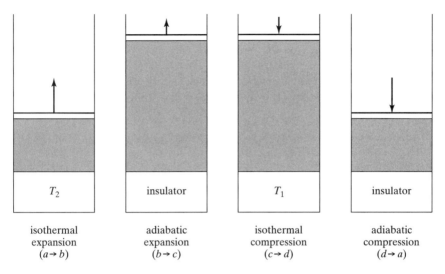

Figure 5.6   The four steps of a Carnot cycle.

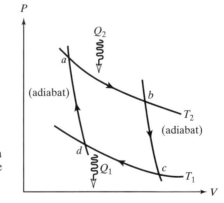

Figure 5.7   *P-V* diagram for a Carnot cycle. The slopes of the isotherms and adiabats are exaggerated.

TABLE 5.1   Heat transfer and work for the four steps of a Carnot cycle.

	Process	Heat	Work
$a \rightarrow b$	Isothermal expansion	$Q_2 > 0$ (in)	$W_2 > 0$ (done by system)
$b \rightarrow c$	Adiabatic expansion	0	$W' > 0$ (done by system)
$c \rightarrow d$	Isothermal compression	$Q_1 < 0$ (out)	$W_1 < 0$ (done on system)
$d \rightarrow a$	Adiabatic compression	0	$W'' < 0$ (done on system)

We can write the first law

$$dU = dQ - dW. \tag{5.14}$$

On the isotherms, for an ideal gas $dU = 0$ since U depends on the temperature alone. Thus $dQ = dW$. Here $Q_2 = W_2$ and $Q_1 = W_1$. Hence

$$Q_2 = W_2 = nRT_2 \ln\left(\frac{V_b}{V_a}\right), \qquad (> 0 \text{ since } V_b > V_a); \tag{5.15}$$

$$Q_1 = W_1 = nRT_1 \ln\left(\frac{V_d}{V_c}\right), \qquad (< 0 \text{ since } V_d < V_c). \tag{5.16}$$

Now b and c are on the same adiabat $PV^\gamma = $ constant. Since $P = nRT/V$, we have

$$\frac{nRT}{V}V^\gamma = nRTV^{\gamma-1},$$

so that $TV^{\gamma-1} = $ constant and

$$T_2 V_b^{\gamma-1} = T_1 V_c^{\gamma-1}. \tag{5.17}$$

Similarly, a and d lie on the same adiabat, so

$$T_2 V_a^{\gamma-1} = T_1 V_d^{\gamma-1}. \tag{5.18}$$

If we divide Equation (5.17) by Equation (5.18) and take the $1/(\gamma - 1)$th root, we obtain

$$\left(\frac{V_b}{V_a}\right) = \left(\frac{V_c}{V_d}\right). \tag{5.19}$$

Combining Equations (5.15), (5.16), and (5.19), we obtain

$$\frac{Q_1}{Q_2} = -\frac{T_1}{T_2}. \tag{5.20}$$

Using this expression in Equation (5.13) gives

$$\eta = 1 - \frac{T_1}{T_2}, \tag{5.21}$$

for an ideal gas. Since $T_1 < T_2$, it follows that $\eta < 1$.

The work performed during the adiabatic steps of the cycle can be calculated using Equation (4.35). However, we note that the use of Equations (5.11), (5.15), and (5.21) yields W, the total work done, which is the quantity of greatest interest. It can be easily shown that this is just the area enveloped by the curves in the P-V diagram delimiting the cyclical process.

It is important to emphasize that a Carnot engine operates between only two reservoirs and that it is reversible. Also, if a working substance other than an ideal gas is used, the shape of the curves in the P-V diagram will, of course, be different. Carnot theorized that the efficiency given by Equation (5.21) is the maximum efficiency for any engine that one might design. We note that the efficiency would be 100 percent if we were able to obtain a low temperature reservoir at absolute zero. However, this is forbidden by the third law.

A Carnot refrigerator is a Carnot engine in reverse (Figure 5.8). Work is done on the engine with the result that heat is removed from a low temperature reservoir (the interior of the refrigerator) and delivered to a high temperature reservoir (the surrounding room). In this case we define a coefficient of performance c as the ratio of the heat Q_1 extracted from the low temperature reservoir to the work done on the system—again, "what you get" divided by "what you pay." Because Q_1 is positive (heat flow *into* the system) and W is negative (work done *on* the system), we introduce a minus sign in order to make the coefficient of performance a positive quantity:

$$c \equiv -\frac{Q_1}{W} = \frac{|Q_1|}{|W|} = \frac{|Q_1|}{|Q_2| - |Q_1|} = \frac{T_1}{T_2 - T_1}. \qquad (5.22)$$

The last step follows from Equation (5.20), which applies to the Carnot refrigerator as well as to the Carnot engine. Since $T_1 < T_2$, the coefficient of performance of a refrigerator, unlike the thermal efficiency of a heat engine, can be made much larger than unity.

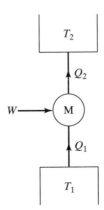

Figure 5.8 The concept of a Carnot refrigerator.

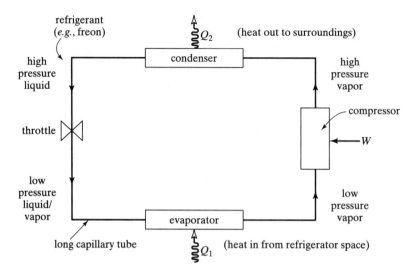

Figure 5.9   Schematic diagram of a typical refrigerator.

A schematic diagram of a typical refrigerator is shown in Figure 5.9. The refrigerant is a substance chosen to be a saturated liquid at the pressure and temperature of the condenser. The liquid undergoes a throttling process in which it is cooled and is partially vaporized. The vaporization is completed in the evaporator; the heat of vaporization is heat absorbed by the refrigerant from the low temperature reservoir (the interior refrigerator space). The low pressure vapor is then adiabatically compressed and isobarically cooled until it becomes a liquid once again.

Refrigerators are designed to extract as much heat as possible from a cold reservoir with small expenditure of work. The coefficient of performance of a household refrigerator is in the range 5 to 10.

PROBLEMS

5-1 The Gay-Lussac–Joule experiments demonstrated that

$$\left(\frac{\partial u}{\partial v}\right)_T = 0$$

for an ideal gas. Show that this result implies that

$$\left(\frac{\partial h}{\partial P}\right)_T = 0.$$

(Hint: Write the differential dh and assume that $h = h(P, T)$ and $u = u(v, T)$.)

5-2 Assume that the specific internal energy of an ideal gas is given by

$$u = u_0 + c_v(T - T_0), \quad u_0 = \text{constant}.$$

(a) Show that the Joule coefficient η is zero.

(b) Show that the Joule–Thomson coefficient μ is zero.

5-3 The specific internal energy of a van der Waals gas is given by

$$u = u_0 + c_v T - \frac{a}{v}, \quad u_0, a \text{ constants}.$$

(a) Find an expression for η. Show that $\eta = 0$ if $a = 0$.

(b) Find an expression for the specific enthalpy h as a function of v and T.

(c) Show that $\mu = \dfrac{\kappa}{c_P} \dfrac{RTv}{(v - b)} - \dfrac{v}{c_P}$.

(d) Calculate the isothermal compressibility κ for the van der Waals gas.

(e) Show that if $a = b = 0$, $\kappa = \dfrac{v}{RT}$, and $\mu = 0$.

5-4 Show that

(a) $\left(\dfrac{\partial h}{\partial T}\right)_v = c_P\left(1 - \dfrac{\beta \mu}{\kappa}\right)$.

(b) $\left(\dfrac{\partial h}{\partial v}\right)_T = \dfrac{\mu c_P}{v\kappa}$.

(c) $\left(\dfrac{\partial T}{\partial v}\right)_h = \dfrac{\mu}{v(\mu\beta - \kappa)}$.

5-5 *Carefully* sketch a Carnot cycle for an ideal gas on

(a) a u-v diagram;

(b) a u-T diagram;

(c) a u-h diagram;

(d) a P-T diagram.

5-6 A Carnot engine is operated between two heat reservoirs at temperatures of 400 K to 300 K.

(a) If the engine receives 1200 kilocalories from the reservoir at 400 K in each cycle, how much heat does it reject to the reservoir at 300 K?

(b) If the engine is operated as a refrigerator (i.e., in reverse) and receives 1200 kilocalories from the reservoir at 300 K, how much heat does it deliver to the reservoir at 400 K?

(c) How much work is done by the engine in each case?

(d) What is the efficiency of the engine in (a) and the coefficient of performance in (b)?

5-7 Fifty kg of water at 0°C must be frozen into ice in a refrigerator. The room temperature is 20°C. The latent heat of fusion of water is 3.33×10^5 J kg^{-1}. What is the minimum power required if the freezing is to take place in one hour?

5-8 Which gives the greater increase in the efficiency of a Carnot engine: increasing the temperature of the hot reservoir or lowering the temperature of the cold reservoir by the same amount?

5-9 An ideal monatomic gas is the working substance of a Carnot engine. During the isothermal expansion the volume doubles. The ratio of the final volume to the initial volume in the adiabatic expansion is 5.7. The work output of the engine is 9×10^5 J in each cycle. Compute the temperature of the reservoirs between which the engine operates. (Assume $n = 1$ kilomole).

5-10 A heat engine having two kilomoles of an ideal monatomic gas as the working substance undergoes a reversible cycle of operations made up of the four steps shown in Figure 5.10. Here $(T_1, V_1) = (300 \text{ K}, 2 \text{ m}^3)$, $(T_2, V_2) = (400 \text{ K}, 2 \text{ m}^3)$, $(T_3, V_3) = (400 \text{ K}, 10 \text{ m}^3)$, $(T_4, V_4) = (300 \text{ K}, 10 \text{ m}^3)$. Find:
(a) the heat flow in each step;
(b) the work done by the gas in the cycle;
(c) the thermal efficiency of the engine.

5-11 During some integral number of cycles, a reversible engine works between *three* heat reservoirs. It absorbs Q_1 joules of heat from a reservoir at T_1, also absorbs Q_2 joules of heat from a reservoir at T_2, delivers Q_3 joules of heat to a reservoir at T_3, and performs W joules of mechanical work. If $T_1 = 400$ K, $T_2 = 200$ K, $T_3 = 300$ K, $Q_1 = 1200$ J, and $W = 200$ J, find Q_2 and Q_3.

5-12 A student claims to have constructed a prototype cyclical engine that operates between two reservoirs at temperatures of 500 K and 350 K, respectively. In each cycle the engine extracts 6000 J of heat from the high temperature reservoir, rejects 3900 J of heat to the low temperature reservoir and performs 2100 J of work on the surroundings. Would you be prepared to give financial support to the development of the engine?

5-13 A heat pump is a device that can be used in winter to warm a house by refrigerating the ground, the outdoor air, or the water in the mains. A Carnot engine, operating in reverse as a heat pump, extracts heat Q_1 from a cold reservoir at temperature T_1 and delivers heat Q_2 to a reservoir at a higher temperature T_2 in each cycle.
(a) Defining the coefficient of performance c as the ratio of "what you get" to "what you pay," derive an expression for c for a Carnot heat pump.
(b) On a day when the outside temperature is 32° F, the interior of a house is maintained at a temperature of 68° F by means of a Carnot heat pump. Calculate its coefficient of performance.

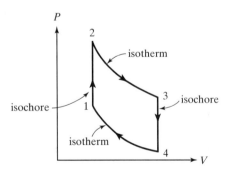

Figure 5.10 *P-V* diagram for a four-step cycle.

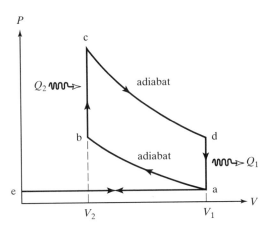

Figure 5.11 *P-V* diagram for
the Otto cycle.

5-14 The behavior of a four-stroke gasoline engine can be approximated by the so-called Otto cycle, shown in Figure 5.11. The processes are as follows:

$e \rightarrow a$	isobaric intake (at atmospheric pressure)
$a \rightarrow b$	adiabatic compression (compression stroke)
$b \rightarrow c$	isochoric increase of temperature during ignition (Gas combustion is an irreversible process, here replaced by a reversible isochoric process in which heat is assumed to flow in from a reservoir.)
$c \rightarrow d$	adiabatic expansion (power stroke)
$d \rightarrow a$	isochoric decrease of temperature (exhaust valve opened)
$a \rightarrow e$	isobaric exhaust (at atmospheric pressure)

(a) Assume that the working substance is an ideal gas and show that the efficiency is given by

$$\eta = 1 - \frac{T_d - T_a}{T_c - T_b} = 1 - \frac{1}{r^{\gamma-1}},$$

where $r = v_a/v_b$ is the compression ratio of the engine.
(b) Calculate η for the realistic values $r = 5$ and $\gamma = 1.5$.

Note: Since $\gamma \geq 1$ we want r as large as possible. The maximum practical obtainable value is approximately 7. For greater values, the rise of temperature upon compression is large enough to cause an explosion *before* the advent of the spark. This is called *preignition*. For diesel engines preignition is not a problem and higher compression ratios are possible. This is partly the reason that diesel engines are inherently more efficient than gasoline engines.

5-15 Consider the Joule cycle, consisting of two isobars and two adiabats (Figure 5.12). Assume that the working substance is an ideal gas with constant specific heat capacities c_P and c_v. Show that the efficiency is

$$\eta = 1 - \left(\frac{P_1}{P_2}\right)^{(\gamma-1)/\gamma}.$$

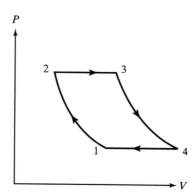

Figure 5.12 *P-V* diagram for the Joule cycle.

Chapter 6

The Second Law of Thermodynamics

6.1 INTRODUCTION

The first law of thermodynamics is remarkable in its generality, simplicity, and utility. Yet our statement of it is in some respects less than satisfactory. Furthermore, in our discussion of heat engines in the previous chapter, we hinted that the first law does not constitute a complete theory because certain processes that it permits do not occur in nature.

The problems are as follows. First, classical thermodynamics is concerned with states of equilibrium and various processes connecting them. We have seen that there is a very substantial advantage in dealing with state variables, changes that are expressible as exact differentials. Then the exact process by which a final state is reached from a given initial state is immaterial; the transition is independent of the particular path taken. However, two of the three quantities appearing in our statement of the first law—the work performed and the heat exchanged—are inexact differentials. The question arises: Is there any way in which we can write the first law in terms of state variables only?

A second concern is that our theory emphasizes *reversible* processes— that is, quasi-static processes in which no dissipative forces such as friction are present. But reversible processes are idealizations whereas real processes are irreversible, and the first law draws no distinction between them. This leads us to ask: Is there some state variable by which we can distinguish between a reversible and an irreversible process?

Finally, in Chapter 5 we noted that certain processes are impossible that are by no means prohibited by the first law. It is not possible to construct a machine that converts all the heat it extracts from a reservoir into useful mechanical work. It is not possible to transfer heat from a cold body to a hot body without supplying energy in the form of work. The first law says nothing about these prohibitions, these "principles of impotency," as they have been called. Is there, then, something missing from our theory? Do we require a second fundamental law for a complete description of our world as it actually is?

The answer is *yes* to all of these questions. We will begin by introducing the concept of entropy as a useful mathematical variable. Phenomenological observations will give the concept its physical meaning.

6.2 THE MATHEMATICAL CONCEPT OF ENTROPY

The first law in its most general form (for a closed system) is

$$dU = \bar{d}Q - \bar{d}W. \tag{6.1}$$

Neither $\bar{d}Q$ nor $\bar{d}W$ is an exact differential; only their difference is exact. We saw that for a reversible process the work is configuration work alone, which can be expressed as

$$\bar{d}W_r = P\,dV,$$

where the subscript denotes a reversible process. Thus

$$\frac{\bar{d}W_r}{P} = dV. \tag{6.2}$$

Now, the volume V is a state variable and dV is an exact differential. By multiplying $\bar{d}W_r$ by the *integrating factor* $1/P$, we have replaced an inexact differential with an exact one.

Is there an integrating factor for $\bar{d}Q$ or $\bar{d}Q_r$ that will lead to an exact differential? We note that P is an intensive state variable while $\bar{d}W_r$ is an extensive variable. The quantity $\bar{d}W_r/P$ is therefore extensive. Also, two of the three fundamental state variables P, V, and T appear in Equation (6.2). This leaves T as the "unused" fundamental state variable and suggests the relation

$$\frac{\bar{d}Q_r}{T} \equiv dS. \tag{6.3}$$

That is, the reciprocal of the absolute temperature is an integrating factor that permits the replacement of the inexact differential $\bar{d}Q_r$ by the exact differential dS. Equation (6.3) is in fact Clausius's definition of the *entropy S*. The term is derived from the Greek word *en-+trepein* for "turning" or "changing."

When we substitute Equations (6.2) and (6.3) in Equation (6.1) we obtain

$$dU = T\,dS - P\,dV \tag{6.4}$$

for a *reversible process*. Thus we have succeeded in writing the first law entirely in terms of state variables.

There are two remaining concerns: (1) we have not proved that dS is an exact differential; and (2) Equation (6.4) appears to hold for reversible processes only. This would seem to indicate that we have replaced a completely general statement of the first law with one that is restricted to a special case. Both of these problems will be addressed later in this chapter.

6.3 IRREVERSIBLE PROCESSES

We have already encountered irreversible processes in discussing dissipative work. Consider the following processes and the results observed in nature:

1. A battery will discharge through a resistor, releasing energy (Figure 6.1). The reverse will not happen: adding energy to the resistor by heating will not cause the battery to charge itself.

2. Two gases, initially in separate adjoining chambers, will mix uniformly when the partition separating the chambers is removed (Figure 6.2). The gases will not separate spontaneously once mixed.

3. A gas in one portion of a chamber is allowed to undergo a free expansion into an evacuated section of the chamber (Figure 6.3). The gas will not compress itself back into its original volume.

4. Heat flows from a high temperature body to a low temperature reservoir in the absence of other effects (Figure 6.4). The reverse process does not take place. If $T_2 > T_1$ initially, the body comes to temperature T_1.

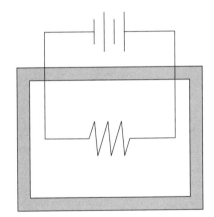

Figure 6.1 Battery discharging current through a resistor in a medium to which heat can be supplied.

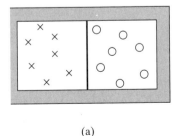

(a)

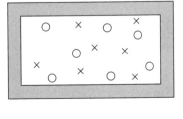

(b)

Figure 6.2 Two different gases, (a) before mixing, and (b) after mixing.

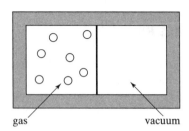

Figure 6.3 Free expansion of a gas into an evacuated chamber.

gas   vacuum

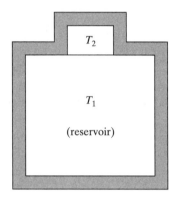

Figure 6.4 Body in contact with a lower temperature reservoir before equilibrium is reached.

T_2

T_1

(reservoir)

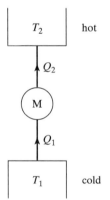

Figure 6.5 Schematic diagram of a device forbidden by the Clausius statement of the second law. If $T_2 > T_1$ then $Q_2 = Q_1$, with $W = 0$, is impossible.

T_2 hot

Q_2

M

Q_1

T_1 cold

In processes such as these, the impossibility of the occurrence of the reverse process was first encountered in two famous statements of the second law:

Clausius statement: It is impossible to construct a device that operates in a cycle and whose sole effect is to transfer heat from a cooler body to a hotter body (Figure 6.5).

Kelvin-Planck statement: It is impossible to construct a device that operates in a cycle and produces no other effect than the performance of work and the exchange of heat with a single reservoir (Figure 6.6).

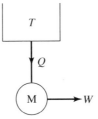

Figure 6.6 Schematic
diagram of a device forbidden
by the Kelvin-Planck
statement of the second law. It
is impossible to have $W = Q$.

6.4 CARNOT'S THEOREM

We have said that Carnot argued that efficient engines must operate as closely
as possible to a Carnot cycle. Using the Clausius statement of the second law,
we can prove Carnot's theorem which states:

> *No engine operating between two reservoirs can be more efficient than a
> Carnot engine operating between those same two reservoirs.*

We consider two engines, M (Carnot) and M′ (hypothetical), and
assume $\eta' > \eta$. That is,

$$\frac{W'}{Q_2'} > \frac{W}{Q_2}. \tag{6.5}$$

The engines are depicted schematically in Figure 6.7.

Since a Carnot engine is reversible, it may be driven backward (as a
refrigerator) by the work from M′. We can arrange the position of the adiabats

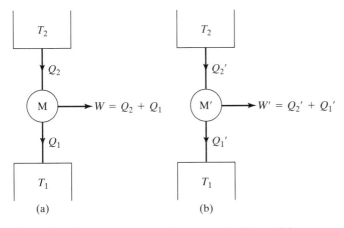

Figure 6.7 Two engines: (a) a Carnot engine M and (b) a
hypothetical engine M′ with efficiency assumed to exceed that
of the Carnot engine.

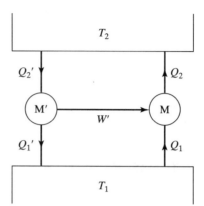

Figure 6.8 The two engines of Figure 6.7 harnessed together, with the hypothetical engine driving the Carnot engine configured as a refrigerator.

so that in one cycle of each engine, M uses exactly as much work as M' produces. Thus $|W| = |W'|$. The "composite" engine is shown in Figure 6.8. Paying careful attention to signs, we note that

$$|W'| = |Q_2'| - |Q_1'|$$

and

$$-|W| = |Q_1| - |Q_2|,$$

or

$$|W| = |Q_2| - |Q_1|.$$

The equality $|W| = |W'|$ therefore gives

$$|Q_2| - |Q_1| = |Q_2'| - |Q_1'|$$

or

$$|Q_2| - |Q_2'| = |Q_1| - |Q_1'|.$$

Taking account of the signs in Equation (6.5) and noting that $|W| = |W'|$, we see that $|Q_2| > |Q_2'|$ and therefore $|Q_1| > |Q_1'|$. We conclude that the composite engine does no work, but extracts heat $|Q_1| - |Q_1'|$ from the cold reservoir and delivers an amount $|Q_2| - |Q_2'|$ to the hot reservoir. This conclusion violates the Clausius statement, so our hypothetical engine cannot exist. It follows that $\eta \geq \eta'$.

We can show that if the Clausius statement is violated, the Kelvin-Planck statement is also violated. In this sense the two statements are equivalent.

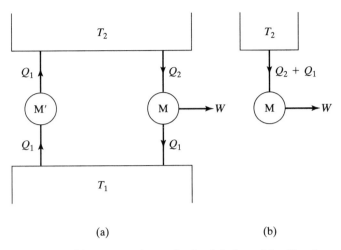

(a)   (b)

Figure 6.9   (a) A composite engine in violation of the Clausius statement; and (b) the equivalent engine, in violation of the Kelvin-Planck statement.

Referring to Figure 6.9, we consider a hypothetical machine M′ that, in violation of the Clausius statement, extracts heat from a cold reservoir and delivers the same amount of heat to the hot reservoir (no work is done on the machine). The Carnot engine M absorbs heat Q_2 from the hot reservoir, does work W, and rejects heat Q_1. The work done is $W = Q_2 + Q_1$. For the *composite* engine, *no* heat is exchanged at the cold reservoir, but heat $Q_2 + Q_1$ is extracted from the hot reservoir and an equal amount of work is done. This violates the Planck-Kelvin statement.

Continuing the proof of Carnot's theorem, we have shown that $\eta \geq \eta'$. Suppose that the hypothetical engine were *reversible.* Then

$$\eta_C \geq \eta_r, \tag{6.6}$$

where the subscript C denotes the Carnot engine and r the reversible hypothetical engine. Since *both* engines are reversible, the Carnot engine could have been used to drive the other engine backward (as a refrigerator), and we would have the result

$$\eta_C \leq \eta_r. \tag{6.7}$$

If both Equations (6.6) and (6.7) are to be satisfied,

$$\eta_C = \eta_r. \tag{6.8}$$

That is to say, *all* reversible engines operating between the same reservoirs have the same efficiency $\eta = 1 - T_1/T_2$. Irreversible engines will have a lesser efficiency. This is Carnot's theorem.

6.5 THE CLAUSIUS INEQUALITY AND THE SECOND LAW

For a Carnot cycle we have seen that (Equation 5.20):

$$\frac{Q_2}{T_2} + \frac{Q_1}{T_1} = 0. \tag{6.9}$$

The quantity Q/T is known as the *Carnot ratio*. From this we deduced the efficiency $\eta = 1 - T_1/T_2$ and proved that the efficiency of all reversible cycles has this value. Applying the Carnot ratio to an infinitely narrow Carnot diagram (finite temperature difference but infinitesimally small quantities of heat extracted and rejected), we obtain

$$\frac{đQ_2}{T_2} + \frac{đQ_1}{T_1} = 0. \tag{6.10}$$

We consider an arbitrary reversible cycle and represent it as a continuous contour in a *P–V* diagram. We replace the process by infinitely narrow Carnot cycles (Figure 6.10).

The fact that the continuous contour is replaced by a sequence of infinitesimal saw-toothed steps is not problematical; the adjacent adiabats cancel each other, leaving only the contributions at the boundary. By an extension of Equation (6.10), we obtain

$$\sum_i \frac{đQ_i}{T_i} \to \oint \frac{đQ_r}{T} = 0, \tag{6.11}$$

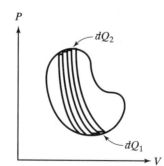

Figure 6.10 *P-V* diagram of a reversible cycle. The process is replaced by infinitesimal Carnot cycles.

where the integration is carried out over the whole contour. The subscript r emphasizes the reversible nature of the cycle. Since the integrand is the differential dS of the entropy, it instantly follows that dS is an *exact* differential and S is a state variable.

Consider next an irreversible cycle. An irreversible Carnot cycle that operates between the same temperatures and produces the same work has a smaller efficiency η' than the efficiency η of a reversible Carnot cycle. That is, $\eta' < \eta$, which implies that

$$\frac{Q_1'}{Q_2'} < \frac{Q_1}{Q_2} = -\frac{T_1}{T_2},$$

or

$$\frac{Q_2'}{T_2} + \frac{Q_1'}{T_1} < 0. \qquad (6.12)$$

For an infinitely narrow Carnot diagram we have

$$\frac{đQ_2'}{T_2} + \frac{đQ_1'}{T_1} < 0$$

instead of Equation (6.10). Following the same reasoning as before, it is clear that for an arbitrary cycle that is partly or wholly irreversible, we must have

$$\oint \frac{đQ'}{T} < 0. \qquad (6.13)$$

Combining Equations (6.11) and (6.13) we obtain the famous *Clausius inequality*

$$\oint \frac{đQ}{T} \leq 0, \qquad (6.14)$$

where the equal sign holds for reversible processes. Equation (6.14) is sometimes taken as a statement of the second law.

Finally, we consider the change in entropy in an irreversible process. Let $1 \rightarrow 2$ be an irreversible change and $2 \rightarrow 1$ be any reversible path connecting the two states in the P-V diagram of Figure 6.11. Then Equation (6.14) gives

$$\oint \frac{đQ}{T} = \underbrace{\int_1^2 \frac{đQ}{T}}_{\text{(irreversible)}} + \int_2^1 \frac{đQ_r}{T} \leq 0.$$

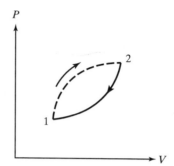

Figure 6.11 An irreversible change (dashed line) and a reversible path connecting states 1 and 2.

Changing the order of integration and the sign of the second term, we have

$$\underbrace{\int_1^2 \frac{đQ}{T}}_{\text{(irreversible)}} \le \int_1^2 \frac{đQ_r}{T} \equiv S_2 - S_1.$$

Thus

$$\Delta S \equiv S_2 - S_1 \ge \int_1^2 \frac{đQ}{T},$$

or, in differential form,

$$dS \ge \frac{đQ}{T}. \tag{6.15}$$

Again, the equality sign holds for a reversible process, and the inequality is appropriate for an irreversible process.

Suppose now that the system is isolated. Then $đQ = 0$ and $dS \ge 0$ or

$$\Delta S \equiv S_2 - S_1 \ge 0 \quad \text{(isolated system)} \tag{6.16}$$

for a finite process. We conclude that

> *The entropy of an isolated system increases in any irreversible process and is unaltered in any reversible process. This is the principle of increasing entropy.*

It is to be noted that this statement refers to *net* entropy changes. It does not say that the entropy of part of the system cannot decrease. If, for example, heat flows from body A to body B at a lower temperature, with both bodies contained in an adiabatic enclosure, ΔS_A is then negative (because of the sign of $đQ_A$), but $\Delta S = \Delta S_A + \Delta S_B$ will still be positive (Figure 6.12).

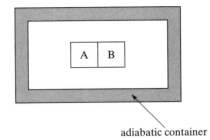

Figure 6.12 Two bodies in thermal contact in an adiabatic enclosure.

adiabatic container

The adiabatic enclosure contains everything that interacts during the process and we can define this assembly as our *universe*. (This is not to be confused with the real universe, which may or may not constitute an isolated system.) If, then, our universe consists of a system and its surroundings, it follows that

$$\Delta S_{\text{universe}} = \Delta S_{\text{system}} + \Delta S_{\text{surroundings}} \geq 0. \tag{6.17}$$

The fact that the entropy of an isolated system can never decrease in a process provides a direction for the sequence of natural events. The laws of mechanics are second-order equations in time t and are unaltered by the replacement of t with $-t$. As far as these equations are concerned, all physical processes can run backward as well as forward. Clearly this is not so; the story of Humpty Dumpty illustrates that processes go only in the direction of increasing entropy. It is for this reason that the law of increasing entropy is often described as providing "the arrow of time" for the evolution of natural processes.

6.6 ENTROPY AND AVAILABLE ENERGY

Another way of expressing the results of the second law is in terms of what is known as *available energy*. It is impossible to utilize all the internal energy of a body for the production of mechanical work, since work can only be obtained by extracting heat from the body and giving it to an engine whose efficiency is less than unity. Suppose that the temperature of a body is T_2 and that a reversible engine works between T_2 and T_1, which is the temperature of a large reservoir to which the engine can give up heat and is also the lowest temperature available (Figure 6.13). Then, if the body gives an amount of heat dQ to the engine, only part of it—namely, $dQ(1 - T_1/T_2)$—can be converted into mechanical work. If the engine is irreversible, still less work can be obtained. The available energy is defined to be $dQ(1 - T_1/T_2)$, while the unavailable energy is $T_1 dQ/T_2$.

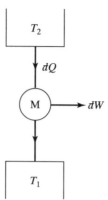

Figure 6.13 Diagram showing the amount of heat available to do mechanical work: $dW = dQ(1 - T_1/T_2)$.

Corresponding to the principle that the entropy always increases in a spontaneous process there is the principle that the available energy always decreases in an irreversible cycle. We could state that

There exists no process that can increase the available energy in the universe.

The property of increasing entropy is equivalent to saying that energy is always being degraded into forms that are more and more difficult to use for the production of work.

The significance of entropy becomes clearer when we consider heat to be the energy of molecular motion. The lack of complete availability of that energy for the production of work is due to the randomness of the molecular motion. It is impossible to reduce the motion of each molecule simultaneously to zero by the action of forces acting on the body as a whole, and so it is impossible to extract all the heat energy from a body. Thus the property of increasing entropy means that the molecular motion of an isolated system always tends to become more random, and the entropy can be thought of as a measure of the "randomness" of the internal motion of a system. This connection between entropy and randomness will be made quantitative by means of statistical thermodynamics.

6.7 ABSOLUTE TEMPERATURE

The Carnot cycle together with Carnot's theorem can serve as the basis for defining an absolute temperature scale. Carnot's theorem shows that the ratio Q_1/Q_2 has the same value for all reversible engines that operate between the same temperatures. The special fact about a reversible Carnot cycle is that the efficiency is independent of the nature of the working substance. It can therefore be used to define an absolute scale of temperature as follows.

Let θ denote an empirical temperature, based on the expansion of an ideal gas. Divide a Carnot cycle into two subcycles, each using the same material (Figures 6.14 and 6.15).

Let the path $a \rightarrow b$ in a θ-V diagram represent an isothermal expansion during which heat Q_2 flows into the system. Similarly, $c \rightarrow d$ is an isothermal compression during which heat Q_1 flows out of the system. For the cycle $abcda$,

$$\frac{Q_1}{Q_2} = f(\theta_2, \theta_1),$$

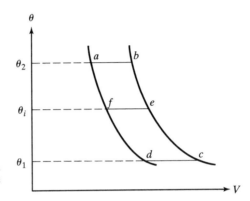

Figure 6.14   Representation in the θ-V plane of a Carnot cycle divided into two sub-cycles.

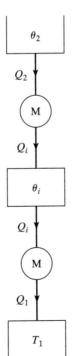

Figure 6.15   The two Carnot engines in series depicted in the θ-V diagram of Figure 6.14.

where f is an unknown function of the two temperatures θ_1 and θ_2. Our challenge is to find f. To do this, consider cycle $abefa$. For this subcycle,

$$\frac{Q_i}{Q_2} = f(\theta_2, \theta_i).$$

Similarly, for the subcycle $fecdf$,

$$\frac{Q_1}{Q_i} = f(\theta_i, \theta_1).$$

Thus

$$\frac{Q_1}{Q_i} \cdot \frac{Q_i}{Q_2} = \frac{Q_1}{Q_2},$$

or

$$f(\theta_i, \theta_1) f(\theta_2, \theta_i) = f(\theta_2, \theta_1).$$

Because the right-hand is a function of θ_1 and θ_2 only, the left-hand side must also depend on these variables alone. This is only possible if

$$f(\theta_i, \theta_1) = \frac{\phi(\theta_1)}{\phi(\theta_i)} \quad \text{and} \quad f(\theta_2, \theta_i) = \frac{\phi(\theta_i)}{\phi(\theta_2)}.$$

Then

$$\frac{Q_1}{Q_2} = f(\theta_2, \theta_1) = \frac{\phi(\theta_1)}{\phi(\theta_2)},$$

independent of the properties of any given substance. Here ϕ is another unknown function. Since the scale of temperature is arbitrary, we can introduce a thermodynamic temperature scale using ϕ itself instead of θ. Kelvin suggested

$$T = A\phi(\theta), \quad A = \text{constant}, \tag{6.18}$$

so that

$$\frac{|Q_1|}{|Q_2|} = \frac{T_1}{T_2}. \tag{6.19}$$

This ratio is the same as the ratio of the temperatures obtained in our analysis of the Carnot cycle, in which the working substance is an ideal gas.

The problem remains to determine the function $\phi(\theta)$. The proof is long, but is worth the effort. Using the definition of the entropy, we can write the first law in the following form (in terms of intensive variables):

$$ds = \frac{1}{T}(du + P\,dv). \tag{6.20}$$

This equation is true in general, as will be discussed in Section 6.8. We take T and v to be the fundamental independent variables and let $u = u(T, v)$. Then

$$du = \left(\frac{\partial u}{\partial T}\right)_v dT + \left(\frac{\partial u}{\partial v}\right)_T dv. \tag{6.21}$$

Substituting this expression in Equation (6.20), we have

$$ds = \frac{1}{T}\left(\frac{\partial u}{\partial T}\right)_v dT + \frac{1}{T}\left[\left(\frac{\partial u}{\partial v}\right)_T + P\right]dv. \tag{6.22}$$

But also, $s = s(T, v)$, so that

$$ds = \left(\frac{\partial s}{\partial T}\right)_v dT + \left(\frac{\partial s}{\partial v}\right)_T dv. \tag{6.23}$$

It follows from Equations (6.22) and (6.23) that

$$\left(\frac{\partial s}{\partial T}\right)_v = \frac{1}{T}\left(\frac{\partial u}{\partial T}\right)_v, \tag{6.24}$$

and

$$\left(\frac{\partial s}{\partial v}\right)_T = \frac{1}{T}\left[\left(\frac{\partial u}{\partial v}\right)_T + P\right]. \tag{6.25}$$

Now

$$\left[\frac{\partial}{\partial v}\left(\frac{\partial s}{\partial T}\right)_v\right]_T = \left[\frac{\partial}{\partial T}\left(\frac{\partial s}{\partial v}\right)_T\right]_v,$$

since the second derivatives are independent of the order of differentiation. We can therefore differentiate Equation (6.24) with respect to v and Equation (6.25) with respect to T and equate the resulting expressions. This leads to the important relation

$$\left(\frac{\partial u}{\partial v}\right)_T = T\left(\frac{\partial P}{\partial T}\right)_v - P. \tag{6.26}$$

Given Equation (6.18), we can write

$$\left(\frac{\partial u}{\partial v}\right)_T = \left(\frac{\partial u}{\partial v}\right)_\theta, \tag{6.27}$$

and, by the chain rule,

$$\left(\frac{\partial P}{\partial T}\right)_v = \left(\frac{\partial P}{\partial \theta}\right)_v \left(\frac{d\theta}{dT}\right).$$

The last derivative is a total derivative, as seen from Equation (6.18). Thus Equation (6.26) becomes

$$\left(\frac{\partial u}{\partial v}\right)_\theta = T\left(\frac{\partial P}{\partial \theta}\right)_v \left(\frac{d\theta}{dT}\right) - P,$$

or

$$\frac{dT}{T} = \frac{\left(\dfrac{\partial P}{\partial \theta}\right)_v d\theta}{\left(\dfrac{\partial u}{\partial v}\right)_\theta + P}. \tag{6.28}$$

For an ideal gas, $Pv/\theta = K$ (a constant), and $u = u(\theta)$. Thus

$$\left(\frac{\partial P}{\partial \theta}\right)_v = \frac{K}{v} = \frac{1}{v}\left(\frac{Pv}{\theta}\right) = \frac{P}{\theta},$$

and

$$\left(\frac{\partial u}{\partial v}\right)_\theta = 0.$$

Substituting these relations in Equation (6.28), we obtain the result

$$\frac{dT}{T} = \frac{d\theta}{\theta},$$

or

$$T = A'\theta, \tag{6.29}$$

where A' is a constant of proportionality. An appropriate choice of A' makes θ *equal* to T; we can then use T to represent both the empirical temperature and the equivalent thermodynamic temperature. The definition of the Kelvin scale is completed by assigning to T_1 in Equation (6.19) the value of 273.16 K, the temperature of the triple point of water. For a Carnot engine operating between temperatures T and T_1, we have

$$T = 273.16 \text{ K} \frac{|Q|}{|Q_1|}. \tag{6.30}$$

The smaller the value of Q, the lower the corresponding value T. The smallest value of Q is zero and the corresponding value of T is zero, called *absolute zero*. In a Carnot cycle, heat is transferred during the isothermal processes. Hence, if a system undergoes a reversible isothermal process without heat transfer, the temperature at which the process takes place is absolute zero. This is a fundamental definition of absolute zero.

6.8 COMBINED FIRST AND SECOND LAWS

We saw early in the chapter that for a reversible process,

$$dU = \bar{d}Q_r - \bar{d}W_r = T\,dS - P\,dV, \tag{6.31}$$

where $\bar{d}Q_r = T\,dS$ and $\bar{d}W_r = P\,dV$. In its most general form, the first law can be written

$$dU = \bar{d}Q - \bar{d}W \quad \text{(general)}. \tag{6.32}$$

The second law states that $T\,dS = \bar{d}Q_r > \bar{d}Q$ for an irreversible process. We can write

$$\bar{d}Q_r = \bar{d}Q + \varepsilon \quad \text{(irreversible)}, \tag{6.33}$$

where ε is a positive quantity. Substituting this in Equation (6.31) gives

$$dU = \bar{d}Q + \varepsilon - \bar{d}W_r \quad \text{(irreversible)}. \tag{6.34}$$

Comparing Equation (6.34) with Equation (6.32) we see that $\bar{d}W < \bar{d}W_r$, or, more specifically,

$$\bar{d}W = \bar{d}W_r - \varepsilon \quad \text{(irreversible)}. \tag{6.35}$$

This is exactly what we expect: the total work done by the system is the reversible configuration work (the useful mechanical work) plus the (negative) dissipative work associated with frictional forces. The latter appears as heat in Equation (6.33). Therefore,

$$dU = đQ - đW = đQ_r - \varepsilon - (đW_r - \varepsilon), \qquad (6.36)$$

or

$$dU = T\,dS - P\,dV \quad \text{(general)}. \qquad (6.37)$$

Evidently Equation (6.37) has the same universality as Equation (6.32). It's just that $đQ$ is identifiable with TdS and $đW$ with PdV only for a *reversible* process. In fact, Equation (6.37) is not restricted to a process at all; it simply expresses a relationship among the state variables of a system and the difference between the values of these variables for two neighboring equilibrium states.

Consider two irreversible processes. For the free expansion of a gas, $đW = 0$ but PdV is finite. Similarly, $đQ = 0$ but TdS has a nonzero value (the entropy increases in the process). For adiabatic stirring, $đQ = 0$ but $TdS \neq 0$ and $dS > 0$. Also, $dV = 0$ but $đW \neq 0$, since stirring work is done.

Equation (6.37) is by far the most important relation in classical thermodynamics.

PROBLEMS

6-1 A Carnot engine operates on 1 kg of methane (CH_4), which we shall consider to be an ideal gas. Take $\gamma = 1.35$. The ratio of the maximum volume to the minimum volume (c and a on the Carnot cycle diagram of Figure 5.7) is 4 and the cycle efficiency is 25 percent. Find the entropy increase of the methane during the isothermal expansion.

6-2 Find the change in entropy of the system during the following processes:
  (a) 1 kg of water is heated reversibly by an electric heating coil from 20°C to 80°C ($c_P = 1$ cal $g^{-1}°C^{-1} = 4.18 \times 10^3$ J kg^{-1} K^{-1}).
  (b) 1 kg of ice at 0°C and 1 atm pressure melts at the same temperature and pressure. (The latent heat of fusion is 3.34×10^5 J kg^{-1}.)
  (c) 1 kg of steam at 100°C and 1 atm pressure condenses to water at the same temperature and pressure. (The latent heat of vaporization is 2.26×10^6 J kg^{-1}.)

6-3 The low temperature specific heat of a diamond varies with temperature according to

$$c_v = 1.88 \times 10^6 \left(\frac{T}{\theta}\right)^3 \text{ J kilomole}^{-1}\,K^{-1},$$

where the Debye temperature $\theta = 2230$ K. What is the entropy change of 1 g of diamond when it is heated at constant volume from 4 K to 300 K? (The atomic weight of carbon is 12.)

6-4 An electric current of 1 A flows for 10 s in a resistor of resistance 25 ohms. The resistor is submerged in a large volume of water, the temperature of which is 280 K. What is the change in the entropy of the resistor? Of the water?

6-5 A thermally insulated resistor of 20 ohms has a current of 2 A passed through it for 1 second. It is initially at 20°C. The mass of the resistor is 5 g; c_P for the resistor is 850 J kg^{-1} K^{-1}.
(a) What is the temperature rise?
(b) What is the entropy change of the resistor and the universe?
(Hint: In the actual process, dissipative work is done on the resistor. Imagine a reversible process taking it between the same equilibrium states.)

6-6 An inventor claims to have developed an engine that takes in 10^8 J at a temperature of 400 K, rejects 4×10^7 J at a temperature of 200 K, and delivers 15 kilowatt hours of mechanical work. Would you advise investing money to put this engine on the market?

6-7 Derive an expression for the entropy of an ideal gas
(a) As a function of T and V.
(b) As a function of T and P.
Assume that the specific heats of the gas are constants.

6-8 An ideal monatomic gas undergoes a reversible expansion from specific volume v_1 to specific volume v_2.
(a) Calculate the change in specific entropy Δs if the expansion is isobaric.
(b) Calculate Δs if the process is isothermal.
(c) Which is larger? By how much?

6-9 A kilomole of an ideal gas undergoes a reversible isothermal expansion from a volume of 5 liters to a volume of 10 liters at a temperature of 20°C.
(a) What is the change in entropy of the gas? Of the universe?
(b) What are the corresponding changes of entropy if the process is a free expansion?

6-10 (a) Show that for reversible changes in temperature at constant volume, $c_v = T(\partial s/\partial T)_v$.
(b) Assume that $c_v = aT + bT^3$ for a metal at low temperatures. Calculate the variation of the specific entropy with temperature.

6.11 Consider a van der Waals gas.
(a) Show that c_v is a function of T only. (Hint: use Equation (6.26)).
(b) Show that the specific internal energy is

$$u = \int c_v dT - a/v + u_o.$$

(c) Show that the specific entropy is

$$s = \int \frac{c_v}{T} dT + R \ln (v - b) + s_o.$$

6-12 When there is a heat flow out of a system during a reversible isothermal process, the entropy of the system decreases. Why doesn't this violate the second law?

6-13 One kilomole of a monatomic ideal gas is carried around the reversible closed cycle shown in Figure 6.16. Here $P_1 = 10$ atm, $V_1 = 2\,\mathrm{m}^3$, and $V_2 = 4\,\mathrm{m}^3$. Calculate the change in entropy for each leg of the cycle and hence show that the entropy change for the complete cycle is zero.

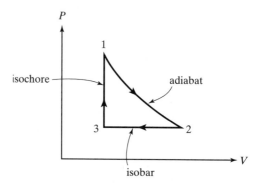

Figure 6.16 *P-V* diagram for Problem 6.13.

Chapter 7

Applications of the Second Law

7.1 ENTROPY CHANGES IN REVERSIBLE PROCESSES

To illustrate the calculation of the entropy change, we first consider reversible processes. In terms of specific quantities, the first law for a reversible process is

$$\bar{d}q_r = du + P dv, \tag{7.1}$$

or

$$\frac{\bar{d}q_r}{T} = \frac{du}{T} + \frac{P}{T} dv = ds. \tag{7.2}$$

We shall examine some special cases.

1. Adiabatic process: $\bar{d}q_r = 0$, $ds = 0$, $s = $ constant. A reversible adiabatic process is an *isentropic* (constant entropy) process. We note in passing that an irreversible adiabatic process is not isentropic.

2. Isothermal process:

$$s_2 - s_1 = \int_1^2 \frac{\bar{d}q_r}{T} = \frac{q_r}{T}. \tag{7.3}$$

3. Isothermal (and isobaric) change of phase:

$$s_2 - s_1 = \frac{l}{T}. \tag{7.4}$$

Here l is the latent heat of transformation.

4. Isochoric process: We assume that $u = u(v, T)$ in general. Since $v = $ constant in an isochoric process, $u = u(T)$, as for an ideal gas, and $du = c_v dT$. Thus

$$s_2 - s_1 = \int_1^2 c_v \frac{dT}{T}.$$

If c_v is constant over the temperature range $T_2 - T_1$, we have

$$s_2 - s_1 = c_v \ln\left(\frac{T_2}{T_1}\right). \tag{7.5}$$

5. Isobaric process: It is convenient to use the specific enthalpy in an isobaric process. Since

$$h = u + Pv,$$

$$dh = du + Pdv + vdP.$$

Substitution in Equation (7.2) yields

$$\frac{dq_r}{T} = \frac{dh}{T} - \frac{v}{T}dP = ds.$$

Assume $h = h(P, T)$ in general. Since $P = $ constant here, $h = h(T)$ only and $dh = c_P dT$. Then

$$s_2 - s_1 = \int_1^2 c_P \frac{dT}{T} = c_P \ln\left(\frac{T_2}{T_1}\right), \tag{7.6}$$

if c_P is constant.

7.2 TEMPERATURE-ENTROPY DIAGRAMS

The total quantity of heat transferred in a reversible process from state 1 to state 2 is given by

$$q_r = \int_1^2 T\,ds.$$

This equals the area under a curve in a T-s diagram. For a reversible isothermal process T is a constant and the curve is a horizontal line. For a reversible adiabatic (and isentropic) process, $ds = 0$ if $T \neq 0$ and the curve is a vertical line.

Consider a T-s diagram for a Carnot cycle (Figure 7.1). Recall that $a \rightarrow b$ is an isothermal expansion, $b \rightarrow c$ an adiabatic expansion, $c \rightarrow d$ an isothermal compression, and $d \rightarrow a$ an adiabatic compression. It follows that the T-s diagram is a simple rectangle for a Carnot cycle. The area under the curve is

$$\oint T\,ds = \oint dq_r = q_{r_2} + q_{r_1}.$$

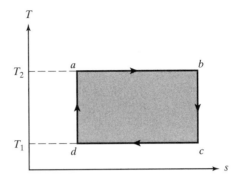

Figure 7.1 *T-s* diagram for a Carnot cycle.

Since $\oint du = 0$, it immediately follows that

$$w = \oint T\,ds = q_{r_2} + q_{r_1}.$$

7.3 ENTROPY CHANGE OF THE SURROUNDINGS FOR A REVERSIBLE PROCESS

In every process in which there is a reversible flow of heat between a system and its surroundings, the temperatures of the system and the surroundings are essentially equal, differing only by dT. The heat flow *out of* the surroundings at every point is equal in magnitude and opposite in sign to the heat flow into the system (Figure 7.2). Thus

$$ds_{\text{system}} + ds_{\text{surroundings}} = ds_{\text{universe}}.$$

But

$$\left(\frac{dq_r}{T + dT}\right)_{\text{surroundings}} \approx -\left(\frac{dq_r}{T}\right)_{\text{surroundings}} = (ds)_{\text{surroundings}}.$$

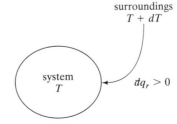

Figure 7.2 Heat flow into a system from its surroundings in a reversible process.

Thus

$$|ds|_{\text{surroundings}} = -|ds|_{\text{system}} \quad \text{and} \quad |ds|_{\text{universe}} = 0.$$

In any reversible process, the entropy change of the universe is always zero. This means that any change in entropy of the system will be accompanied by an entropy change in the surroundings equal in magnitude but opposite in sign. Entropy is conserved in a reversible process. Of course, only idealized processes are reversible; all natural processes are irreversible and entropy is not conserved in general.

7.4 ENTROPY CHANGE FOR AN IDEAL GAS

With $du = c_v dT$ in Equation (7.2), we have

$$\frac{\bar{d}q_r}{T} = \frac{c_v dT}{T} + \frac{P}{T} dv = ds \tag{7.7}$$

for a reversible process. For an ideal gas, $P/T = R/v$, so

$$ds = c_v \frac{dT}{T} + R \frac{dv}{v}.$$

Integrating, we have

$$s_2 - s_1 = c_v \ln\left(\frac{T_2}{T_1}\right) + R \ln\left(\frac{v_2}{v_1}\right), \tag{7.8}$$

if c_v is constant. In any thermal process, only the *change* in entropy is important. In chemistry, however, it is important to be able to assign an absolute value to the entropy. Statistical thermodynamics will make such an assignment possible.

Although Equation (7.8) holds only for an ideal gas, it has characteristics typical of practically all solids, liquids, and gases: (1) if only the temperature is varied, the higher the temperature rise, the greater the increase in entropy; (2) if only the volume is varied, the larger the volume expansion, the greater the entropy increase.

The implications of these results are important. Consider the *isentropic* expansion of a gas. An increase of volume provides a positive contribution to an entropy change. From a molecular point of view, the available energy levels of the system become more closely spaced as the volume increases and the molecules tend to become more randomly distributed among them. The entropy tends to increase. To keep the entropy constant, there must be a compensating negative contribution to the entropy change, which is provided by a decrease in temperature. The drop in temperature that takes place when a gas expands adiabatically (and reversibly) may therefore be regarded as an effect that is needed to offset the volume increase, in order to keep the entropy constant.

7.5 THE *Tds* EQUATIONS

From the combined first and second laws, expressed as

$$T\,ds = du + P\,dv, \tag{7.9}$$

we can obtain some powerful results known as the "*Tds* equations." They involve writing the specific entropy as a function of two independent coordinates, that is, two of the three fundamental state variables P, v, and T. The equations are:

$$T\,ds = c_v dT + T\left(\frac{\partial P}{\partial T}\right)_v dv = c_v dT + \frac{T\beta}{\kappa}dv, \ (s = s(T,v)) \tag{7.10}$$

$$T\,ds = c_P dT - T\left(\frac{\partial v}{\partial T}\right)_P dP = c_P dT - Tv\beta\,dP, \ (s = s(T,P)) \tag{7.11}$$

$$T\,ds = c_P\left(\frac{\partial T}{\partial v}\right)_P dv + c_v\left(\frac{\partial T}{\partial P}\right)_v dP = \frac{c_P}{\beta v}dv + \frac{c_v \kappa}{\beta}dP. \ (s = s(v,P)) \tag{7.12}$$

The *Tds* equations have a variety of uses: (1) they give the heat transferred in a reversible process ($đq_r = T\,ds$); (2) the entropy can be obtained by dividing by T and integrating; (3) the equations express the heat flow or entropy in terms of measurable properties such as c_P, β, κ, T, etc; (4) they can be used to determine the difference in the specific heat capacities c_P and c_v; and (5) the equations can provide relations between pairs of coordinates in a reversible adiabatic process in which $ds = 0$.

The derivation of the *Tds* equations is straightforward. The key is noting that the entropy is a state variable whose partial derivatives satisfy the condition for an exact differential. The first *Tds* equation was essentially derived in Section 6.7 in the steps leading to Equation (6.26), from which the final result can easily be found. Here we shall derive the second *Tds* equation and leave the derivation of the third as a problem.

Let T and P be the independent variables. Then

$$T\,ds = du + P\,dv.$$

The enthalpy is $h \equiv u + Pv$, so that

$$du = dh - Pdv - vdP.$$

Thus

$$T ds = dh - vdP = \left(\frac{\partial h}{\partial T}\right)_P dT + \left(\frac{\partial h}{\partial P}\right)_T dP - vdP,$$

or

$$ds = \frac{1}{T}\left(\frac{\partial h}{\partial T}\right)_P dT + \frac{1}{T}\left[\left(\frac{\partial h}{\partial P}\right)_T - v\right]dP. \tag{7.13}$$

With $s = s(T, P)$, we have

$$ds = \left(\frac{\partial s}{\partial T}\right)_P dT + \left(\frac{\partial s}{\partial P}\right)_T dP. \tag{7.14}$$

Since T and P are independent, it follows that

$$\left(\frac{\partial s}{\partial T}\right)_P = \frac{1}{T}\left(\frac{\partial h}{\partial T}\right)_P \tag{7.15}$$

and

$$\left(\frac{\partial s}{\partial P}\right)_T = \frac{1}{T}\left[\left(\frac{\partial h}{\partial P}\right)_T - v\right]. \tag{7.16}$$

The differential ds is exact. Therefore we can equate the mixed second-order partial derivatives of s:

$$\left[\frac{\partial}{\partial P}\left(\frac{\partial s}{\partial T}\right)_P\right]_T = \frac{\partial^2 s}{\partial P\,\partial T} = \frac{\partial^2 s}{\partial T\,\partial P} = \left[\frac{\partial}{\partial T}\left(\frac{\partial s}{\partial P}\right)_T\right]_P.$$

Substituting Equations (7.15) and (7.16) and carrying out the differentiation, we get

$$\frac{1}{T}\frac{\partial^2 h}{\partial P\,\partial T} = \frac{1}{T}\left[\frac{\partial^2 h}{\partial T\,\partial P} - \left(\frac{\partial v}{\partial T}\right)_P\right] - \frac{1}{T^2}\left[\left(\frac{\partial h}{\partial P}\right)_T - v\right].$$

Two of the terms cancel, and we have

$$\left(\frac{\partial h}{\partial P}\right)_T = -T\left(\frac{\partial v}{\partial T}\right)_P + v, \tag{7.17}$$

analogous, incidentally, to Equation (6.26).

For a reversible process, $(\partial h/\partial T)_P = c_P$. Using this and Equation (7.17) in Equation (7.13), we obtain the result

$$Tds = c_P dT - T\left(\frac{\partial v}{\partial T}\right)_P dP.$$

Finally, since $(\partial v/\partial T)_P = v\beta$, we have

$$Tds = c_P dT - Tv\beta \, dP.$$

As an example, suppose that one kilomole of an ideal gas undergoes a reversible isothermal change in pressure from P_1 to P_2. We wish to find the quantity of heat transferred in the process. For an ideal gas, $\beta = 1/T$. Thus

$$Tds = c_P \underbrace{dT}_{0} - vdP = -\frac{RT}{P} dP.$$

Then

$$q_r = \int Tds = -RT \int_{P_1}^{P_2} \frac{dP}{P} = -RT \ln\left(\frac{P_2}{P_1}\right).$$

Heat is absorbed if $P_2 < P_1$ and evolved if $P_2 > P_1$.

An important application of the *Tds* equations is the determination of $c_P - c_v$, the difference in the specific heat capacities of a given substance. Equating the first and second *Tds* equations, we have

$$c_P \, dT - Tv\beta \, dP = c_v dT + \frac{T\beta}{\kappa} dv. \tag{7.18}$$

Solving for dT, we get

$$dT = \frac{T\beta}{\kappa(c_P - c_v)} dv + \frac{Tv\beta}{(c_P - c_v)} dP = \left(\frac{\partial T}{\partial v}\right)_P dv + \left(\frac{\partial T}{\partial P}\right)_v dP.$$

Thus

$$\left(\frac{\partial T}{\partial v}\right)_P = \frac{T\beta}{\kappa(c_P - c_v)},$$

and

$$\left(\frac{\partial T}{\partial P}\right)_v = \frac{Tv\beta}{(c_P - c_v)}.$$

Solving the first equation for $c_P - c_v$ and using the reciprocal relation gives

$$c_P - c_v = \frac{T\beta}{\kappa}\left(\frac{\partial v}{\partial T}\right)_P,$$

or

$$c_P - c_v = \frac{Tv\beta^2}{\kappa}. \tag{7.19}$$

The equation for $(\partial T/\partial P)_v$ gives the same result. This equation is noteworthy since c_P is measured in experiments, whereas c_v is difficult to measure but can be calculated from theory; the equation affords a comparison of the two. Also, the quantities on the right-hand side of the equation are all positive, showing that c_P is always greater than c_v. It is left to the student to show that $c_P - c_v = R$ for an ideal gas.

Equation (7.19) can be used to calculate the specific heat capacity c_v of a solid. As an example, let us calculate c_v for copper at 1000 K. At atmospheric pressure, $c_P = 29 \times 10^3$ J kilomole^{-1} K^{-1}, $\beta = 6.5 \times 10^{-5}$ K^{-1}, and $\kappa = 9.5 \times 10^{-12}$ Pa^{-1}. The specific volume in m^3 kilomole^{-1} can be found from the atomic weight of copper and its density:

$$v = \frac{V}{n} = \frac{m}{n\rho} = \frac{63.6 \text{ kg kilomole}^{-1}}{8.96 \times 10^3 \text{ kg m}^{-3}} = 7.1 \times 10^{-3} \text{ m}^3 \text{ kilomole}^{-1}.$$

Hence

$$c_v = c_P - \frac{Tv\beta^2}{\kappa} = 29 \times 10^3 - \frac{(1000)(7.1 \times 10^{-3})(6.5 \times 10^{-5})^2}{9.5 \times 10^{-12}}$$

$$= 29 \times 10^3 - 3.2 \times 10^3 \approx 26 \times 10^3 \text{ J kilomole}^{-1} \text{ K}^{-1}.$$

This high-temperature value is very nearly equal to $3R$, a result that will be explored in detail in Chapter 16.

Finally, we consider a reversible adiabatic process $ds = 0$. Equation (7.12) gives

$$\frac{c_P}{\beta v} dv = -\frac{c_v \kappa}{\beta} dP, \tag{7.20}$$

or

$$-\frac{1}{v}\left(\frac{\partial v}{\partial P}\right)_s = \kappa \frac{c_v}{c_P} = \frac{\kappa}{\gamma}.$$

In analogy with the isothermal compressibility κ we introduce the adiabatic compressibility

$$\kappa_s \equiv -\frac{1}{v}\left(\frac{\partial v}{\partial P}\right)_s = \frac{\kappa}{\gamma}. \tag{7.21}$$

Since $\gamma > 1$, $\kappa_s < \kappa$; the adiabatic compressibility is less than the isothermal compressibility because the increase in pressure causes the temperature to rise. The temperature increase, in turn, results in an expansion that partially offsets the compression associated with the pressure increase. It follows that in the adiabatic case the volume change is less and κ_s is smaller.

It is interesting to note that the speed of sound in a gas is given by

$$c = \sqrt{\frac{1}{\rho \kappa_s}} = \sqrt{\frac{\gamma}{\rho \kappa}}, \tag{7.22}$$

where ρ is the density of the gas. The adiabatic compressibility is used since the process of compression and rarefaction takes place very fast; there is no time for heat to be exchanged with the surroundings.

Using Equation (7.22), we can estimate the speed of sound in air. If air is regarded as an ideal diatomic gas, then $\gamma \approx 1.4$, $\kappa \approx 1/P \approx 10^{-5}\,\text{Pa}^{-1}$ for atmospheric pressure, and $\rho \approx 1.2\,\text{kg m}^{-3}$. Thus

$$c = \left[\frac{1.4}{(1.2)(10^{-5})}\right]^{1/2} \approx 340\,\text{ms}^{-1}.$$

This compares favorably with the measured value of $343\,\text{ms}^{-1}$ at 20°C.

7.6 ENTROPY CHANGE IN IRREVERSIBLE PROCESSES

How do we calculate the entropy change for an irreversible process when the entropy is only defined in terms of reversible heat flow? We resort again to an earlier argument. The entropy is a state variable, and the entropy difference is the same between any two equilibrium states regardless of the nature of the process. Thus, we can find Δs for an irreversible process by choosing any convenient reversible path from the initial to the final state and be certain that it is equal to the change produced by the actual, irreversible path.

Consider the example of Figure 7.3. A body at temperature T_1 is in thermal contact with a single reservoir at temperature T_2 with $T_2 > T_1$. The system is allowed to come to thermal equilibrium with the reservoir. The process is *irreversible* since there is a finite temperature difference; the process cannot be reversed by an infinitesimal change. We assume that the process is *isobaric*. Then

$$\dd q_r = c_P \, dT - \underbrace{v dP}_{=0}.$$

Here we are using an equation representing the first law for a *reversible* process. That is, we are effecting a simple heat exchange by bringing up a whole series of reservoirs between T_1 and T_2 (keeping the pressure constant) in such a way that the body passes through a series of equilibrium states. Then

$$ds = \frac{\dd q_r}{T} = c_P \frac{dT}{T},$$

so that

$$(\Delta s)_{\text{body}} = c_P \ln\!\left(\frac{T_2}{T_1}\right),$$

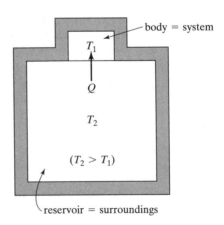

body = system

T_1

Q

T_2

$(T_2 > T_1)$

reservoir = surroundings

Figure 7.3 Body initially at temperature T_1 in thermal contact with a reservoir at temperature T_2.

while

$$(\Delta s)_{\text{reservoir}} = -\frac{|q_r|}{T_2} = -c_P\frac{(T_2 - T_1)}{T_2}.$$

As far as the reservoir is concerned, the process is isothermal (the temperature of the reservoir remains constant) as well as isobaric. The heat flow out of the reservoir is $-c_P(T_2 - T_1)$.

Now

$$(\Delta s)_{\text{universe}} = (\Delta s)_{\text{body}} + (\Delta s)_{\text{reservoir}},$$

or

$$(\Delta s)_{\text{universe}} = c_P\left[\ln\left(\frac{T_2}{T_1}\right) - \frac{T_2 - T_1}{T_2}\right]. \tag{7.23}$$

It is easily seen that the entropy of the universe will always increase, whether $T_2 > T_1$ or $T_2 < T_1$. Let $x = T_2/T_1$. Then the term in brackets can be written

$$\left[\ln\left(\frac{T_2}{T_1}\right) - \frac{T_2 - T_1}{T_2}\right] = \ln x - \left(1 - \frac{1}{x}\right) \equiv f(x).$$

Taking the derivative of $f(x)$ and setting it equal to zero shows that the function has a maximum or a minimum at $x = 1$, where the function is zero. At $x = 1$ the second derivative is positive, so the extreme value is a minimum. Hence $f(x)$ is always positive except when $T_2 = T_1$, in which case there is no heat exchange at all.

In the example, the increase in entropy of the body is larger than the decrease in entropy of the reservoir and the entropy of the universe is greater at the end of the process than at the beginning. In all real changes, the entropy will always increase; in other words, entropy is *created* in the process. The additional entropy is an onus that the actual universe must bear. For an indefinitely expanding universe, there is no known process that can cause the entropy to decrease.*

To illustrate the calculation of entropy changes in irreversible processes, we consider the following examples.

EXAMPLE I

Suppose that 0.5 kg of water at 90°C is cooled to 20°C, the temperature of the surrounding room. The specific heat capacity at constant pressure of water is

---

*However, a universe that expands without limit is different from a finite, closed system. There is no final equilibrium state and energy conservation may not apply.

$4180 \text{ J kg}^{-1} \text{ K}^{-1}$. The change in entropy of the system (the water) is the change that would occur if the water were cooled reversibly from $T_1 = 363\text{K}$ to $T_2 = 293\text{K}$:

$$(\Delta S)_{\text{system}} = \int \frac{dQ_r}{T} = mc_P \int_{T_1}^{T_2} \frac{dT}{T} = mc_P \ln\left(\frac{T_2}{T_1}\right)$$

$$= (0.5)(4180)\ln\left(\frac{293}{363}\right) = -448 \text{ JK}^{-1}.$$

The entropy of the system decreases. The change in entropy of the surroundings is equal to the heat transferred to the room divided by the temperature of the room, which remains constant. The sign is positive since the heat flows *into* the room:

$$(\Delta s)_{\text{surroundings}} = -\frac{mc_P}{T_2} \int_{T_1}^{T_2} dT = mc_P\left(\frac{T_1 - T_2}{T_2}\right)$$

$$= (0.5)(4180)\frac{(363 - 293)}{293} = +499 \text{ J K}^{-1}.$$

Therefore

$$(\Delta s)_{\text{universe}} = (\Delta s)_{\text{system}} + (\Delta s)_{\text{surroundings}} = -448+499 = +51 \text{ J K}^{-1}.$$

The entropy of the universe has increased.   ∎

EXAMPLE 2

Heat $Q = 4 \times 10^5$ J flows through a diathermal wall separating a high temperature heat reservoir at $T_H = 500$ K from a low temperature heat reservoir at $T_L = 200$ K. The pair of reservoirs constituting the system is thermally insulated from its surroundings. Then

$$(\Delta S)_{\text{universe}} = (\Delta S)_{\text{system}} = (\Delta S)_{\text{reservoir } H} + (\Delta S)_{\text{reservoir } L}$$

$$= -\frac{Q_r}{T_H} + \frac{Q_r}{T_L} = -\frac{4 \times 10^5}{500} + \frac{4 \times 10^5}{200} = +1200 \text{ J K}^{-1}.$$

Note that if the two reservoirs are at the same temperature, no heat is exchanged and $(\Delta S)_{\text{universe}} = 0$.   ∎

EXAMPLE 3

A 5 kg mass falls to the ground from a height of 50 m. The temperature is constant at 20°C. The process is irreversible but we can imagine the mass being slowly and reversibly lowered by a string and pulley arrangement. Since no heat exchange is involved, $(\Delta S)_{\text{system}} = 0$, but

$$(\Delta S)_{\text{universe}} = (\Delta S)_{\text{surroundings}} = \frac{W_r}{T} = \frac{mgh}{T}$$

$$= \frac{(5)(9.8)(50)}{293} = +8.36 \text{ J K}^{-1}.$$

We note that for the surroundings, which are unchanged if the mass is small,

$$T(\Delta S) = Q_r = \Delta U + W_r = 0 + mgh. \qquad \blacksquare$$

7.7 FREE EXPANSION OF AN IDEAL GAS

We have seen that in the free expansion of an ideal gas, $du = 0$ and $\bar{d}q = 0$ so that $\bar{d}w = 0$ also. Referring to Figure 7.4, we observe that the equations describing the equilibrium end states are

$$P_0 = \frac{RT_0}{v_0} \quad \text{and} \quad P_1 = \frac{RT_1}{v_1}.$$

A reversible isothermal process would be described by the equation $Pv = \text{constant}$ or $P_0v_0 = P_1v_1$. However, the latter equation does *not* describe the free expansion because P_1 is initially zero *and* the process is irreversible.

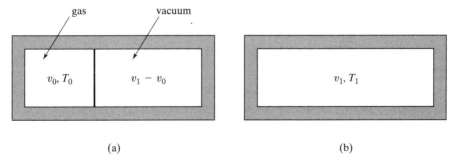

(a)   (b)

Figure 7.4   Free expansion of an ideal gas: (a) initial state; and (b) final state.

Nonetheless, the entropy change can be calculated by assuming that a reversible, isothermal expansion takes place between the initial and final states of the system. Thus,

$$ds = \underbrace{c_v \frac{dT}{T}}_{= 0} + R\frac{dv}{v},$$

so

$$(\Delta s)_{\text{system}} = R \ln\left(\frac{v_1}{v_0}\right), \qquad (7.24)$$

and

$$(\Delta s)_{\text{universe}} = (\Delta s)_{\text{system}} > 0, \qquad (7.25)$$

since the system is isolated (adiabatically insulated). In an irreversible free expansion, the available energy levels become more closely spaced, leading to greater randomness and increased entropy.

We note that in a reversible, isothermal expansion, the work done would be

$$w_r = RT_0 \ln\left(\frac{v_1}{v_0}\right).$$

Since $du = 0$, $w_r = q_r$, so that

$$\Delta s = \frac{q_r}{T_0} = R \ln\left(\frac{v_1}{v_0}\right).$$

In a free expansion no work is done but the change in entropy in the irreversible process is as if work *were* done in a reversible, isothermal process between the same end points.

If the situation involves the mixing of gases ($P_1 \neq 0$), the problem is quite different and will be treated in Section 9.5.

7.8 ENTROPY CHANGE FOR A LIQUID OR SOLID

The equation of state of a liquid or solid is, to a first approximation,

$$v = v_0[1 + \beta(T - T_0) - \kappa(P - P_0)] \qquad (7.26)$$

(see Section 2.5). The second Tds equation is:

$$Tds = c_P dT - T\left(\frac{\partial v}{\partial T}\right)_P dP. \tag{7.27}$$

Setting

$$\left(\frac{\partial v}{\partial T}\right)_P = v_0 \beta,$$

we obtain

$$ds = c_P \frac{dT}{T} - v_0 \beta \, dP.$$

Integrating, we have

$$s - s_0 = c_P \ln\left(\frac{T}{T_0}\right) - v_0 \beta (P - P_0). \tag{7.28}$$

The entropy increases if the temperature increases and decreases if the pressure increases.

PROBLEMS

7-1 The latent heat of fusion of ice at a pressure of 1 atm and 0°C is 3.348×10^5 J kg^{-1}. The density of ice under these conditions is 917 kg m^{-3} and the density of water is 999.8 kg m^{-3}. If 1 kilomole of ice is melted, what will be
(a) the work done?
(b) the change in internal energy?
(c) the change in entropy?

7-2 Ten kg of water at 20°C is converted to ice at $-10°C$ by being put in contact with a reservoir at $-10°C$. The process takes place at constant pressure. The heat capacities at constant pressure of water and ice are 4180 and 2090 kg^{-1} K^{-1}, respectively. The heat of fusion of water is 3.35×10^5 J kg^{-1}. Calculate the change in entropy of the universe.

7-3 Calculate the change in the entropy of the universe as a result of each of the following processes:
(a) A copper block of mass 0.4 kg and heat capacity 150 JK^{-1} at 100°C is placed in a lake at 10°C.

(b) The same block at 10°C is dropped from a height of 100 m into the lake.

(c) Two similar blocks at 100°C and 10°C are joined together. (Hint: See Problem 7-8.)

(d) One kilomole of a gas at 0°C is expanded reversibly and isothermally to twice its initial volume.

(e) One kilomole of a gas at 0°C is expanded reversibly and adiabatically to twice its initial volume.

7-4 Suppose that the specific heat capacity c_P of the body discussed in Section 7.6 of the text is 10 J kg^{-1} K^{-1} and $T_1 = 200$ K. Assume that the mass of the body is 1 kg.

(a) Calculate the change in entropy of the body and of the reservoir if $T_2 = 400$ K.

(b) Make the same calculations for $T_2 = 100$ K.

(c) Find the entropy change of the universe in both cases.

7-5 One kilomole of an ideal gas undergoes a free expansion, tripling its volume. What is the entropy change of

(a) the gas?

(b) the universe?

7-6 An ideal gas has a specific heat given by $c_v = A + BT$, where A and B are constants. Show that the change in entropy per kilomole in going from state (v_1, T_1) to state (v_2, T_2) is

$$\Delta s = A \ln\left(\frac{T_2}{T_1}\right) + B(T_2 - T_1) + R \ln\left(\frac{v_2}{v_1}\right).$$

7-7 A 50 kg bag of sand at 25°C falls 10 m onto the pavement and comes to an abrupt stop. Neglect any transfer of heat between the sand and the surroundings and assume that the thermal capacity of the sand is so large that its temperature is unchanged.

(a) What is the dissipative work done on the sand?

(b) What is the change in the internal energy of the sand?

(c) What is the entropy change associated with this ΔU at constant T?

(The sand does no work as it deforms when it hits the pavement; only its shape changes, not its volume.)

7-8 Two equal quantities of water, each of mass m and at temperatures T_1 and T_2, are adiabatically mixed together, the pressure remaining constant.

(a) Show that the entropy change of the universe is

$$\Delta S = 2mc_P \ln\left(\frac{T_1 + T_2}{2\sqrt{T_1 T_2}}\right),$$

where c_P is the specific heat capacity of the water at constant pressure.

(b) Show that $\Delta S > 0$ for any finite temperatures T_1 and T_2. (Hint: $(a - b)^2 > 0$ for all real a and b.)

7-9 Two identical blocks of copper are held at constant volume with a constant heat capacity $C_V = 380$ J K^{-1}. One is at an initial temperature of 320 K, the other at a temperature of 280 K. The two blocks are thermally isolated and placed in contact with each other. What is the entropy change of the system?

7-10 (a) Assuming that P and v are independent, show that in a reversible process,

$$\left(\frac{\partial s}{\partial P}\right)_v = \frac{c_v}{T}\left(\frac{\partial T}{\partial P}\right)_v$$

and

$$\left(\frac{\partial s}{\partial v}\right)_P = \frac{c_P}{T}\left(\frac{\partial T}{\partial v}\right)_P.$$

(b) Use these results to derive the third Tds equation (Equation (7.12)).

7-11 Use the first Tds equation to calculate the entropy for a van der Waals gas.

7-12 Calculate the difference in the specific heat capacities $c_P - c_v$ of copper at 300 K. Take $\beta = 4.9 \times 10^{-5}\ \text{K}^{-1}$, $\kappa = 7.7 \times 10^{-12}\ \text{Pa}^{-1}$, and $v = 7.1 \times 10^{-3}$ $\text{m}^3\ \text{kilomole}^{-1}$. How does your answer differ from the result of Section 7.5?

7-13 (a) Show that the difference between the isothermal and adiabatic compressibilities is

$$\kappa - \kappa_s = \frac{Tv\beta^2}{c_P}.$$

(b) What is this difference for a monatomic ideal gas?

7-14 Using the fact that dv/v is an exact differential, prove that

$$\left(\frac{\partial \beta}{\partial P}\right)_T = -\left(\frac{\partial \kappa}{\partial T}\right)_P.$$

7-15 (a) Show that the Joule coefficient may be written

$$\eta \equiv \left(\frac{\partial T}{\partial v}\right)_u = \frac{1}{c_v}\left(P - \frac{T\beta}{\kappa}\right).$$

(b) Show that the Joule–Thomson coefficient may be written

$$\mu \equiv \left(\frac{\partial T}{\partial P}\right)_h = \frac{v}{c_P}(T\beta - 1).$$

(c) Using these results, find η and μ for a van der Waals gas and show that both are zero for an ideal gas.

7-16 Calculate the velocity of sound in liquid He^4 using Equation (7.22). Take $\gamma = 1.48$, $\rho = 162\text{kg m}^{-3}$, and $\kappa = 9.43 \times 10^{-8}\ \text{Pa}^{-1}$.

7-17 (a) The temperature of a block of copper is increased from T_0 to T without any appreciable change in its volume. Show that the change in its specific entropy is

$$\Delta s = c_P \ln\left(\frac{T}{T_0}\right) - \frac{v_0\beta^2}{\kappa}(T - T_0).$$

(b) Calculate Δs in units of $\text{J kg}^{-1}\ \text{K}^{-1}$ if the temperature increases from 300 K to 310 K. Take $c_P = 390\ \text{J kg}^{-1}\ \text{K}^{-1}$, $\beta = 4.9 \times 10^{-5}\ \text{K}^{-1}$, and $\kappa = 7.7 \times 10^{-12}$ Pa^{-1}. The density of copper is $9.85 \times 10^3\ \text{kg m}^{-3}$.

Chapter 8

Thermodynamic Potentials

8.1 INTRODUCTION

In the context of the first law, we have defined two functions of the state variables with the dimensions of energy: the internal energy U and the enthalpy H. Since neither of these is well suited to the analysis of certain processes, it will be convenient to introduce two additional functions—the Helmholtz function F and the Gibbs function G. Because of their role in determining the equilibrium states of systems under prescribed constraints, they are known as *thermodynamic potentials,* in analogy with the potential energy in mechanics.

The enthalpy, Helmholtz function, and Gibbs function, are all related to the internal energy and can be derived from it using a procedure known as a Legendre differential transformation. To see how this is done, we consider the combined first and second laws, written as

$$dU = T dS - P dV. \tag{8.1}$$

The two independent variables S and V are intrinsically extensive quantities. The two *intensive* variables, T and $-P$, are said to be *canonically conjugate* to them. That is, the canonically conjugate pairs are

$$T, S \quad \text{and} \quad -P, V.$$

Note that T and S are thermal variables, whereas P and V are by nature mechanical variables.

In Equation (8.1) we assume $U = U(S, V)$ so that

$$dU = \left(\frac{\partial U}{\partial S}\right)_V dS + \left(\frac{\partial U}{\partial V}\right)_S dV. \tag{8.2}$$

Comparing Equations (8.1) and (8.2), we see that

$$\left(\frac{\partial U}{\partial S}\right)_V = T, \quad \left(\frac{\partial U}{\partial V}\right)_S = -P. \tag{8.3}$$

However, the selection of the two independent variables is a matter of choice. There are four possible ways in which a thermal variable can be paired with a mechanical variable:

$$S, V \quad S, P \quad T, V \quad T, P. \tag{8.4}$$

We consider the feasibility of defining a thermodynamic potential for each of the four pairs.

8.2 THE LEGENDRE TRANSFORMATION

Consider the function $Z = Z(x, y)$ and write the differential

$$dZ(x, y) = X \, dx + Y \, dy, \tag{8.5}$$

where x, X and y, Y are, by definition, canonically conjugate pairs. We wish to replace (x, y) by (X, Y) as independent variables. To do this we transform the function Z to a function M. The equation of transformation is

$$M(X, Y) \equiv Z - xX - yY. \tag{8.6}$$

Then

$$dM = dZ - X \, dx - Y \, dy - x \, dX - y \, dY.$$

Since, by Equation (8.5) the first three terms on the right-hand side sum to zero, we have

$$dM = -x \, dX - y \, dY. \tag{8.7}$$

Thus Equation (8.6) is the transformation that takes us from a function of one pair of variables to the other. Equations (8.5) and (8.7) give *reciprocity* relations as follows:

$$\frac{\partial Z}{\partial x} = X, \quad \frac{\partial Z}{\partial y} = Y, \tag{8.8}$$

$$\frac{\partial M}{\partial X} = -x, \quad \frac{\partial M}{\partial Y} = -y. \tag{8.9}$$

If we wish to replace only *one* of the variables, say y, by its canonically conjugate variable Y, we must consider the function

$$N(x, Y) = Z - yY. \tag{8.10}$$

Then

$$dN = dZ - Y \, dy - y \, dY,$$

and, using Equation (8.5), we obtain

$$dN = X \, dx - y \, dY, \qquad (8.11)$$

with reciprocity relations

$$\frac{\partial N}{\partial x} = X, \qquad \frac{\partial N}{\partial Y} = -y.* \qquad (8.12)$$

The equations of transformation, Equations (8.6) and (8.10), may look mysterious at first glance. However, they are based on a very simple idea: a curve in a plane can be equally well represented by pairs of coordinates (point geometry), or as the envelope of a family of tangent lines (line geometry), as depicted in Figure 8.1.

Suppose that the curve is given by the relation $Z = Z(y)$, where y is the independent variable. Consider a tangent line that goes through the point (y, Z) and has a slope $dZ/dy \equiv Y$. If the Z intercept is N, the equation of the line is

$$Y = \frac{Z - N}{y - 0},$$

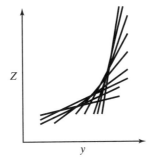

Figure 8.1   A curve represented as the envelope of a family of tangent lines.

---

*It is noteworthy that in classical mechanics, the transformation from the Lagrangian to the Hamiltonian is a Legendre transformation:

$$H(q_k, p_k) = \sum_k p_k \dot{q}_k - L(q_k, \dot{q}_k).$$

Here the $q_k, \dot{q}_k$, and p_k are generalized coordinates, velocities, and momenta, respectively.

or

$$N = Z - yY,$$

which is Equation (8.10). If now we differentiate this expression, we have

$$dN = dZ - y\, dY - Y\, dy.$$

But $dZ = Y\, dy$, so

$$dN = -y\, dY,$$

which is the reciprocal relation

$$y = -\frac{dN}{dY}.$$

Thus the Legendre transformation is a mapping from a $(y,\ Z)$ space to a (Y, N) space, from a point representation of a curve to a tangent line representation. The extension of the transformation to functions of more than one independent variable is straightforward, as shown.

8.3 DEFINITION OF THE THERMODYNAMIC POTENTIALS

1. To transform the internal energy $U(S, V)$ to the enthalpy $H(S, P)$, we replace V by its conjugate variable $-P$. Thus

$$U(S, V) \rightarrow H(S, P).$$

Since

$$dU = T\, dS + (-P)\, dV,$$

we make the following associations with the variables of Equations (8.5) and (8.10):

$$Z = U, \quad X = T, \quad x = S, \quad Y = -P, \quad y = V, \quad N = H.$$

Equations (8.10), (8.11), and (8.12) immediately give

$$H = U + PV,$$

$$dH = T\,dS + V\,dP, \tag{8.13}$$

$$\left(\frac{\partial H}{\partial S}\right)_P = T, \quad \left(\frac{\partial H}{\partial P}\right)_S = V.$$

These are the fundamental relations involving the enthalpy.

2. Next we replace the thermal variable S in the function $U(S, V)$ by its conjugate variable T. This leads to a new potential, the Helmholtz function F. The Legendre transformation is

$$U(S, V) \to F(T, V).$$

With the identification

$$Z = U, \quad X = -P, \quad x = V, \quad Y = T, \quad y = S, \quad N = F,$$

we obtain the expressions

$$F = U - ST,$$

$$dF = -P\,dV - S\,dT, \tag{8.14}$$

$$\left(\frac{\partial F}{\partial V}\right)_T = -P, \quad \left(\frac{\partial F}{\partial T}\right)_V = -S.$$

3. Finally, in $U(S, V)$ we replace S by its conjugate T and V by its conjugate $-P$, yielding the Gibbs function G, the last thermodynamic potential required by the pairing of thermal and mechanical variables. The transformation

$$U(S, V) \to G(T, P)$$

is effected by using Equations (8.5) and (8.6) and setting

$$Z = U, \quad X = T, \quad x = S, \quad Y = -P, \quad y = V, \quad M = G.$$

The resulting expressions are

$$G = U - ST + PV,$$

$$dG = -S\,dT + V\,dP, \tag{8.15}$$

$$\left(\frac{\partial G}{\partial T}\right)_P = -S, \quad \left(\frac{\partial G}{\partial P}\right)_T = V.$$

These potentials have interesting properties, which will be discussed later in this chapter.

8.4 THE MAXWELL RELATIONS

Each of the four thermodynamic potentials is a state variable whose differential is exact. We can use the condition for exactness discussed in Appendix A, which states that the value of a mixed second partial derivative is independent of the order in which the differentiation is applied. As an example, we consider

$$dU = T dS + (-P) dV = \left(\frac{\partial U}{\partial S}\right)_V dS + \left(\frac{\partial U}{\partial V}\right)_S dV.$$

The exactness of dU immediately gives

$$\frac{\partial^2 U}{\partial V \, \partial S} = \left(\frac{\partial T}{\partial V}\right)_S = \frac{\partial^2 U}{\partial S \, \partial V} = -\left(\frac{\partial P}{\partial S}\right)_V.$$

The equality of the first derivatives,

$$\left(\frac{\partial T}{\partial V}\right)_S = -\left(\frac{\partial P}{\partial S}\right)_V,$$

is known as a *Maxwell relation*. Its utility lies in the fact that each of the partials is a state variable that can be integrated along any convenient reversible path to obtain differences in values of the fundamental state variables between given equilibrium states.

A summary of relationships involving the thermodynamic potentials, including the Maxwell relations, is presented in Table 8.1.

8.5 THE HELMHOLTZ FUNCTION

We have seen that a key property of the internal energy is that the change in U is the heat flow in an isochoric reversible process. Similarly, the change in the enthalpy H is the heat flow in an isobaric reversible process. We ask: With what quantities can we associate changes in F and G as a result of reversible processes?

Suppose that a system is in thermal contact with a reservoir environment that is at a constant temperature T. Let the system undergo a process from some initial state to a final state. According to the second law,

$$\Delta S + \Delta S_0 \geq 0,$$

TABLE 8.1   Relationships involving the thermodynamic potentials.

Thermodynamic Potential	Independent Variables	Reciprocity Relations	Maxwell Relations
Internal energy	S, V	$T = \left(\dfrac{\partial U}{\partial S}\right)_V$	$\left(\dfrac{\partial T}{\partial V}\right)_S = -\left(\dfrac{\partial P}{\partial S}\right)_V$
U	$dU = T\,dS - P\,dV$	$-P = \left(\dfrac{\partial U}{\partial V}\right)_S$	$= \dfrac{\partial^2 U}{\partial V\,\partial S}$
Enthalpy	S, P	$T = \left(\dfrac{\partial H}{\partial S}\right)_P$	$\left(\dfrac{\partial T}{\partial P}\right)_S = \left(\dfrac{\partial V}{\partial S}\right)_P$
$H = U + PV$	$dH = T\,dS + V\,dP$	$V = \left(\dfrac{\partial H}{\partial P}\right)_S$	$= \dfrac{\partial^2 H}{\partial P\,\partial S}$
Helmholtz function	T, V	$S = -\left(\dfrac{\partial F}{\partial T}\right)_V$	$\left(\dfrac{\partial S}{\partial V}\right)_T = \left(\dfrac{\partial P}{\partial T}\right)_V$
$F = U - TS$	$dF = -S\,dT - P\,dV$	$-P = \left(\dfrac{\partial F}{\partial V}\right)_T$	$= -\dfrac{\partial^2 F}{\partial V\,\partial T}$
Gibbs function	T, P	$S = -\left(\dfrac{\partial G}{\partial T}\right)_P$	$\left(\dfrac{\partial S}{\partial P}\right)_T = -\left(\dfrac{\partial V}{\partial T}\right)_P$
$G = U - TS + PV$ $= H - TS$	$dG = -S\,dT + V\,dP$	$V = \left(\dfrac{\partial G}{\partial P}\right)_T$	$= -\dfrac{\partial^2 G}{\partial P\,\partial T}$

where ΔS is the entropy change of the system and ΔS_0 is the entropy change of the reservoir. If the reservoir transfers heat to the system, then $\Delta S_0 = -Q/T$ and therefore,

$$Q \leq T\,\Delta S,$$

where the equality sign applies if the process is reversible. If we substitute this in the first law expressed in the form

$$W = -\Delta U + Q,$$

we obtain

$$W \leq -\Delta U + T\,\Delta S,$$

or, since $F = U - TS$,

$$W \leq -\Delta F \quad \text{(no change in } T). \tag{8.16}$$

Thus the change in the Helmholtz function in an isothermal reversible process is the work done on or by the system. More generally, the decrease in F equals the maximum energy that can be freed in an isothermal process and made available for work.* The function is therefore often called the Helmholtz free energy. (Elsewhere it is frequently given the symbol A, an abbreviation for the German word for work, *Arbeit.*)

If the system held in thermal contact with the reservoir has a uniform pressure throughout its volume and the latter is also held constant, then the work performed will be zero and

$$(\Delta F)_{T,V} \leq 0, \tag{8.17}$$

or

$$F_f \leq F_i, \tag{8.18}$$

where i and f denote the initial and final values, respectively. Hence, if the Helmholtz function is a minimum, any change in the state of the system would increase F, which would be contradictory to Equation (8.18). It follows that the condition for *equilibrium* in a system in thermal contact with a reservoir and kept at constant volume is

$$dF = 0, \tag{8.19}$$

with F a *minimum.* Since T and V are the most convenient fundamental state variables for the purposes of statistical mechanics, as we shall see, the Helmholtz function becomes a very significant means of specifying a system's properties.

8.6 THE GIBBS FUNCTION

The Gibbs function is useful in problems in which T and P are the fundamental variables. Consider a system in a surrounding environment that constitutes a temperature and pressure reservoir. By this we mean that the reservoir is so large that its temperature and pressure remain unchanged. Most chemical reactions and some phase changes (ice melting in a beaker exposed to the atmosphere, for example) take place in this way. As before,

$$Q \leq T \, \Delta S,$$

with

$$Q = \Delta U + P \Delta V,$$

---

* Actually, it is not necessary that the temperature remain constant during the process; it is only required that the two end points be at temperature T.

where it is assumed that the work done by the system on the surroundings is reversible.

Combining these expressions, we have

$$\Delta U + P \, \Delta V - T \, \Delta S \leq 0.$$

Since

$$G = U + PV - TS,$$

and T and P are constant here, we obtain

$$(\Delta G)_{T,P} \leq 0, \tag{8.20}$$

or

$$G_f \leq G_i. \tag{8.21}$$

Spontaneous processes occur in the direction of decreasing G. A system in thermal contact with a heat and pressure reservoir moves to a state of *minimum G* for which

$$dG = 0. \tag{8.22}$$

It can be shown that if nonmechanical forces (electric forces, magnetic forces, etc.) act on a system, doing work W_{nm}, then

$$W_{nm} \leq -\Delta G \quad \text{(no change in } T \text{ or } P\text{).} \tag{8.23}$$

The decrease in the Gibbs function is equal to the maximum energy that can be freed in an isothermal, isobaric process and made available for nonmechanical work. Again, T and P need not be fixed throughout the process; they need only have the same initial and final values.

Table 8.2 lists the applicable conditions for various system states or processes.

8.7 APPLICATION OF THE GIBBS FUNCTION TO PHASE TRANSITIONS

Consider a system consisting of the liquid and vapor phases of some substance in equilibrium at temperature T and pressure P. Let n_1'' be the number of kilomoles in the liquid phase and n_1''' the number of kilomoles in the vapor

TABLE 8.2* Conditions on thermodynamic variables for different systems or processes.

State of System or Type of Process	Valid Equation	Valid Inequality	Equilibrium Condition
S and V constant	$dS = dV = 0$	$dU \leq 0$	Minimum U
S and P constant	$dS = dP = 0$	$dH \leq 0$	Minimum H
T and V constant	$dT = dV = 0$	$dF \leq 0$	Minimum F
T and P constant	$dT = dP = 0$	$dG \leq 0$	Minimum G
Adiabatic	$đQ = dU + P\,dV = 0$	$dS \geq 0$	Maximum S

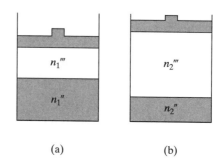

Figure 8.2 Liquid and vapor phases of a substance in equilibrium at temperature T and pressure P; (a) initial state; (b) final state.

(a)   (b)

phase.[†] The state of the system is defined in terms of the variables (T, P, n_1'', n_1'''). Consider a second state differing from the first only in the number of kilomoles of liquid and vapor and defined by (T, P, n_2'', n_2''') (Figure 8.2). Mass is conserved so that

$$n_1'' + n_1''' = n_2'' + n_2'''. \tag{8.24}$$

We define g'' and g''' as the specific Gibbs functions of the liquid and vapor, respectively, associated with the particular substance under investigation. Noting that the Gibbs function is an extensive variable, we have for the two states:

$$G_1 = n_1'' g'' + n_1''' g''', \tag{8.25}$$

$$G_2 = n_2'' g'' + n_2''' g'''. \tag{8.26}$$

---

*Adapted from Table 7.2 in *Thermodynamics and Statistical Mechanics* by P. L. Landsberg, Dover Publications, New York, 1990.

[†] The notation is that used in section 4.3: one, two, and three primes denote the solid, liquid, and vapor phases, respectively. Here 1 refers to the initial state and 2 to the final state.

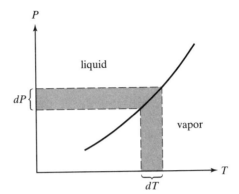

Figure 8.3 Relationship between temperature and pressure for a liquid and vapor in equilibrium. The derivative dP/dT is the slope of the vaporization curve.

Suppose that a reversible transition takes place from state 1 to state 2. Since $(\Delta G)_{T,P} = 0$ for a reversible process, it follows that $G_1 = G_2$. Equating Equations (8.25) and (8.26) and using Equation (8.24), we find that

$$g'' = g'''. \tag{8.27}$$

The specific Gibbs function is the same for the two phases. This is true for all phases in equilibrium, that is, for all points on the curve of the phase transformation (Figure 8.3).

Since at a temperature $T + dT$ and a pressure $P + dP$ we still have equilibrium, it follows that $g'' + dg'' = g''' + dg'''$. Combining this with Equation (8.27), we have

$$dg'' = dg'''.$$

Using the expression for the differential previously derived, we can write

$$-s''\, dT + v''\, dP = -s'''\, dT + v'''\, dP,$$

or

$$(s''' - s'')dT = (v''' - v'')dP.$$

Thus

$$\frac{dP}{dT} = \frac{s''' - s''}{v''' - v''}. \tag{8.28}$$

From the definition of entropy,

$$s''' - s'' = \frac{\ell_{23}}{T}, \tag{8.29}$$

where ℓ_{23} is the latent heat of vaporization. Since heat is absorbed as a liquid becomes a vapor, ℓ_{23} is positive and $s''' > s''$. Substituting Equation (8.29) in Equation (8.28) gives

$$\left(\frac{dP}{dT}\right)_{23} = \frac{\ell_{23}}{T(v''' - v'')} \quad \text{(liquid-vapor)}. \tag{8.30}$$

This is the famous Clausius-Clapeyron equation. It gives the *slope* of the curve denoting the boundary between the liquid and vapor phases, that is, the vaporization curve. Similar expressions hold for the sublimation and fusion curves:

$$\left(\frac{dP}{dT}\right)_{13} = \frac{\ell_{13}}{T(v''' - v')} \quad \text{(solid-vapor)}, \tag{8.31}$$

$$\left(\frac{dP}{dT}\right)_{12} = \frac{\ell_{12}}{T(v'' - v')} \quad \text{(solid-liquid)}. \tag{8.32}$$

The latent heats in these expressions are positive, and the slopes are all positive for substances that expand on melting. A notable exception is water, which contracts when ice melts into liquid; for this case $(dP/dT)_{12} < 0$.

The Clausius-Clapeyron equation, combined with the appropriate equations of state, can in principle yield equations for the phase transformation curves. A simple example is the vaporization curve describing, say, the conversion of liquid water to steam. Here $v''' \gg v''$ (see Chapter 2), and so

$$\left(\frac{dP}{dT}\right)_{23} \approx \frac{\ell_{23}}{Tv'''}.$$

If we treat the vapor as an ideal gas,

$$v''' \approx \frac{RT}{P},$$

so that

$$\frac{dP}{dT} \approx \frac{\ell_{23}}{R}\frac{P}{T^2}.$$

If ℓ_{23} is assumed to be temperature-independent, upon integrating we obtain

$$P = P_0 \exp\left[-\frac{\ell_{23}}{R}\left(\frac{1}{T} - \frac{1}{T_0}\right)\right], \qquad (8.33)$$

where (T_0, P_0) denotes some fixed point on the curve.*

The Clausius-Clapeyron equation can also help us understand why the ice point of water (273.15 K) is 0.01 K below the triple point (273.16 K). Solving for dT in Equation (8.32), we have

$$dT = \frac{T(v'' - v')}{\ell_{12}} dP.$$

For the small temperature change involved, we can write

$$\Delta T = \frac{T(v'' - v')}{\ell_{12}} \Delta P$$

and assume that the specific volumes v'' and v' are constant. This equation gives the change in temperature as the pressure is increased from the low value at the triple point ($P_{TP} = 4.58$ Torr) to atmospheric pressure, the pressure at which the normal melting point of water is defined. The density of pure liquid water is 1000 kg m^{-3} and the density of ice is 916 kg m^{-3}. Noting that the specific volume in m^3 kg^{-1} is just the reciprocal of the density, we have $v'' = 1.00 \times 10^{-3}$ m^3 kg^{-1} and $v' = 1.09 \times 10^{-3}$ m^3 kg^{-1}, to three significant figures. The latent heat of melting for water is $\ell_{12} = 3.34 \times 10^5$ J kg^{-1}. Writing P_{TP} in Pascals, we have

$$\Delta P = P_{atm} - P_{TP} = 1.01 \times 10^5 \text{ Pa} - 610 \text{ Pa} \approx 1.01 \times 10^5 \text{ Pa}.$$

Then

$$\Delta T = \frac{273(1.00 \times 10^{-3} - 1.09 \times 10^{-3})(1.01 \times 10^5)}{3.34 \times 10^5} = -0.0074 \text{ K}.$$

The presence of dissolved air in a mixture of ice and water further lowers the temperature at which ice melts by 0.0023 K. The total reduction of temperature below the triple point of water is therefore 0.0097 K, or approximately 0.01 K.

*The more general solution is

$$P = P_0 \exp\left[\frac{1}{R}\int \frac{\ell_{23}(T)}{T^2} dT\right].$$

8.8 AN APPLICATION OF THE MAXWELL RELATIONS

As an example of the use of a Maxwell relation, consider an ideal gas that undergoes an isothermal reversible change from pressure P_0 to pressure P. By definition of the entropy,

$$đQ_r = T \, dS.$$

We express S as a function of the fundamental state variables of the problem—namely, T and P:

$$S = S(T, P).$$

Then

$$dS = \left(\frac{\partial S}{\partial T}\right)_P dT + \left(\frac{\partial S}{\partial P}\right)_T dP.$$

Since the process is isothermal, $dT = 0$, and

$$đQ_r = T\left(\frac{\partial S}{\partial P}\right)_T dP.$$

From Table 7.1, we select the Maxwell relation

$$\left(\frac{\partial S}{\partial P}\right)_T = -\left(\frac{\partial V}{\partial T}\right)_P.$$

Thus

$$đQ_r = -T\left(\frac{\partial V}{\partial T}\right)_P dP.$$

For an ideal gas,

$$PV = nRT,$$

$$\left(\frac{\partial V}{\partial T}\right)_P = \frac{nR}{P},$$

and

$$Q_r = -nRT \int_{P_0}^{P} \frac{dP}{P} = -nRT \ln\left(\frac{P}{P_0}\right).$$

If $P > P_0$, the negative sign indicates that heat flows *out* of the system in the process.

8.9 CONDITIONS OF STABLE EQUILIBRIUM

The second law states that in spontaneous processes the entropy increases. Maximum entropy corresponds to thermodynamic equilibrium. Furthermore, maximum entropy is a point of *stable* equilibrium: if the system is perturbed slightly from its equilibrium state, it will return to that state spontaneously. We wish to examine the implications of this fundamental law of nature.

Since $U = U(S, V)$, it follows that $S = S(U, V)$. The maximization of S is a problem in multivariate calculus.* A theorem states that the function $f(x, y)$ has an absolute maximum if, for all points in the domain,

1. $f_{xx} < 0$,
2. $D > 0$,

where

$$D \equiv \begin{vmatrix} f_{xx} & f_{xy} \\ f_{yx} & f_{yy} \end{vmatrix} = f_{xx}f_{yy} - (f_{xy})^2.$$

(In this section, the subscripts indicate differentiation; for example, $f_{xy} = \partial^2 f / \partial x \, \partial y$.) Thus the conditions on S are

1. $S_{UU} < 0$,
2. $D = S_{UU} S_{VV} - (S_{UV})^2 > 0$.

From the combined first and second laws,

$$dS = S_U \, dU + S_V \, dV = \frac{1}{T} dU + \frac{P}{T} dV. \tag{8.34}$$

Thus

$$S_U = \frac{1}{T}, \quad S_V = \frac{P}{T}. \tag{8.35}$$

Taking the differential of the first of these equations, we obtain

$$S_{UU} \, dU + S_{VU} \, dV = -\frac{dT}{T^2}.$$

---

*See, for example, *Calculus: One and Several Variables,* 4th edition, by S. L. Salas and E. Hille, Wiley, New York, 1982, pp. 792 ff.

Then

$$S_{UU}\left(\frac{\partial U}{\partial T}\right)_V = -\frac{1}{T^2}.$$

Since

$$C_V = \left(\frac{\partial U}{\partial T}\right)_V,$$

it follows that

$$S_{UU} = -\frac{1}{C_V T^2}. \tag{8.36}$$

The temperature T is always positive. Therefore the first condition for stability, $S_{UU} < 0$, implies that

$$C_V > 0. \tag{8.37}$$

Now let's take the differentials of the *two* equations labeled Equation (8.35):

$$S_{UU}\,dU + S_{VU}\,dV = -\frac{dT}{T^2},$$

$$S_{VU}\,dU + S_{VV}\,dV = -\frac{P}{T^2}dT + \frac{dP}{T}.$$

We can solve for dV (or dU) using Cramer's rule:

$$dV = \frac{\begin{vmatrix} S_{UU} & -\dfrac{dT}{T^2} \\[2mm] S_{VU} & -\dfrac{P}{T^2}dT + \dfrac{dP}{T} \end{vmatrix}}{D}, \tag{8.38}$$

where $D \equiv S_{UU}\,S_{VV} - (S_{UV})^2$. Expanding the determinant in the numerator and rearranging terms, we obtain

$$D\,dV = \frac{S_{UU}}{T}dP + (S_{VU} - S_{UU}\,P)\frac{dT}{T^2}.$$

Then

$$D\left(\frac{\partial V}{\partial P}\right)_T = \frac{S_{UU}}{T},$$

or

$$D = \frac{S_{UU}}{T}\left(\frac{\partial P}{\partial V}\right)_T.$$

Now

$$\kappa \equiv -\frac{1}{V}\left(\frac{\partial V}{\partial P}\right)_T, \quad \text{and} \quad S_{UU} = -\frac{1}{C_V T^2}$$

from Equation (8.36), so that

$$D = \frac{1}{C_V \kappa T^2 V}. \tag{8.39}$$

Since C_V, T, and V are all positive, the second condition for stability, $D > 0$, implies that

$$\kappa > 0. \tag{8.40}$$

We conclude that for *thermal* stability, the addition of heat to a body causes the temperature to increase:

$$C_V \equiv \left(\frac{đQ}{dT}\right)_V > 0.$$

For *mechanical* stability, a decrease of pressure causes an increase in volume:

$$\kappa \equiv -\frac{1}{V}\left(\frac{\partial V}{\partial P}\right)_T > 0.$$

That is, for a closed system, stable *thermodynamic* equilibrium consists of thermal and mechanical equilibrium.

There are a few exceptions to this otherwise general result. At critical points the heat capacities and the compressibility can diverge and the stability conditions are then violated. It is also interesting to note that long-range forces can give rise to negative heat capacities. Such modifications have by no means been fully investigated as yet.

PROBLEMS

8-1 A van der Waals gas and an ideal gas are originally at the same pressure, temperature, and volume. If each gas undergoes a reversible isothermal compression, which gas will experience the greater change in entropy?

8-2 Show that for the an ideal gas

(a) $f = c_v(T - T_0) - c_v T \ln\left(\dfrac{T}{T_0}\right) - RT \ln\left(\dfrac{v}{v_0}\right) - s_0 T.$

(b) $g = c_P(T - T_0) - c_P T \ln\left(\dfrac{T}{T_0}\right) + RT \ln\left(\dfrac{P}{P_0}\right) - s_0 T.$

8-3 A cylinder contains a piston on each side of which is one kilomole of an ideal gas. The walls of the cylinder are diathermal and the system is in contact with a heat reservoir at a temperature of 0°C. The initial volumes of the gaseous subsystems on either side of the piston are 12 liters and 2 liters, respectively. The piston is now moved reversibly so that the final volumes are each 7 liters. What is the change in the Hemholtz potential? (Note that this is the work delivered to the system by the reservoir.)

8-4 Derive the following equations:

(a) $F = U + T\left(\dfrac{\partial F}{\partial T}\right)_V$;

(b) $C_V = -T\left(\dfrac{\partial^2 F}{\partial T^2}\right)_V$ (reversible process);

(c) $H = G - T\left(\dfrac{\partial G}{\partial T}\right)_P$;

(d) $C_p = -T\left(\dfrac{\partial^2 G}{\partial T^2}\right)_P$ (reversible process);

(The third relation is the Gibbs-Helmholtz equation, alluded to in Chapter 10.)

8-5 The Helmholtz function of a certain gas is

$$F = -\frac{n^2 a}{V} - nRT \ln(V - nb) + J(T),$$

where J is a function of T only. Derive an expression for the pressure of the gas.

8-6 The Gibbs function of a certain gas is

$$G = nRT \ln P + A + BP + \frac{CP^2}{2} + \frac{DP^3}{3},$$

where $A, B, C,$ and D are constants. Find the equation of state of the gas.

8-7 The specific Gibbs function of a gas is given by

$$g = RT \ln\left(\frac{P}{P_0}\right) - AP,$$

where A is a function of T. Find expressions for:
(a) the equation of state;
(b) the specific entropy;
(c) the specific Helmholtz function.

8-8 (a) A van der Waals gas undergoes an isothermal expansion from specific volume v_1 to specific volume v_2. Calculate the change in the specific Helmholtz function.
(b) Calculate the change in the specific internal energy in terms of v_1 and v_2.

8-9 Start with the first Maxwell relation listed in Table 8.1 and derive the second by using the cyclical and recipirocal relations (Appendix A) and the identity

$$\left(\frac{\partial x}{\partial y}\right)_f \left(\frac{\partial y}{\partial z}\right)_f \left(\frac{\partial z}{\partial x}\right)_f = 1.$$

(The remaining Maxwell relations can be derived in a similar manner.)

8-10 (a) Prove that

$$c_P = T\left(\frac{\partial s}{\partial T}\right)_P,$$

and use the result to show that

$$\left(\frac{\partial c_P}{\partial P}\right)_T = -T\left(\frac{\partial^2 v}{\partial T^2}\right)_P.$$

(b) Prove that c_P for an ideal gas is a function of T only.

8-11 In Figure 2.2 depicting the isotherms for a van der Waals gas, the shaded areas were shown by Maxwell to be equal. The path bcd is the segment of an isotherm along which liquid and vapor are in equilibrium. Prove Maxwell's result by noting that $\Delta g = 0$ for a phase change (P and T constant) and calculating Δf.

8-12 The P-v diagram in Figure 8.4 shows two neighboring isotherms in the region of a liquid-gas phase transition. By considering a Carnot cycle between temperatures T and $T + dT$ in the region shown, derive the Clausius-Clapeyron equation $dP/dT = \ell_{23}/T(\Delta v)$. Here Δv is the specific volume change between gas and liquid.

8-13 The equations of the sublimation and the vaporization curves of a particular material are given by

$$\ln P = 0.04 - 6/T \quad \text{(sublimation)},$$
$$\ln P = 0.03 - 4/T \quad \text{(vaporization)},$$

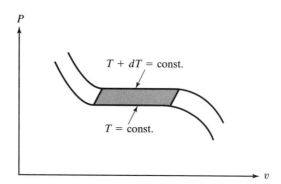

Figure 8.4 P-v diagram for
an infinitesimal Carnot cycle.

where P is in atmospheres.

(a) Find the temperature of the triple point.

(b) Show that the specific latent heats of vaporization and sublimation are $4R$
and $6R$, respectively. (You may assume that the specific volume in the vapor
phase is much larger than the specific volume in the liquid and solid
phases.)

(c) Find the latent heat of fusion.

8-14 (a) Calculate the slope of the fusion curve of ice in PaK^{-1} at the normal melting
point. At this temperature, the heat of fusion is 3.34×10^5 Jkg^{-1} and the
change in specific volume on melting is -9.05×10^{-5} m^3kg^{-1}.

(b) Ice at $-2°C$ and atmospheric pressure is compressed isothermally. Find the
pressure at which the ice starts to melt (in atmospheres).

Chapter 9

The Chemical Potential and Open Systems

9.1 THE CHEMICAL POTENTIAL

Until now we have confined our discussion to closed physical systems, which cannot exchange matter with their surroundings. We turn our attention in this chapter to *open* systems, in which the quantity of matter is not fixed.

Suppose that *dn* kilomoles of matter are introduced into a system. Each kilomole of added matter has its own internal energy that is released to the rest of the system, possibly in a chemical reaction. The added energy is proportional to *dn* and may be written as μdn. The quantity μ is called the *chemical potential.*

The chemical potential is associated with intermolecular forces. An electrically polarized molecule experiences a Coulomb attraction when it is brought into the vicinity of another such molecule.* This force is expressed as a negative potential energy, a sort of "potential well." As the new particle approaches its neighbor, it gains kinetic energy while losing potential energy. The kinetic energy is imparted to other particles through collisions, so the system gains internal energy in the process.

Consider a motionless molecule infinitely distant from other molecules. Its kinetic energy and potential energy are both zero. The molecule is moved into the force field of a second molecule. This can, in principle, be done slowly so that the kinetic energy is negligibly small. Left by itself, however, the molecule picks up kinetic energy equal in magnitude to the depth of the potential well (Figure 9.1). Quantitatively,

$$E = K + V(r),$$

where E is the total energy, K is the kinetic energy, and $V(r)$ is the potential energy; r is the distance between the molecule and its neighbor. At $r = \infty$, $K = 0$, and $V = 0$, so $E = 0$ everywhere. At $r = r_o$,

$$K - |V(r_o)| = 0, \quad \text{so} \quad K = |V(r_o)| > 0.$$

---

*See Section 11.1 for a discussion of the molecular interaction potential.

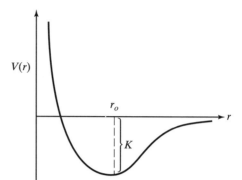

Figure 9.1 Schematic diagram of a potential well due to intermolecular forces. A more descriptive graph is given in Section 11.1.

Energy is conserved, but a conversion from potential energy to kinetic energy takes place. The kinetic energy is added to the internal energy of the system.

It is reasonable to ask what the magnitude of μ is. In a standard laboratory experiment, sulfuric acid is added to water, producing an increase in temperature. Imagine that 10^{-5} kilomoles of acid at room temperature are added to a liter of water (5.56×10^{-2} kilomoles), also at room temperature. The temperature is observed to rise 0.1°C. We wish to determine the chemical potential of acid in water.

We shall assume that the interaction among the acid molecules is small compared with their interaction with the water molecules (both acid and water molecules are electrically polarized). The specific heat capacity of water is 7.52×10^4 J kilomole^{-1} K^{-1}. Thus the heat gained by the water through the addition of the acid is

$$Q = mc_p\Delta T = (5.56 \times 10^{-2})(7.52 \times 10^4)(0.1) = 418 \text{ J}.$$

Since the mass of the water is more than 10^3 times the mass of the acid, we can ignore the heat capacity of the latter. The chemical potential, then, is

$$\mu = -\frac{Q}{\Delta n} = -\frac{418}{10^{-5}} = -4.18 \times 10^7 \text{ J kilomole}^{-1}.$$

The sign is negative because heat is transferred *from* the acid to the water.

We are interested in the chemical energy *per particle,* which is essentially the depth of the potential well. Since a kilomole has 6.02×10^{26} molecules, this value is -6.94×10^{-20} J or -0.43 eV. (One electron volt is equal to 1.6×10^{-19} joules.) Most chemical potentials are of this order of magnitude.

To account for the effect of adding mass to a system, we need to add a term to our fundamental equation of thermodynamics:

$$dU = T\,dS - P\,dV + \mu\,dn. \tag{9.1}$$

Here dn is the increment of mass added (in kilomoles) and μ is the chemical potential in joules per kilomole. If, in an open system, $U = U(S, V, n)$ and

$$dU = \left(\frac{\partial U}{\partial S}\right)_{V,n} dS + \left(\frac{\partial U}{\partial V}\right)_{S,n} dV + \left(\frac{\partial U}{\partial n}\right)_{S,V} dn, \tag{9.2}$$

then

$$\mu = \left(\frac{\partial U}{\partial n}\right)_{S,V}. \tag{9.3}$$

That is, the chemical potential is defined as the internal energy per kilomole added under conditions of constant entropy and volume.

If there is more than one type of particle added to the system (say m types), then Equation (9.1) becomes

$$dU = TdS - PdV + \sum_{j=1}^{m} \mu_j \, dn_j, \tag{9.4}$$

with

$$\mu_j = \left(\frac{\partial U}{\partial n_j}\right)_{S,V,n_k}. \tag{9.5}$$

The subscript n_k means that all other n's except n_j are held constant.

Another way of seeing the relationship between U and μ_j is to integrate Equation (9.4). This can be done by using Euler's theorem for homogenous functions. Euler's theorem states that if

$$\lambda f(x, y, z) = f(\lambda x, \lambda y, \lambda z), \tag{9.6}$$

then

$$f = x\frac{\partial f}{\partial x} + y\frac{\partial f}{\partial y} + z\frac{\partial f}{\partial z}. \tag{9.7}$$

The theorem can be easily proved by differentiating Equation (9.6) with respect to λ and then setting λ equal to unity. Now $U = U(S, V, n_1 \ldots n_m)$. Suppose the amounts of all the types of substance, called *constituents*, in the system were doubled or halved or, more generally, changed by the factor λ without changing any of the fundamental state variables. Then the extensive variable U would be changed by λ and all the independent, extensive state variables would also be changed by the factor λ. Thus U is a homogeneous function and Euler's theorem can be applied to it:

$$U = S\left(\frac{\partial U}{\partial S}\right)_{V,n_k} + V\left(\frac{\partial U}{\partial V}\right)_{S,n_k} + \sum_{j=1}^{m} n_j\left(\frac{\partial U}{\partial n_j}\right)_{S,V,n_k}. \tag{9.8}$$

From the differential of U we know that

$$\left(\frac{\partial U}{\partial S}\right)_{V,n_k} = T, \quad \left(\frac{\partial U}{\partial V}\right)_{S,n_k} = -P, \quad \left(\frac{\partial U}{\partial n_j}\right)_{S,V,n_k} = \mu_j. \tag{9.9}$$

Substituting the relations of Equation (9.9) in Equation (9.8), we have

$$U = ST - PV + \sum_{j=1}^{m} \mu_j n_j. \tag{9.10}$$

Recall that the Gibbs function is defined as $G = U - ST + PV$. It is therefore immediately clear that

$$G = \sum_{j=1}^{m} \mu_j n_j. \tag{9.11}$$

If only one constituent is present, then $G = \mu n$ or $\mu = G/n$; so μ in this case is simply the Gibbs function per kilomole of the substance.

Finally, if we take the differential of Equation (9.10), we obtain

$$dU = TdS + SdT - PdV - VdP + \sum_{j}\mu_j dn_j + \sum_{j} n_j d\mu_j. \tag{9.12}$$

Equating Equations (9.4) and (9.12), we have

$$SdT - VdP + \sum_{j=1}^{m} n_j d\mu_j = 0, \tag{9.13}$$

a relation known as the Gibbs-Duhem equation. Taking the differential of Equation (9.11) gives

$$dG = \sum_{j}\mu_j dn_j + \sum_{j} n_j d\mu_j. \tag{9.14}$$

Note that $G = G(T, P, n_1 \ldots n_m)$. Thus if we have a process that takes place at constant temperature and pressure, the first two terms of Equation (9.13) vanish so the sum also vanishes. Equation (9.14) then yields the important result

$$(dG)_{T,P} = \sum_j \mu_j dn_j. \tag{9.15}$$

We now have the mathematical tools to look at some applications.

9.2 PHASE EQUILIBRIUM

We wish to find the condition of equilibrium for two subsystems under particle exchange where the subsystems are two phases of the same substance. An example would be ice melting in liquid water; an exchange of H_2O molecules occurs between the two subsystems. We assume that the total system is enclosed in a rigid adiabatic wall (Figure 9.2). Since there can be heat and particle flow across the boundary between the two phases, the boundary will move as the process evolves. The various interactions are subject to the following conditions:

$$n_A + n_B = n = \text{constant} \quad \text{(conservation of mass)}, \tag{9.16}$$

$$V_A + V_B = V = \text{constant} \quad \text{(conservation of volume)}, \tag{9.17}$$

$$U_A + U_B = U = \text{constant} \quad \text{(conservation of energy)}. \tag{9.18}$$

The total internal energy is constant, according to the first law, since no heat flows into or out of the combined system and no work is done. At equilibrium, the entropy of the combined system will be a maximum:

$$S_A + S_B = S \quad \text{(maximum)}. \tag{9.19}$$

Let all the quantities n, V, U, and S change by infinitesimal amounts. Then

$$dS = dS_A + dS_B = 0. \tag{9.20}$$

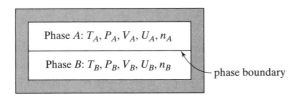

Figure 9.2   Two phases of a substance in thermal equilibrium.

But

$$dS_A = \frac{1}{T_A}(dU_A + P_A \, dV_A - \mu_A dn_A), \qquad (9.21)$$

and

$$dS_B = \frac{1}{T_B}(dU_B + P_B \, dV_B - \mu_B dn_B). \qquad (9.22)$$

We can eliminate three infinitesimals by invoking Equations (9.16), (9.17), and (9.18):

$$dn_B = -dn_A, \quad dV_B = -dV_A, \quad dU_B = -dU_A.$$

Combining these expressions with Equations (9.21) and (9.22), and substituting in Equation (9.20) gives the result

$$\left(\frac{1}{T_A} - \frac{1}{T_B}\right)dU_A + \left(\frac{P_A}{T_A} - \frac{P_B}{T_B}\right)dV_A - \left(\frac{\mu_A}{T_A} - \frac{\mu_B}{T_B}\right)dn_A = 0. \qquad (9.23)$$

Since Equation (9.23) must be true for arbitrary increments dU_A, dV_A, dn_A, all of the coefficients must be zero:

$$T_A = T_B \quad \text{(thermal equilibrium)}, \qquad (9.24)$$
$$P_A = P_B \quad \text{(mechanical equilibrium)}, \qquad (9.25)$$
$$\mu_A = \mu_B \quad \text{(diffusive equilibrium)}. \qquad (9.26)$$

We conclude that if we have two phases or subsystems in thermal and mechanical equilibrium, then they will also be in *diffusive* equilibrium (equilibrium against particle exchange) if their chemical potentials are equal.

Suppose that the two interacting systems are not yet in equilibrium. Then the entropy of the combined system will not have reached its maximum value but will still be increasing:

$$\Delta S = \Delta S_A + \Delta S_B > 0. \qquad (9.27)$$

Consider a state very close to an equilibrium state and differing from it only in that there is a small excess of mass $\Delta n_A (>0)$ in subsystem A over the equilibrium value n_A, and a corresponding deficit of mass $-\Delta n_A$ in subsystem B. As the combined system moves toward equilibrium, subsystem A will give up its excess mass with an accompanying entropy change

$$\Delta S_A = -\frac{\mu_A}{T}(-\Delta n_A). \tag{9.28}$$

Simultaneously, the entropy change of subsystem B as a result of acquiring the additional mass will be

$$\Delta S_B = -\frac{\mu_B}{T}(+\Delta n_A). \tag{9.29}$$

The temperature is the same in both expressions. Substituting Equations (9.28) and (9.29) in Equation (9.27) gives

$$\frac{\Delta n_A}{T}(\mu_A - \mu_B) > 0,$$

or

$$\mu_A > \mu_B. \tag{9.30}$$

Thus, in the approach to equilibrium, heat energy must flow from the hotter to the cooler body, volume will be gained by the body under lower pressure at the expense of the other, and mass will tend to flow from the body of higher chemical potential toward the one of lower chemical potential. All of these tendencies follow from the second law of thermodynamics.

9.3 THE GIBBS PHASE RULE

Consider a system in equilibrium with k constituents in π phases.* We note the following:

1. Only one gaseous phase can exist because of diffusion.
2. Several liquids can coexist in equilibrium if they are immiscible.
3. Several solids can coexist.
4. Only rarely do more than three phases of a given constituent coexist.

---

*The symbol π, traditionally used to denote the number of phases, is not to be confused here with 3.14. . . .

Let μ_i^γ be the chemical potential of the ith constituent in the γth phase. According to Equation (9.11) the Gibbs function is

$$G = \sum_{i=1}^{k} \sum_{\gamma=1}^{\pi} \mu_i^\gamma n_i^\gamma, \tag{9.31}$$

where n_i^γ is the number of kilomoles of the ith constituent in the γth phase. For equilibrium, $(dG)_{T,P} = 0$, so Equation (9.15) gives

$$\sum_{i=1}^{k} \sum_{\gamma=1}^{\pi} \mu_i^\gamma \, dn_i^\gamma = 0, \quad T \text{ and } P \text{ fixed.} \tag{9.32}$$

It also follows from the Gibbs-Duhem equation (Equation (9.13)) that for constant temperature and pressure,

$$\sum_{i=1}^{k} \sum_{\gamma=1}^{\pi} n_i^\gamma \, d\mu_i^\gamma = 0. \tag{9.33}$$

If we consider the special case of a closed system in which mass dn_i of the ith constituent is transferred from one phase to another with the total mass of the constituent unchanged, then

$$\sum_{\gamma=1}^{\pi} dn_i^\gamma = 0, \quad i = 1, 2, \ldots k. \tag{9.34}$$

This equation is simply a statement of the conservation of mass: No constituent is created or destroyed.

Suppose that there are two constituents $(i = 1, 2)$ and two phases $(\gamma = \alpha, \beta)$. Then Equation (9.32) becomes

$$\mu_1^\alpha \, dn_1^\alpha + \mu_2^\alpha \, dn_2^\alpha + \mu_1^\beta \, dn_1^\beta + \mu_2^\beta \, dn_2^\beta = 0, \tag{9.35}$$

and Equation (9.34) gives

$$dn_1^\alpha + dn_1^\beta = 0, \quad dn_2^\alpha + dn_2^\beta = 0. \tag{9.36}$$

Combining these equations gives

$$(\mu_1^\alpha - \mu_1^\beta)dn_1^\alpha + (\mu_2^\alpha - \mu_2^\beta)dn_2^\alpha = 0. \tag{9.37}$$

Since the differentials dn_1^α and dn_2^α are arbitrary, the coefficients must vanish. Hence

$$\mu_1^\alpha = \mu_1^\beta \quad \text{and} \quad \mu_2^\alpha = \mu_2^\beta. \tag{9.38}$$

In other words, the equality of the chemical potentials for the two phases of each of the constituents is the condition for equilibrium with T and P fixed.

From the conservation of mass it follows that the number of kilomoles of each constituent is fixed provided no chemical reaction occurs. Thus

$$n_1^\alpha + n_1^\beta = n_1 \quad \text{and} \quad n_2^\alpha + n_2^\beta = n_2, \tag{9.39}$$

where n_1 and n_2 are the given number of kilomoles of constituents 1 and 2, respectively. It is convenient to introduce for each constituent the concept of the kilomole fraction, the ratio of the number of kilomoles of a given phase to the total number of kilomoles of all constituents in that phase:

$$x_1^\alpha = \frac{n_1^\alpha}{n_1}, \quad x_1^\beta = \frac{n_1^\beta}{n_1}, \tag{9.40}$$

$$x_2^\alpha = \frac{n_2^\alpha}{n_2}, \quad x_2^\beta = \frac{n_2^\beta}{n_2}.$$

From Equations (9.39) and (9.40), it is evident that

$$x_1^\beta = 1 - x_1^\alpha, \quad \text{and} \quad x_2^\beta = 1 - x_2^\alpha. \tag{9.41}$$

Therefore, in our example of $k = 2$, $\pi = 2$, the state of the system is determined by four independent variables: T, P, x_1^α, and x_2^α. However, since there are two equilibrium conditions (Equation (9.38)), the number of variables is reduced to two.

The generalization of this special case leads to the so-called Gibbs phase rule. For a multiconstituent, multiphase system with T and P fixed, the condition for equilibrium is

$$\mu_i^\alpha = \mu_i^\beta = \mu_i^\gamma \ldots = \mu_i^\pi, \quad i = 1, 2, \ldots k. \tag{9.42}$$

The kilomole fractions are:

$$x_i^\gamma = \frac{n_i^\gamma}{\displaystyle\sum_{i=1}^{k} n_i^\gamma}, \quad i = 1, 2 \ldots k; \quad \gamma = 1, 2 \ldots \pi. \tag{9.43}$$

Thus, there would be $k\pi$ kilomole fractions were it not for the identity

$$\sum_{i=1}^{k} x_i^\gamma = 1, \tag{9.44}$$

which subtracts π kilomole fractions from the total. Counting the two prescribed quantities, T and P, there are therefore $2 + k\pi - \pi$ independent variables and $k(\pi - 1)$ equilibrium conditions (Equation (9.42)). The number of remaining "degrees of freedom," called the variance f, is

$$f = [2 + k\pi - \pi] - [k(\pi - 1)],$$

or

$$f = k - \pi + 2 \quad \text{(no chemical reaction).} \tag{9.45}$$

This is the usual form of the Gibbs phase rule.

If $f = 0$, the system has zero variance and is completely determined. If $f = 1$, the system is monovariant and is not determined until one additional variable is specified.

Consider the following examples:

1. A homogeneous fluid with one constituent and one phase. Here $k = 1$, $\pi = 1, f = 2$. Thus we can choose T and P arbitrarily. This is the case of an ideal gas, for instance.

2. A homogeneous system consisting of two chemically different gases in the same phase. Here $k = 2, \pi = 1, f = 3$. Here we can choose T, P, and one kilomole fraction.

3. Water in equilibrium with its saturated vapor. In this case $k = 1, \pi = 2$, $f = 1$. Thus we can choose only the temperature (*or* the pressure) arbitrarily. Equilibrium is defined by a point on the vaporization curve for water; specification of either T or P gives the value of the other variable.

4. Ice, liquid water, and water vapor in equilibrium. Here $k = 1, \pi = 3$, $f = 0$. The three phases can coexist in equilibrium only for the fixed set of values T and P defining the triple point.

9.4 CHEMICAL REACTIONS

As we have noted, the chemical potential is a measure of the chemical energy per kilomole that a substance can generate in a reaction. Chemical thermodynamics involves, in part, the measurement of the chemical potentials of compounds.

A chemical reaction is usually described by an equation, such as

$$2H_2 + O_2 \rightarrow 2H_2O. \tag{9.46}$$

Such an equation can be generalized as

$$\sum_{j=1}^{m} \nu_j M_j = 0, \tag{9.47}$$

which states that a certain number of molecules of the initial reactants will combine to yield a certain number of final product molecules. Here the M_j denote chemical symbols and the ν_j are so-called *stoichiometric coefficients*. The coefficients are positive or negative integers; negative values represent initial reactants and positive values correspond to final products. Thus, in the previous example,

$$\nu_{H_2} = -2, \quad \nu_{O_2} = -1 \quad \text{and} \quad \nu_{H_2O} = +2.$$

For a chemical reaction to occur, more than one substance must be present either initially or finally. The number of kilomoles n_j of each of the constituents will change in a manner consistent with the conservation of atomic species. In the example, for every two kilomoles of hydrogen and one kilomole of oxygen that disappear, two kilomoles of water will appear. The change in the number of kilomoles will therefore be proportional to the stoichiometric coefficients:

$$dn_{H_2}: dn_{O_2}: dn_{H_2O} = \nu_{H_2}: \nu_{O_2}: \nu_{H_2O} = -2: -1: +2.$$

Suppose that a chemical reaction takes place under conditions of constant temperature and pressure. This would occur when, say, the reactants are immersed in a water bath at atmospheric pressure. However, it is not necessary that the temperature and pressure be constant throughout the process. It is sufficient that they return to their initial values after the reaction happens, which is virtually always the case. Under these conditions, the Gibbs-Duhem equation (Equation (9.13)) states that the chemical potentials of the reactants are fixed. Then, at equilibrium $(dG)_{T,P} = 0$, so that

$$\sum_j \mu_j dn_j = 0,$$

where the μ_j's are constants. Finally, since $dn_j \propto \nu_j$, we can write

$$\sum_j \mu_j \nu_j = 0 \qquad (9.48)$$

for *chemical equilibrium* in a reaction. Once initiated, most chemical reactions proceed spontaneously either until some reactant runs out or until the reaction arrives at equilibrium. If the reaction is being used to determine the chemical potentials of the constituents, the experiment will be designed to reach equilibrium. In the reaction of Equation (9.46), we obtain the useful result

$$\mu_{H_2O} = \frac{1}{2}(2\mu_{H_2} + \mu_{O_2}). \qquad (9.49)$$

This relation can serve as a check on the measured values of the individual μ's.

9.5 MIXING PROCESSES

In Section 9.2 we considered the problem of particle exchange between two phases of a given substance. Here we shall address the phenomenon of the mixing or *interdiffusion* of two different gases. We assume that the mixing takes place at constant temperature and pressure, so that the Gibbs function is again a central concept in the description of the process.

Dalton's law of partial pressures states that the pressure P_j of the jth constituent gas of a mixture of gases is given by

$$P_j = x_j P, \tag{9.50}$$

where x_j is the kilomole fraction of the jth constituent gas defined in Section 9.3, and P is the pressure of the mixture.* Equation (9.50) is a definition of the partial pressure; the relation follows immediately from the ideal gas law for which

$$P_j = \frac{n_j RT}{V}, \quad P = \frac{nRT}{V},$$

since then

$$\sum_j P_j = \frac{P}{n} \sum_j n_j = P \sum_j x_j = P.$$

The Gibbs function for a mixture is

$$G = \sum_j n_j g_j, \tag{9.51}$$

where g_j is the specific Gibbs function for the jth constituent. To calculate the specific Gibbs function for an ideal gas we start with the first law expressed in the form

$$T ds = c_P dT - v dP.$$

With $v = RT/P$, we obtain

$$ds = c_P \frac{dT}{T} - R \frac{dP}{P}.$$

---

*We omit the superscript on the kilomole fraction because there is only one phase in this process. Here $\sum_j n_j = n$, the total number of kilomoles of the system.

Integration gives

$$s = c_P \ln T - R \ln P + s_0.$$

Since

$$g = u + Pv - Ts = h - Ts,$$

and

$$h = c_P T + h_0,$$

it follows that

$$g = c_P T - c_P T \ln T + RT \ln P - T s_0 + h_0,$$

or

$$g = RT (\ln P + \phi), \tag{9.52}$$

where ϕ is a function that depends on T only.

We can now calculate the change in the Gibbs function as a result of the mixing of two ideal gases (Figure 9.3). The initial Gibbs function (before mixing) is

$$G_i = n_1 g_{1i} + n_2 g_{2i}, \tag{9.53}$$

where

$$g_{1i} = RT (\ln P + \phi_1), \quad g_{2i} = RT (\ln P + \phi_2),$$

so that

$$G_i = n_1 RT (\ln P + \phi_1) + n_2 RT (\ln P + \phi_2). \tag{9.54}$$

Figure 9.3 The mixing of two ideal gases. (a) The initial state; the gases are at the same temperature and pressure and are separated by a diaphragm. (b) The final state; the diaphragm has been removed, allowing the gases to mix. The final temperature is T and the final pressure is the sum of the partial pressures.

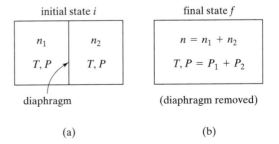

Although ϕ_1 and ϕ_2 are functions of the same temperature T, the functions themselves are different because c_P will in general not have the same value for the two gases.

The final value of the Gibbs function is

$$G_f = n_1 g_{1f} + n_2 g_{2f},$$

with

$$g_{1f} = RT(\ln P_1 + \phi_1), \quad g_{2f} = RT(\ln P_2 + \phi_2).$$

Since T doesn't change in the mixing process, ϕ_1 and ϕ_2 will be the same for the final state as for the initial state. Using $P_1 = x_1 P$ and $P_2 = x_2 P$, we obtain

$$g_{1f} = RT(\ln P + \phi_1 + \ln x_1), \quad g_{2f} = RT(\ln P + \phi_2 + \ln x_2),$$

so

$$G_f = n_1 RT(\ln P + \phi_1 + \ln x_1) + n_2 RT(\ln P + \phi_2 + \ln x_2). \quad (9.55)$$

Suppose that we define

$$\mu \equiv RT(\ln P + \phi + \ln x)$$

$$= g + RT \ln x.$$

Then

$$G_f = n_1 \mu_1 + n_2 \mu_2.$$

Comparing this with Equation (9.11) we see that μ is the chemical potential! Since g_1 and g_2 are characteristics of the constituents, they are the same for the initial and final states and we can drop the subscripts on g in Equation (9.53), which becomes

$$G_i = n_1 g_1 + n_2 g_2. \quad (9.56)$$

Our final result is therefore

$$\Delta G = G_f - G_i = n_1(\mu_1 - g_1) + n_2(\mu_2 - g_2)$$

$$= RT(n_1 \ln x_1 + n_2 \ln x_2),$$

or

$$\Delta G = nRT(x_1 \ln x_1 + x_2 \ln x_2). \quad (9.57)$$

We note that since x_1 and x_2 are both positive numbers less than unity, $\Delta G < 0$. Also, since $\mu = g + RT \ln x, \mu = g$ for one constituent ($x = 1$).

Using Equation (9.57), one can immediately calculate the entropy change for the mixing process. From the reciprocity relation $S = -(\partial G/\partial T)_P$, it is also true that

$$\Delta S = -\left[\frac{\partial(\Delta G)}{\partial T}\right]_P.$$

Hence

$$\Delta S = -nR(x_1 \ln x_1 + x_2 \ln x_2). \tag{9.58}$$

Clearly $\Delta S > 0$, as must be the case.

Entropy increase is to be expected when two different gases are mixed. But what if the two gases are the same? Does the removal of a diaphragm separating the two halves of a volume V filled with one kind of gas change its entropy or not? Then

$$x_1 = \frac{n_1}{n_1 + n_2}, \quad x_2 = \frac{n_2}{n_1 + n_2},$$

with

$$n_1 = \frac{P(V/2)}{RT} = n_2,$$

so that

$$x_1 = \frac{1}{2} = x_2$$

and

$$\Delta S = nR \ln 2 > 0. \tag{9.59}$$

Suppose that the properties of gas 1 approach those of gas 2. The previous result gives a finite entropy increase. However, if the two gases are identical there can be no change in entropy when the diaphragm is removed ($\Delta S = 0$). This is the *Gibbs paradox*.

The resolution will be fully discussed when we treat statistical thermodynamics. Suffice it to say now that there is a positive change in entropy when different, *distinguishable* gases are mixed. The transition from different to identical gases is not a continuous change, so Equation (9.59) does not apply to the case where the gases are the same.

PROBLEMS

9-1 For a one-component open system,

$$dU = T\,dS - P\,dV + \mu\,dn.$$

(a) Use Euler's theorem for homogeneous functions to show that

$$U = TS - PV + \mu n.$$

(b) Prove that $(dG)_{T,P} = \mu\,dn$ or, equivalently,

$$\mu = \left(\frac{\partial G}{\partial n}\right)_{T,P}.$$

(c) Prove the alternative definition of the chemical potential

$$\mu = \left(\frac{\partial F}{\partial n}\right)_{T,V}.$$

9-2 (a) Express the chemical potential of an ideal gas in terms of the temperature T and the volume V:

$$\mu = c_p T - c_v T \ln T - RT \ln V - s_0 T + \text{constant}.$$

(Hint: find the entropy $S = S(T, V)$; use $G = U - TS + PV$ and write $\mu = G/n$.)

(b) Similarly, find μ in terms of T and P. Show that the chemical potential at the fixed temperature T varies with pressure as

$$\mu = \mu_0 + RT \ln\left(\frac{P}{P_0}\right),$$

where μ_0 is the value of μ at the reference point (P_0, T). This expression is of great use in chemistry.

9-3 A container of volume V is divided by partitions into three parts containing one kilomole of helium gas, two kilomoles of neon gas, and three kilomoles of argon gas, respectively. The temperature of each gas is initially 300 K and the pressure is 2 atm. The partitions are removed and the gases diffuse into one another.

(a) Calculate the kilomole fraction and the partial pressure of each gas in the mixture.

(b) Calculate the change of the Gibbs function and the change of the entropy of the system in the mixing process.

9-4 Show that for a closed system consisting of two phases coexisting in equilibrium at a temperature T and under a pressure P,

$$\left(\frac{\partial P}{\partial V}\right)_s = -\frac{T}{C_v}\left(\frac{dP}{dT}\right)^2.$$

Here dP/dT is the slope of the phase equilibrium curve.

9-5 A mixture of gold and thallium can exist in equilibrium with four phases present: solution, vapor, solid gold, and solid thallium. What is the variance? (This point is known as the *eutectic point.*)

9-6 The Gibbs phase rule can be generalized to systems in which there occur r chemical reactions:

$$f = k - \pi - r + 2.$$

Determine the number of degrees of freedom at equilibrium of a chemically reactive system containing solid sulfur S and the three gases O_2, SO_2, and SO_3. The elements S and O_2 appear in the reactions

$$S + O_2 \rightarrow SO_2 \quad \text{and} \quad S + \frac{3}{2}O_2 \rightarrow SO_3.$$

9-7 (a) Show that

$$\left(\frac{\partial S}{\partial P}\right)_{H,n} = \frac{-V}{T}.$$

(b) For an ideal gas, the entropy $S(H, P, n)$ has the form

$$S(H, P, n) = nR \ln\left[\frac{\Phi(H,n)}{P}\right],$$

where $\Phi(H, n)$ is a function of the enthalpy and the number of kilomoles. Use the result of part (a) to derive the equation of state.

9-8 A container is initially separated by a diathermal wall into two compartments of equal volume. The left compartment is filled with 1 kilomole of neon gas at a pressure of 4 atmospheres and the right with argon gas at 1 atmosphere. The gases may be considered ideal. The whole system is initially at temperature $T = 300$K, and is thermally insulated from the outside world. Suppose that the diathermal wall is removed.
(a) What is the new temperature of the system? The new pressure?
(b) What is the change in the Gibbs function of the system?
(c) What is the change in the entropy of the system?

9-9 (a) Show that for an open system with one component,

$$dG = -SdT + VdP + \mu dn.$$

(b) Using this result, calculate G for a van der Waals gas, assuming a fixed amount of material at a given, fixed temperature. Show that

$$G = -nRT \ln(V - nb) + \frac{n^2 bR}{V - nb} - \frac{2n^2 a}{V} + C(T),$$

where the integration constant $C(T)$ is, in general, different for different temperatures.

9-10 Show that, during a first-order phase transition:
(a) The change of entropy of the system undergoing the transition is a linear function of the volume change.

(b) The change of internal energy is given by

$$\Delta U = L\left(1 - \frac{d \ln T}{d \ln P}\right),$$

where L is the latent heat of transformation.

9-11 (a) Show that the Gibbs-Duhem equation leads to the relation

$$\sum_i x_i d\mu_i = 0 \quad (T, P \text{ constant}),$$

where x_i is the kilomole fraction of the ith component.

(b) Show that for a two-component system

$$\left(\frac{d\mu_1}{d \ln x_1}\right)_{T,P} = \left(\frac{d\mu_2}{d \ln x_2}\right)_{T,P}.$$

(c) Show that for a two-component liquid phase whose vapor can be treated as an ideal gas mixture,

$$\left(\frac{d \ln P_1}{d \ln x_1}\right)_{T,P} = \left(\frac{d \ln P_2}{d \ln x_2}\right)_{T,P},$$

where P_1 and P_2 are the partial pressures of the components of the vapor, and x_1 and x_2 are the kilomole fractions of the liquid. (This is known as the Duhem-Margules equation.)

Chapter *10*

The Third Law of Thermodynamics

10.1 STATEMENTS OF THE THIRD LAW

The third law of thermodynamics is concerned with the behavior of systems in equilibrium as their temperature approaches zero. The definition of entropy given by

$$S = \int_0^T \frac{đQ_r}{T} + S_0 \qquad (10.1)$$

is incomplete because of the undetermined additive constant S_0, the entropy at absolute zero. In this short chapter we shall introduce a principle that will enable us to determine S_0. The principle, which was discovered by Nernst in 1906, is often referred to as Nernst's postulate or the third law of thermodynamics.

As long as we deal only with differences of the entropy, a knowledge of S_0 is unnecessary. However, the absolute entropy is by no means unimportant. Suppose, for example, that we wish to determine the change in the Gibbs function. Since $dG = -SdT + PdV$, it is not sufficient to know dS; we must know S itself. In chemistry, moreover, knowing the absolute entropy makes it possible to calculate the equilibrium constants of a chemical reaction from the thermal properties of the reactants.

We have seen that a spontaneous process can occur in a system at constant temperature and pressure if the Gibbs function decreases. The Gibbs function is related to the enthalpy by the equation $G = H - TS$. Using the reciprocity relation $S = -(\partial G/\partial T)_P$, we obtain an expression known as the Gibbs-Helmholtz equation:

$$G = H + T\left(\frac{\partial G}{\partial T}\right)_P. \qquad (10.2)$$

If this relation is applied to the initial and final states of a system undergoing an isothermal process, it takes the form

$$\Delta G = \Delta H + T\left[\frac{\partial(\Delta G)}{\partial T}\right]_P, \qquad (10.3)$$

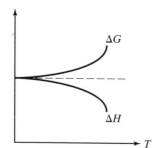

Figure 10.1   Variation of ΔG and ΔH in the vicinity of absolute zero.

which shows that the change in enthalpy and the change in the Gibbs function are equal at $T = 0$ for an isobaric process. Based on the results of experiments, Nernst postulated that as the temperature tends to zero, not only do ΔG and ΔH become equal, but their temperature rates of change approach zero as well:

$$\lim_{T \to 0}\left[\frac{\partial(\Delta G)}{\partial T}\right]_P = 0, \qquad \lim_{T \to 0}\left[\frac{\partial(\Delta H)}{\partial T}\right]_P = 0. \tag{10.4}$$

This behavior is illustrated in Figure 10.1.

We may write the first of the two equations as

$$\lim_{T \to 0}\left[\frac{\partial(G_2 - G_1)}{\partial T}\right]_P = \lim_{T \to 0}\left[\left(\frac{\partial G_2}{\partial T}\right)_P - \left(\frac{\partial G_1}{\partial T}\right)_P\right] = 0, \tag{10.5}$$

where the subscripts 1 and 2 refer to the initial and final states, respectively. From the reciprocity relation, we have

$$\lim_{T \to 0}(S_1 - S_2) = 0. \tag{10.6}$$

This is the Nernst formulation of the third law:

All reactions in a liquid or solid in thermal equilibrium take place with no change of entropy in the neighborhood of absolute zero.

Although the Nernst theorem is quoted for solids and liquids, it is believed to hold for all systems in equilibrium states. (There are certain quantum systems that constitute gas-like aggregations at absolute zero.)

In 1911 Planck extended Nernst's hypothesis by assuming that it holds for G_1 and G_2 separately. Specifically, he proposed that

$$\lim_{T \to 0} G(T) = \lim_{T \to 0} H(T), \tag{10.7}$$

and

$$\lim_{T \to 0} \left(\frac{\partial G}{\partial T} \right)_P = \lim_{T \to 0} \left(\frac{\partial H}{\partial T} \right)_P. \tag{10.8}$$

These statements lead to an important conclusion. For convenience, we temporarily introduce a variable $\Phi \equiv G - H$. Equation (10.2) then becomes

$$T \left(\frac{\partial G}{\partial T} \right)_P - \Phi = 0. \tag{10.9}$$

By Equations (10.7) and (10.8), we have in the limit when $T = 0$,

$$\Phi = 0 \quad \text{and} \quad \left(\frac{\partial \Phi}{\partial T} \right)_P = 0. \tag{10.10}$$

Adding the term $-T(\partial H/\partial T)_P$ to both sides of Equation (10.9), we get

$$T \left(\frac{\partial \Phi}{\partial T} \right)_P - \Phi = -T \left(\frac{\partial H}{\partial T} \right)_P,$$

or

$$\left(\frac{\partial \Phi}{\partial T} \right)_P - \frac{\Phi}{T} = - \left(\frac{\partial H}{\partial T} \right)_P. \tag{10.11}$$

By L'Hôpital's rule, the limit of Φ/T for $T = 0$ is the same as the limit of $(\partial \Phi/\partial T)_P$, which is zero, according to Equation (10.10). Therefore,

$$\lim_{T \to 0} \left(\frac{\partial H}{\partial T} \right)_P = 0. \tag{10.12}$$

Finally, by Equation (10.8), we obtain the result

$$\lim_{T \to 0} \left(\frac{\partial G}{\partial T} \right)_P = 0. \tag{10.13}$$

Since $(\partial G/\partial T)_P = -S$, it follows that

$$\lim_{T \to 0} S = 0. \tag{10.14}$$

This is Planck's statement of the third law. In words:

The entropy of a true equilibrium state of a system at absolute zero is zero.

This statement, stronger than the Nernst theorem, needs qualification. Certain glasses have a nonvanishing entropy as the temperature tends to absolute zero. This makes sense, since statistical thermodynamics associates entropy with disorder, and glass is a disordered structure. The Planck statement can be shown to be valid for every pure crystalline solid. Quantum statistics is necessary for a complete understanding of the absolute entropy of a system.

Still another statement of the third law is:

> *It is impossible to reduce the temperature of a system to absolute zero using a finite number of processes.*

In Section 10.3 this so-called unattainability principle will be shown to be equivalent to Nernst's postulate.

10.2 METHODS OF COOLING

A widely used method for cooling a substance is to isolate it from its surroundings (with no heat flow in or out) and reduce its temperature in an adiabatic reversible process. Then work is done by the system solely at the expense of its internal energy. This can be accomplished by varying some parameter such as the magnetic field. The process is called adiabatic demagnetization and is discussed in Chapter 17.

In the 1970s *laser cooling* was proposed, a process in which the deceleration of a beam of atoms is accomplished by directing a laser beam so as to oppose the atomic beam. Under the action of the laser beam, momentum is transferred to an atom by absorption of a photon. The laser light frequency is chosen to excite the lowest order atomic transition. If the laser frequency is lower than the atomic resonant frequency and the atom is moving against the laser beam, the laser frequency in the rest frame of the atom is Doppler shifted toward resonance. The key point is that the laser light "pushes on" only those atoms that are moving into the laser beam.

Following absorption, the atom spontaneously emits the photon in a random direction. This absorption and isotropic reradiation result in an average driving force in the direction of the incident light. Since the force is proportional to the velocity, the atomic motion is slowed and the atoms are thereby cooled. In this process, the atomic vapor is cooled without becoming a liquid or a solid.

Various methods, including laser detuning, have been used to compensate the changing Doppler shift as the atoms decelerate so as to keep them near resonance. Three-dimensional cooling results if orthogonal pairs of laser beams are used in a so-called "optical molasses" configuration.

A thorough understanding of the cooling mechanisms involved has made it possible to conduct novel experiments on cooled atoms and to achieve temperatures of a fraction of a microkelvin, the lowest ever measured.

10.3 EQUIVALENCE OF THE STATEMENTS

By considering the adiabatic reversible process, we can show in a few steps the equivalence of two statements of the third law. We assume that absolute zero is unattainable and prove that in a change from equilibrium state 1 to equilibrium state 2,

$$\Delta S_0 \equiv S_{02} - S_{01} = 0, \tag{10.15}$$

where the naught subscript refers to $T = 0$. We can examine the process in the entropy-temperature plane. The shape of the S versus T curve can be found from the relation

$$S = S_0 + \int_0^T \frac{dQ_r}{T}. \tag{10.16}$$

Now $C_v = (dQ/dT)_V$ and $C_P = (dQ/dT)_P$. We assume that either P or V is held constant and drop the subscript. Thus

$$S = S_0 + \int_0^T C\frac{dT}{T}. \tag{10.17}$$

We note that the Debye law for the heat capacity of a solid gives $C_V \propto T^3$. In this case, S increases with T according to some power law, as Figure 10.2 reflects. The curve S_1 applies to an initial value of the variable parameter and S_2 to some final value.

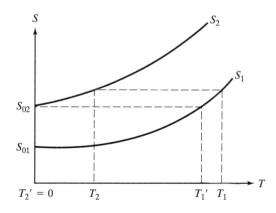

Figure 10.2 Entropy-temperature diagram for a hypothetical cooling process in which $S_1 < S_2$. (The entropy curve for the initial value of the varied parameter lies below that of the final value.)

Cooling from T_1 to T_2 by an adiabatic (and isentropic) reversible process occurs along the horizontal line connecting the S_1 and S_2 curves. Then $S_2 = S_1$ gives

$$S_{02} + \int_0^{T_2} C_2 \frac{dT}{T} = S_{01} + \int_0^{T_1} C_1 \frac{dT}{T}. \qquad (10.18)$$

Transposing terms, we have

$$\int_0^{T_2} C_2 \frac{dT}{T} = -(S_{02} - S_{01}) + \int_0^{T_1} C_1 \frac{dT}{T}. \qquad (10.19)$$

Now, C is always positive; if it were negative, a heat gain would result in a lowering of the temperature, and the attainment of absolute zero would be trivially easy!

If $(S_{02} - S_{01}) > 0$, as shown in Figure 10.2, and because $C > 0$, there must be some temperature T_1' for which the right-hand side of Equation (10.19) is zero. Then

$$\int_0^{T_2'} C_2 \frac{dT}{T} = 0, \qquad (10.20)$$

where T_2' is the final temperature corresponding to T_1'. But T_2' must be zero if Equation (10.20) is to hold. Thus absolute zero could be achieved through the adiabatic change from state 1 to state 2 if $S_{02} > S_{01}$. This violates the unattainability principle. We conclude that

$$S_{02} \le S_{01}. \qquad (10.21)$$

Thus we need to redraw the S-T diagram with $S_{02} < S_{01}$ (Figure 10.3). The adiabat $S_2 = S_1$ gives Equation (10.18) as before, which can be rewritten as

$$-(S_{01} - S_{02}) + \int_0^{T_2} C_2 \frac{dT}{T} = \int_0^{T_1} C_1 \frac{dT}{T}. \qquad (10.22)$$

If $(S_{01} - S_{02}) > 0$, there is some value of T_2 — say T_2'' — such that the left-hand side of Equation (10.22) is zero. Then

$$\int_0^{T_1''} C_1 \frac{dT}{T} = 0,$$

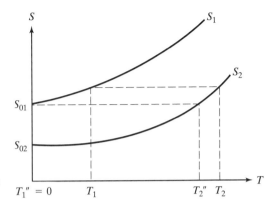

Figure 10.3 Entropy-temperature diagram for a hypothetical cooling process in which $S_1 > S_2$.

requiring that $T_1'' = 0$. This also violates the unattainability principle and we conclude that for this case

$$S_{02} \geq S_{01}. \qquad (10.23)$$

The only way that Equations (10.21) and (10.23) can both be true is if

$$S_{02} = S_{01}, \qquad (10.24)$$

which is the Nernst postulate. According to Equation (10.24), the curves of S_2 and S_1 versus T must come together at $T = 0$. The final sketch, Figure 10.4, nicely illustrates the equivalence of the two statements of the third law. The curves have been drawn to reflect the Planck statement, though the proof does not place the entropy at absolute zero.

A series of isothermal and adiabatic processes (demagnetizations) are represented by the zigzag path in the figure. Each successive process reduces the temperature. It is clear, however, that because the curves intersect at

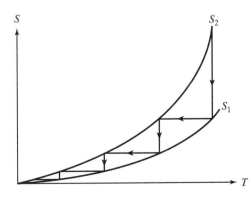

Figure 10.4 Entropy-temperature diagram for an actual cooling process. The unattainability of absolute zero is illustrated by the indefinitely increasing number of steps required to achieve a given temperature reduction as absolute zero is approached.

$T = 0$, an infinite number of steps would be required to reach absolute zero. We conclude that it is impossible to achieve a temperature of absolute zero by any practical method. This fundamental fact of nature should not be especially disquieting, however, given that cryogenics has enabled us to get extremely close to $T = 0$.

10.4 CONSEQUENCES OF THE THIRD LAW

1. *Expansivity*
   The expansivity (coefficient of volume expansion) is defined as

$$\beta = \frac{1}{V}\left(\frac{\partial V}{\partial T}\right)_P.$$

From Section 8.5 we have the Maxwell relation

$$\left(\frac{\partial V}{\partial T}\right)_P = -\left(\frac{\partial S}{\partial P}\right)_T.$$

The Nernst postulate implies that the change in entropy is zero in any process at absolute zero; that is,

$$\lim_{T\to 0}\left(\frac{\partial S}{\partial P}\right)_T = 0.$$

Since the volume V remains finite as $T \to 0$, it is therefore true that

$$\lim_{T\to 0}\beta = 0. \tag{10.25}$$

We conclude that in the case of all known solids, the expansivity approaches zero as the temperature approaches absolute zero.

2. *Slope of the phase transformation curves*
   In connection with any phase change that takes place at low temperature we can invoke the Clausius-Clapeyron equation (Section 8.7):

$$\frac{dP}{dT} = \frac{\Delta S}{\Delta V}.$$

The Nernst theorem states that

$$\lim_{T\to 0}\Delta S = 0.$$

Since ΔV is not zero for so-called first-order transitions, it follows that

$$\lim_{T \to 0} \frac{dP}{dT} = 0. \tag{10.26}$$

The slope of the boundary between two phases is zero at absolute zero. This has been verified for all known sublimation curves, for the vaporization curve of Helium II, and for the fusion curve of solid helium. Experimental results show that for solid helium, $dP/dT \propto T^7$, which approaches zero at a very fast rate indeed.

3. *Heat Capacity*
From the definitions of the heat capacities at constant volume and constant pressure, we have

$$C_V = T \left(\frac{\partial S}{\partial T} \right)_V, \qquad C_P = T \left(\frac{\partial S}{\partial T} \right)_P.$$

Integrating, we obtain

$$S - S_0 = \int_0^T C_V \frac{dT}{T}, \qquad S - S_0 = \int_0^T C_P \frac{dT}{T}.$$

The left-hand sides of the equations will remain finite down to absolute zero, where, by the third law, they will vanish. The right-hand sides must behave in the same way. The integrals show that the heat capacities must decrease to zero at least as rapidly as T. Otherwise the integrals would diverge because of the lower limit of $T = 0$. We conclude that

$$\lim_{T \to 0} C_V = 0, \qquad \lim_{T \to 0} C_P = 0. \tag{10.27}$$

That is, the heat capacities approach zero as the temperature approaches absolute zero. With respect to this behavior, extensive experiments provide a full confirmation of the third law.

PROBLEMS

10-1 Show that the heat capacity at constant volume can be written

$$C_V = \left(\frac{\partial S}{\partial \ln T} \right)_V.$$

It follows that as $T \to 0$, $\ln T \to -\infty$ and $S \to 0$, so that $C_V \to 0$, in confirmation of Equation (10.18). By the same argument, $C_P \to 0$ as $T \to 0$.

10-2 Show that, in view of the third law, neither the ideal gas equation nor the van der Waals equation can hold at $T = 0$.

10-3 **(a)** Show that the third law cannot predict the value of the isothermal compressibility κ at absolute zero. (It remains finite.)
(b) The isothermal bulk modulus B is the reciprocal of κ.

$$B \equiv -V\left(\frac{\partial P}{\partial V}\right)_T.$$

Show that

$$\lim_{T \to 0}\left(\frac{\partial B}{\partial T}\right)_V = 0.$$

10-4 According to Debye's theory of heat capacities, to be discussed in Chapter 16, the specific heat capacity of an insulating solid at very low temperatures is given by

$$c_v = \frac{12}{5}\pi^4 R\left(\frac{T}{\theta_D}\right)^3,$$

where c_v is in the usual units of J kilomole^{-1} K^{-1}. Find the specific entropy of such a solid at $T = 0.01$K and $T = 10$K, given that $\theta_D = 300$K.

10-5 **(a)** Show that if $C_V = bT^\alpha$ at low temperatures, the third law requires that $\alpha > 0$.
(b) If $C_V = aT + bT^3$ at low temperatures, calculate the variation of the entropy with temperature.

10-6 Consider a solid whose equation of state is

$$PV + f(V) = AU,$$

where $f(V)$ is a function of the volume only and A is a constant. Show that $C_V \to 0$ as $T \to 0$.

10-7 A low-temperature physicist wishes to publish his experimental result that the heat capacity of a nonmagnetic dielectric material between 0.05K and 0.5K varies as $AT^{1/2} + BT^3$. As editor of the journal, should you accept the paper for publication?

Chapter 11

The Kinetic Theory of Gases

11.1 BASIC ASSUMPTIONS

In the nineteenth century, scientists realized that in order to go beyond the limitations of classical thermodynamics, we must take into account the structure of matter. The recognition that bulk matter is composed of particles led to the kinetic theory, which assumes that atoms and molecules obey the same laws of mechanics that apply to macroscopic systems. This assumption was ultimately challenged by quantum theory and, indeed, statistical thermodynamics is best treated from the quantum mechanical point of view.

Nonetheless, kinetic theory yields far deeper insight into such concepts as pressure, internal energy, and specific heat than classical thermodynamics provides, and is therefore well worth investigating. In addition, it gives an explanation of transport processes such as viscosity, heat conduction, and diffusion that is quite satisfactory for most purposes.

Kinetic theory, as the name implies, is concerned with the motion of molecules rather than the forces between them. It is therefore especially suitable for describing the behavior of gases at relatively low pressures.

The fundamental assumptions of the kinetic theory of gases are:

1. *A macroscopic volume contains a large number of molecules.*
   What do we mean by this statement? We know that Avogadro's number is

   $$N_A = 6.02 \times 10^{26} \text{ molecules kilomole}^{-1}$$

   and that at standard temperature and pressure (STP), one kilomole of a gas occupies a volume of 22.4 m^3. This gives a molecular density of about 3×10^{25} molecules m^{-3} or 3×10^{19} molecules cm^{-3}. Even at high vacuums down to 10^{-11} atmospheres of pressure, at ordinary temperatures there are still many millions of molecules in each cubic centimeter.

2. *The separation of molecules is large compared with molecular dimensions and with the range of intermolecular forces.*

Take a volume of 22.4 m³, divide it into cubical cells with one molecule in each cell. The volume of each cell (at STP) is $22.4 \text{ m}^3/6.02 \times 10^{26} \approx 30 \times 10^{-27} \text{ m}^3$. Thus the distance between molecules in this case is equal to the cube root of this volume, or roughly 3×10^{-9} m. This is a factor of about 10 greater than the diameter of a typical gas molecule.

We must also consider the range of the forces of interaction between molecules, which are electrical in nature. The force between two molecules is strong and repulsive at separations approximately equal to the molecular diameter and is associated with the overlap of electron clouds. At greater distances the force is attractive and is due to induced electrical dipoles. The behavior is often characterized by the total potential energy

$$V(r) = 4(\Delta E)\left[\left(\frac{d}{r}\right)^{12} - \left(\frac{d}{r}\right)^6\right], \tag{11.1}$$

where ΔE and d are empirically determined parameters. Equation (11.1) is known as the Lennard-Jones, or "6-12" potential, and is plotted in Figure 11.1. The curve has a minimum at $r/d = 2^{1/6} = 1.12$, where $V(r) = -\Delta E$. To the left of the minimum, the curve is very steep; to the right it is comparatively flat. Typically, ΔE lies in the range of 2×10^{-4} eV to 5×10^{-2} eV, and d varies between 2.2×10^{-10} m and 4.5×10^{-10} m. For molecular separations of an order of magnitude greater, the interaction force is evidently negligibly small.

In a liquid or solid, the molecules have nearest neighbors at a distance r only slightly greater than d. Thus there is substantial interaction between them. In contrast, the molecules of a gas under normal conditions are most of the time traveling freely through space. An ideal gas is a good

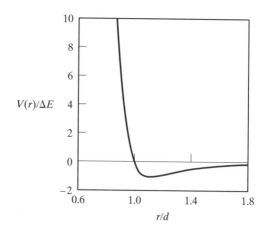

Figure 11.1   Variation of the intermolecular potential $V(r)$ with the intermolecular distance r for a pair of gas molecules.

approximation at these low densities; dry air at STP deviates from the ideal gas law by less than 0.1 percent. For high density gases, on the other hand, it is essential to include the intermolecular forces, as is done in the van der Waals model.

3. *No forces exist between molecules except those associated with collisions.*
That is, molecules move in straight-line paths between collisions with constant velocities. Each molecule moves forever in a zigzag path, for it collides many million times per second with other molecules of the gas or with the molecules that make up the walls of the container. These collisions are so brief that in our analysis we entirely ignore their duration. Furthermore, the distance a molecule moves between collisions can lie between zero and thousands of times the molecular diameter. However, the very short and very long distances are unusual. The *average* distance between collisions is called the *mean free path*. In oxygen at STP, the mean free path is about 10^{-7} m.

4. *The collisions are elastic.*
Both momentum and mechanical energy are conserved in the collision process.

5. *The molecules are uniformly distributed within a container.*

6. *The directions of the velocities of the molecules are uniformly distributed.*

The last two assumptions characterize the randomness of the motion. The basic idea of kinetic theory is that the molecules of the gas are moving rapidly and randomly, colliding with each other and with the walls of the container, thereby creating a pressure due to their countless impacts. We are concerned with a gas that is in thermal equilibrium at a fixed temperature and in mechanical equilibrium with the walls. It follows from these assumptions that the average molecular speed and the probability distribution of speeds are unchanging.

Let $f(v)dv$ be the fractional number of molecules in the speed range v to $v + dv$. The probability density function $f(v)$ has the dimensions of v^{-1} since

$$\int_0^\infty f(v)dv = 1.$$

The mean speed is given by

$$\bar{v} \equiv \int_0^\infty vf(v)dv, \tag{11.2}$$

and the mean square speed is

$$\overline{v^2} \equiv \int_0^\infty v^2 f(v)dv. \tag{11.3}$$

The square root of $\overline{v^2}$ is called the *root mean square* or *rms* speed:

$$v_{rms} \equiv \sqrt{\overline{v^2}}. \tag{11.4}$$

The nth moment of the distribution is defined as

$$\overline{v^n} \equiv \int_0^\infty v^n f(v)\,dv.$$

We shall also be interested in the most probable speed, obtained by differentiating $f(v)$ and setting the derivative equal to zero. The speed distribution will be derived in Section 11.6. We can develop the fundamental aspects of the kinetic theory, however, without knowing the exact form of the distribution function.

11.2 MOLECULAR FLUX

In many applications of kinetic theory, we need to know the molecular flux, the number of molecules striking a unit area per unit time. Consider a slanting cylinder of slant height $v\,dt$ at angle θ measured with respect to the normal to the elementary area dA (Figure 11.2). We want to find the number of molecules that strike the area dA in time dt.

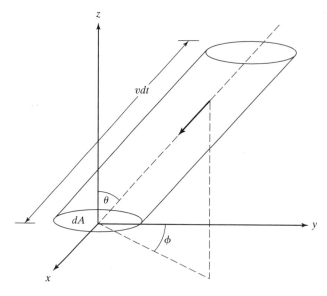

Figure 11.2 Slant cylinder geometry used to calculate the number of molecules that strike the area dA in time dt.

We define $n \equiv N/V$, the number of molecules per unit volume. Now, the number of molecules with speeds v to $v + dv$ in the cylinder is $nf(v)dv$ × (volume of the cylinder). Since the volume of the cylinder is $(dA \cos\theta)(vdt)$, the number of molecules with speeds v to $v + dv$ in the cylinder is $n v f(v) dv \cos\theta dA dt$. Only a few of these molecules are traveling toward dA. Let the velocity of each molecule be represented by a vector, and imagine that all the velocity vectors are moved parallel to themselves to a common origin (Figure 11.3). Assumption 6 states that the head ends are distributed uniformly in a spherical shell of radius v and thickness dv. The element of area of the surface of this sphere is

$$d\sigma = v^2 \sin\theta \, d\theta \, d\phi,$$

with

$$\int_{\text{sphere}} d\sigma = 4\pi v^2.$$

Therefore, the fraction of the total number of molecules with angular distributions θ to $\theta + d\theta$, ϕ to $\phi + d\phi$, is

$$\frac{d\sigma}{\displaystyle\int_{\text{sphere}} d\sigma} = \frac{\sin\theta \, d\theta \, d\phi}{4\pi}.$$

The number of molecules in the cylinder at time t traveling with speeds v to $v + dv$ *toward dA* (i.e., the θ, ϕ molecules) is

$$dN = n v f(v) dv \cos\theta dA dt \frac{\sin\theta \, d\theta \, d\phi}{4\pi}. \tag{11.5}$$

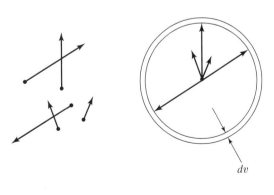

Figure 11.3 (a) Randomly oriented molecular velocities; (b) the same vectors referred to a common origin.

(a)   (b)

The corresponding particle flux is, by definition,

$$d\Phi = \frac{dN}{dA\,dt}.$$ (11.6)

Substituting Equation (11.5) in Equation (11.6) and integrating, we obtain

$$\Phi = \frac{n}{4\pi} \int_0^\infty vf(v)\,dv \int_0^{\pi/2} \sin\theta\,\cos\theta\,d\theta \int_0^{2\pi} d\phi,$$ (11.7)

or

$$\Phi = \frac{1}{4}n\bar{v},$$ (11.8)

where $\bar{v}$ is the mean speed, defined in Equation (11.2).

11.3 GAS PRESSURE AND THE IDEAL GAS LAW

Next we consider a single molecule colliding with a surface. The normal component of the change of momentum is (Figure 11.4)

$$mv\cos\theta - (-mv\cos\theta) = 2mv\cos\theta.$$

The change of momentum dp produced by all the molecules with speeds v to $v + dv$ traveling toward an element of the surface dA in time dt is

$$\begin{aligned}
dp &= \int 2mv\cos\theta\,dN \\
&= \int \int \int 2\,mv\cos\theta \cdot nvf(v)\,dv\cos\theta\,dA\,dt \cdot \frac{\sin\theta\,d\theta\,d\phi}{4\pi} \\
&= \frac{nm\,dA\,dt}{2\pi} \int_0^\infty v^2 f(v)\,dv \int_0^{\pi/2} \sin\theta\cos^2\theta\,d\theta \int_0^{2\pi} d\phi \\
&= \frac{1}{3}nm\overline{v^2}\,dA\,dt.
\end{aligned}$$

From Newton's second law and the definition of the pressure P, we know that

$$\frac{dp}{dA\,dt} = P,$$

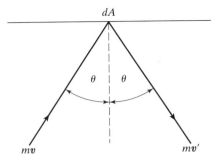

Figure 11.4 Change in momentum of a molecule striking a surface element dA.

so that

$$P = \frac{1}{3}nm\overline{v^2}. \tag{11.9}$$

We note that $nm = Nm/V$. Then

$$PV = \frac{1}{3} Nm\overline{v^2} = \frac{2}{3}\left(\frac{1}{2}Nm\overline{v^2}\right). \tag{11.10}$$

The right-hand side is two-thirds of the average translational kinetic energy of the molecules of the gas.

Equation (11.10) is highly suggestive of the ideal gas law

$$PV = nRT. \tag{11.11}$$

The number of kilomoles is simply related to N:

$$n = \text{no. of kilomoles} = \frac{\text{no. of molecules}}{\text{no. of molecules kilomole}^{-1}} = \frac{N}{N_A},$$

where N_A is Avogadro's number. Thus

$$PV = NkT, \tag{11.12}$$

where

$$k = \frac{R}{N_A} = \frac{8.31 \times 10^3 \text{ J kilomole}^{-1} \text{ K}^{-1}}{6.02 \times 10^{26} \text{ molecules kilomole}^{-1}} = 1.38 \times 10^{-23} \text{ JK}^{-1}.$$

Here k is Boltzmann's constant; it is the ratio of two universal constants and is therefore also a universal constant. We see that Equation (11.12) will agree

with Equation (11.10), the result we derived from kinetic theory, if we make
the association

$$NkT = \frac{1}{3}Nm\overline{v^2},$$

or

$$\frac{3}{2}kT = \frac{1}{2}m\overline{v^2}. \tag{11.13}$$

Thus, the theory has provided a statistical interpretation of the absolute temperature of a dilute gas—namely, the temperature is proportional to the average kinetic energy of a molecule. The result is one of the great insights given to
us by the kinetic theory.

From Equation (11.13) it is apparent that the mean kinetic energy is
independent of the pressure, the volume, or the molecular species. At a given
temperature, the mean kinetic energies of different gases are all the same.

It is instructive to note that at room temperature (293 K),

$$\frac{3}{2}kT = 6.06 \times 10^{-21}\text{J} = 0.038\text{eV}. \tag{11.14}$$

We can compare this energy with ΔE, the depth of the potential "well" discussed at the beginning of the chapter. There we observed that ΔE has a magnitude ranging from about 0.0002 eV to 0.05 eV.

11.4 EQUIPARTITION OF ENERGY

Let the molecular velocity have components (v_x, v_y, v_z). Then

$$\overline{v^2} = \overline{v_x^2} + \overline{v_y^2} + \overline{v_z^2},$$

since taking an average is a linear operation. By assumption, there are no pre-
ferred directions, so

$$\overline{v_x^2} = \overline{v_y^2} = \overline{v_z^2} = \frac{1}{3}\overline{v^2}.$$

Thus the mean kinetic energy per molecule associated with any one component—
the x-component, for example—is

$$\frac{1}{2}m\overline{v_x^2} = \frac{1}{6}m\overline{v^2} = \frac{1}{2}kT,$$

since

$$\frac{1}{2}m\overline{v^2} = \frac{3}{2}kT. \tag{11.15}$$

It follows that the average translational kinetic energy associated with each component of velocity is one-third the total and has a magnitude of $kT/2$. Each component represents a *degree of freedom* f of a mechanical system. An atom or a monatomic molecule, for example, has three translational degrees of freedom so that $f = 3$.

A diatomic molecule is modeled as two masses connected by a spring. As such, it has seven degrees of freedom: three translational, two rotational, and two vibrational. There are two rotational degrees of freedom, because two angles are needed to specify the orientation of an axially symmetric molecule. The vibrational motion is represented by a one-dimensional harmonic oscillator. For simple harmonic motion, the energy is partly kinetic and partly potential; the average values of the two are equal, and both are expressible as quadratic forms. Thus these are two vibrational degrees of freedom.

The equipartition theorem (proved in Chapter 14) states that (1) there is one degree of freedom associated with every form of energy expressible in quadratic form, and (2) the mean value of the energy of a molecule corresponding to each degree of freedom is $kT/2$. Thus the total average energy of a molecule $\overline{\varepsilon}$ is $f\,kT/2$. It follows that the internal energy of a gas must be the total average energy of its N molecules. That is,

$$U = N\overline{\varepsilon} = \frac{f}{2}NkT = \frac{f}{2}nRT, \tag{11.16}$$

or

$$u = \frac{U}{n} = \frac{f}{2}RT. \tag{11.17}$$

11.5 SPECIFIC HEAT CAPACITY OF AN IDEAL GAS

For a reversible process, the specific heat capacity at constant volume is given by

$$c_v = \left(\frac{\partial u}{\partial T}\right)_v.$$

From Equation (11.17), this gives

$$c_v = \frac{d}{dT}\left(\frac{f}{2}RT\right) = \frac{f}{2}R.$$

TABLE 11.1   Ratio of specific heat capacities of various
gases at temperatures close to room temperature. The values
are obtained by measuring the velocity of sound in the gas.

Gas	γ	Gas	γ	Gas	γ
He	1.66	H_2	1.40	CO_2	1.29
Ne	1.64	O_2	1.40	NH_3	1.33
Ar	1.67	N_2	1.40	CH_4	1.30
Kr	1.69	CO	1.42	Air	1.40

Mayer's equation for an ideal gas is $c_P = c_v + R$. Hence,

$$c_P = \frac{f}{2}R + R = \left(\frac{f+2}{2}\right)R,$$

and the ratio of specific heats is

$$\gamma = \frac{c_P}{c_v} = \frac{f+2}{f}.$$

For a monatomic gas,

$$f = 3, \quad c_v = \frac{3}{2}R, \quad c_P = \frac{5}{2}R, \quad \gamma = \frac{5}{3} = 1.67.$$

These results are in good agreement with measured values at temperatures in
the vicinity of room temperature (Table 11.1).
　　For a diatomic gas,

$$f = 7, \quad c_v = \frac{7}{2}R, \quad c_P = \frac{9}{2}R, \quad \gamma = \frac{9}{7} = 1.28.$$

Here the agreement is poor; the number of degrees of freedom is too large.
We try

$$f = 5, \quad c_v = \frac{5}{2}R, \quad c_P = \frac{7}{2}R, \quad \gamma = \frac{7}{5} = 1.40.$$

Good agreement is obtained with this adjustment. Evidently, near room temperature the rotational degrees of freedom *or* the vibrational degrees of freedom are excited, but *not both*. Maxwell considered the resolution of this
question the most pressing challenge confronting the kinetic theory. It took
statistical thermodynamics to provide the final answer (see Chapter 15).

Two other observations are noteworthy. For large molecules, f will be large and $\gamma \rightarrow 1$. In general,

$$1 \leq \gamma \leq 1.67,$$

in pretty fair agreement with experimental results.

The kinetic theory predicts that c_v and c_P are constants for an ideal gas. We have seen that this is not true at low temperatures when both must approach zero. Kinetic theory, which is a classical theory, is incapable of accounting for this behavior. A quantum mechanical description is required.

11.6 DISTRIBUTION OF MOLECULAR SPEEDS

We have yet to derive the probability distribution of molecular speeds from which various averages can then be found. In Chapter 14, the so-called Maxwell-Boltzman distribution will be determined from a knowledge of the distribution of particles among energy levels. We include here a more physically satisfying derivation that makes reference to the molecular collisions through which the speed distribution is maintained.*

In a dilute gas, it is reasonable to assume that collisions are binary events, involving only two molecules. Simultaneous collisions of more than two molecules are exceedingly rare. In a binary collision, the velocity of particle 1 is changed from v_1 to v_1' and the velocity of particle 2 is changed from v_2 to v_2'. (Here we use boldface symbols to denote vectors.) For identical point masses, conservation of momentum implies coplanar collisions, with

$$v_1 + v_2 = v_1' + v_2'. \tag{11.18}$$

The assumptions of kinetic theory ensure that under equilibrium conditions, for each binary collision there is an inverse collision in which the original velocities are restored (Figure 11.5). The velocity vectors are collinear with the original set.

If we let $F(v)\ dv$ be the number of molecules with velocities in the range v to $v + dv$, then the number of collisions occurring between molecules 1 and 2 per unit time will be $aF(v_1)\ F(v_2)$, where a is a constant of proportionality.

For the inverse collisions, the number will be $a'F(v_1')F(v_2')$. Because the velocity distribution is unchanged by the total of all the elastic collisions, the two rates must be the same. Also, the collisions will be completely equivalent when viewed in a center of mass frame of reference, so $a = a'$. Thus

*See, for example, *Thermal Physics,* by M. Sprackling, American Institute of Physics, New York, 1991.

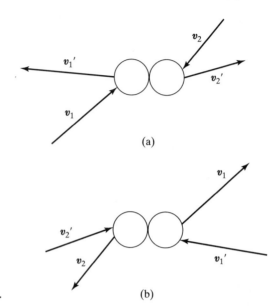

Figure 11.5 (a) A binary
molecular collision; (b) the
corresponding inverse collision.

$$F(\mathbf{v}_1)F(\mathbf{v}_2) = F(\mathbf{v'}_1)F(\mathbf{v'}_2),$$

or

$$\ln F(\mathbf{v}_1) + \ln F(\mathbf{v}_2) = \ln F(\mathbf{v'}_1) + \ln F(\mathbf{v'}_2). \tag{11.19}$$

Since kinetic energy is conserved and $\mathbf{v} \cdot \mathbf{v} = |v|^2 = v^2$, we also have

$$v_1^2 + v_2^2 = v_1'^2 + v_2'^2. \tag{11.20}$$

The functional relation of Equation (11.19), with the constraint of Equation
(11.20), can be solved by inspection.* A solution is

$$F(\mathbf{v}) = A \exp(-\alpha \mathbf{v} \cdot \mathbf{v}), \tag{11.21}$$

where A and α are constants. This is the Maxwell-Boltzman *velocity* distribution. To obtain a *speed* distribution, we let $N(v)dv$ be the number of molecules whose speeds lie in the range v to $v + dv$, irrespective of direction. As in
Figure 11.3, we can visualize the velocity vectors as displaced parallel to themselves with their ends at a common point. Since the directions are random, the
velocity distribution has spherical symmetry. Therefore $N(v)dv$ is the number

---

*The solution can also be found by the method of Lagrange multipliers, discussed in
Chapter 13.

of velocity vectors whose tips lie within the spherical shell between radii v and $v + dv$. It follows that

$$N(v)dv = 4\pi v^2 F(v)dv = 4\pi v^2 A e^{-\alpha v^2} dv. \tag{11.22}$$

The constants A and α can be found from two conditions. The total number of particles N is given by

$$N = \int_0^\infty N(v)dv, \tag{11.23}$$

and the total energy of the molecules in the container is

$$\frac{3}{2}NkT = \frac{1}{2}m \int_0^\infty v^2 N(v)dv. \tag{11.24}$$

To determine α we can divide Equation (11.24) by Equation (11.23) and use Equation (11.22). The result is

$$\frac{3kT}{m} = \frac{\displaystyle\int_0^\infty v^2 N(v)dv}{\displaystyle\int_0^\infty N(v)dv} = \frac{\displaystyle\int_0^\infty v^4 e^{-\alpha v^2}dv}{\displaystyle\int_0^\infty v^2 e^{-\alpha v^2}dv} = \frac{\dfrac{3\sqrt{\pi}}{8\alpha^{5/2}}}{\dfrac{\sqrt{\pi}}{4\alpha^{3/2}}} = \frac{3}{2\alpha}. \tag{11.25}$$

(The values of the definite integrals are given in Appendix D.) Thus, $\alpha = m/2kT$. The constant A can be then found from Equation (11.23):

$$N = \int_0^\infty N(v)dv = 4\pi A \int_0^\infty v^2 e^{-\alpha v^2}dv = 4\pi A \left(\frac{\sqrt{\pi}}{4\alpha^{3/2}} \right). \tag{11.26}$$

Hence

$$A = \left(\frac{\alpha}{\pi} \right)^{3/2} N = \left(\frac{m}{2\pi kT} \right)^{3/2} N. \tag{11.27}$$

Finally,

$$N(v)dv = 4\pi N \left(\frac{m}{2\pi kT} \right)^{3/2} v^2 e^{-mv^2/2kT} dv. \tag{11.28}$$

The function $N(v)$ is shown schematically in Figure 11.6.

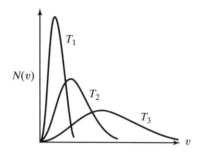

Figure 11.6 Graph of the Maxwell speed distribution for three temperatures: $T_3 > T_2 > T_1$.

If we divide Equation (11.28) by N, we normalize $N(v)$ and obtain a probability density function $f(v) = N(v)/N$ that can then be used to obtain some useful average values. As noted in Section 11.1, the definition of the mean value of v^n is

$$\overline{v^n} = \int_0^\infty v^n f(v)\,dv,$$

where $f(v)$ is the probability density function. Three speeds that characterize the Maxwell-Boltzmann distribution are of special importance:

1. *Mean speed:*

$$\overline{v} = \frac{1}{N}\int_0^\infty vN(v)\,dv$$

$$= 4\pi\left(\frac{m}{2\pi kT}\right)^{3/2}\int_0^\infty v^3 e^{-\alpha v^2}\,dv,$$

$$= \left(\frac{8kT}{\pi m}\right)^{1/2} = 1.596\left(\frac{kT}{m}\right)^{1/2}. \tag{11.29}$$

2. *Root mean square speed:*
The mean square speed is given by

$$\overline{v^2} = \frac{1}{N}\int_0^\infty v^2 N(v)\,dv$$

$$= 4\pi\left(\frac{m}{2\pi kT}\right)^{3/2}\int_0^\infty v^4 e^{-\alpha v^2}\,dv$$

$$= \left(\frac{3kT}{m}\right).$$

The root mean square (rms) speed is, then,

$$v_{rms} \equiv \sqrt{\overline{v^2}} = \left(\frac{3kT}{m}\right)^{1/2} = 1.732\left(\frac{kT}{m}\right)^{1/2}. \tag{11.30}$$

3. *Most probable speed*:
   This is the value v_m for which $N(v)/N$ is a maximum, or for which

$$\frac{d}{dv}(v^2 e^{-\alpha v^2}) = 0, \quad \alpha = \frac{m}{2kT}.$$

The derivative is zero for $v = (1/\alpha)^{1/2}$, which yields

$$v_m = \left(\frac{2kT}{m}\right)^{1/2} = 1.414\left(\frac{kT}{m}\right)^{1/2}. \tag{11.31}$$

It is evident that (Figure 11.7)

$$v_m : \overline{v} : v_{rms} = 1.414 : 1.596 : 1.732$$
$$= 1 : 1.128 : 1.225. \tag{11.32}$$

For a nitrogen molecule ($m = 4.65 \times 10^{-26}$ kg) at $T = 273$ K, $v_{rms} = 493$ ms^{-1}, which is slightly greater than the speed of a sound wave in nitrogen gas. Numerical values of the mean speed and root mean square speed for various gases are given in Table 11.2.

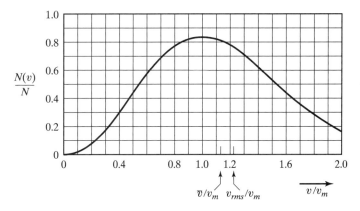

Figure 11.7 The Maxwell-Boltzmann speed distribution function at a particular temperature showing the most probable speed v_m, the mean speed $\overline{v}$, and the root mean square speed v_{rms}.

TABLE 11.2 Molecular
Speeds at 273 K, in ms^{-1}

Gas	$\bar{v}$	v_{rms}
H$_2$	1690	1840
He	1210	1310
H$_2$O	570	620
Ne	530	580
N$_2$	450	490
O$_2$	420	460
Ar	400	430

There are other speed distributions of less interest to us; one is the number of particles whose x-component of velocity lies in the range v_x to $v_x + dv_x$ regardless of the values of the other components:

$$N(v_x)dv_x = N\left(\frac{m}{2\pi kT}\right)^{1/2} e^{-mv_x^2/2kT} dv_x; \tag{11.33}$$

(Figure 11.8). Clearly, $\bar{v}_x = \bar{v}_y = \bar{v}_z$ since the velocity *directions* are assumed to be uniformly distributed over all angles in space.

11.7 MEAN FREE PATH AND COLLISION FREQUENCY

We are interested in how far the molecules of a gas travel between collisions and how often they undergo collisions. These concepts are called, respectively, the mean free path ℓ and the collision frequency f_c.

Suppose that the radius of a molecule of the gas is R. Then a collision occurs when one molecule approaches another within a center-to-center distance $2R$. In time t, a moving molecule sweeps out a cylindrical "exclusion" volume of length $\bar{v}t$ and cross-sectional area $\sigma = \pi(2R)^2$ (Figure 11.9). If the neighboring molecules were at rest, the number of molecules with centers in this volume would be $\underline{\underline{n}}\sigma\bar{v}t$ and the mean free path would therefore be

$$\ell = \frac{\bar{v}t}{\underline{\underline{n}}\sigma\bar{v}t} = \frac{1}{\underline{\underline{n}}\sigma}.$$

This answer is only approximately correct because we have used the mean speed $\bar{v}$ for all the molecules instead of performing an integration over the Maxwell-Boltzman speed distribution. If that is done, and the most probable

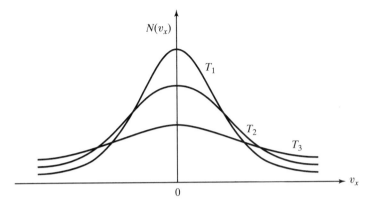

Figure 11.8   The Maxwell-Boltzman distribution function for the x-component of velocity for three temperatures: $T_3 > T_2 > T_1$. (The distributions for the y- and z-components are identical with this distribution.)

speed is used in place of $\overline{v}$, then

$$\ell = \frac{1}{\sqrt{\dfrac{8}{\pi}n\sigma}}, \qquad (11.34)$$

which is the correct expression for the mean free path. The corresponding collision frequency is

$$f_c = \frac{v_m}{\ell} = \sqrt{\frac{8}{\pi}}v_m n\sigma. \qquad (11.35)$$

Note that the mean free path is independent of the molecular speed and therefore of temperature. Since the number of molecules per unit volume n is directly proportional to the pressure, the mean free path increases as the pressure decreases. For an oxygen molecule, $2R \approx 3.6 \times 10^{-10}$ m. Under standard temperature and pressure, $n = 2.7 \times 10^{25}$ m^{-3} and $v_m = 370$ ms^{-1}. Thus

$$\ell \approx 5.7 \times 10^{-8} \text{ m}$$

and

$$f_c \approx 6.5 \times 10^9 \text{ s}^{-1}.$$

The average time between collisions, the reciprocal of the collision frequency, is approximately 1.5×10^{-10} s. Note that ℓ is the order of $300R$, validating our kinetic theory assumption that the separation of molecules is large compared with molecular dimensions.

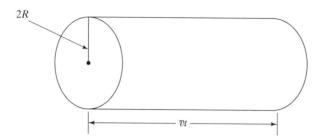

2R

$\overline{v}t$

Figure 11.9   Exclusion volume swept out in time t by a
particle of radius R traveling with average speed $\overline{v}$.

The actual distance that a molecule will travel between collisions is subject to a large statistical variation, independent of time. Let the probability
that a collision will occur in a distance dx be αdx. Then the probability that a
collision will *not* occur when the molecule travels a distance dx will be 1
$- \alpha dx$. If $p(x)$ is the probability that a molecule will travel a finite distance x
before making a collision, then $p(x)(1 - \alpha dx)$ is the probability that a molecule will travel a distance $x + dx$ before experiencing a collision. Thus

$$p(x + dx) = p(x)(1 - \alpha dx).$$

Expanding $p(x + dx)$ in a Taylor series and retaining the first two terms, we
have

$$p(x) + \frac{dp(x)}{dx}dx = p(x) - \alpha p(x)dx,$$

or

$$\frac{dp(x)}{dx} = -\alpha p(x).$$

Integrating, we obtain

$$p(x) = Ae^{-\alpha x},$$

where A is a constant. Since $p(0)$ must be unity, $A = 1$ and

$$p(x) = e^{-\alpha x}.$$

To determine the mean free path, we need the probability density function
$f(x)$ associated with the probability $p(x)$. Because $p(x) = \int f(x)dx$, it follows
that $f(x) = |dp(x)/dx| = \alpha e^{-\alpha x}$, since $f(x)$ must be positive. Then

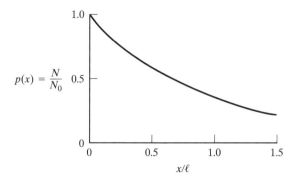

Figure 11.10   Graph of the survival equation.

$$\ell \equiv \bar{x} = \int_0^\infty xf(x)dx = \alpha \int_0^\infty xe^{-\alpha x}dx = \frac{1}{\alpha},$$

so that

$$p(x) = e^{-x/\ell}. \tag{11.36}$$

The probability $p(x)$ can be interpreted as the ratio N/N_0, where N_0 is the initial number of particles in a molecular beam, and N is the number of particles remaining in the beam after a distance x. Figure 11.10 is a graph of this so-called "survival equation." We see that the fraction of molecules with free paths longer than ℓ is e^{-1} or 37 percent, while the fraction with free paths shorter than the mean is 63 percent. Equation (11.36) shows how a beam of particles is attenuated due to collisions as it passes through a low density gas. Use of this curve is independent of the starting point: the probability of a collision in a length ℓ is independent of where the last collision occurred.

11.8 EFFUSION

If a container of gas, surrounded by a vacuum, has a small hole in its wall, the contained gas will leak through the hole into the surrounding space. If the hole is sufficiently small, the equilibrium state of the gas in the container will not be appreciably affected. In that case, the number of molecules escaping through the hole will be the same as the number that would impinge upon the area of the hole if the latter were not present. The emergence of molecules through the hole is a process known as *effusion*.

How small must the hole be? Molecules move unimpeded through a distance approximately equal to the mean free path ℓ. Thus the density of molecules at a given point is maintained by the flow of molecules from the surface of an imagined sphere of radius ℓ and area $4\pi\ell^2$. If the flow from the area A of a circular hole is to be cut off without making much of a difference to the molecular density at the center of the sphere, then $A \ll 4\pi\ell^2$. Let the diameter of the hole be D. Then $\pi D^2/4 \ll 4\pi\ell^2$, or $D^2 \ll 16\ell^2$. We may interpret this somewhat arbitrarily as $D^2 < 0.16\ell^2$ or $D < 0.4\ell$. At room temperature and atmospheric pressure, $\ell \approx 10^{-7}$ m. Thus, the opening must be extremely small if the state of the gas in the container is to remain virtually undisturbed. Since the mean free path is inversely proportional to the pressure, it follows that there is an upper limit to the pressure for which effusion occurs. Effusion is essentially a low pressure phenomenon.

In Section 11.2 we found that the number of molecules with speeds in the range v to $v + dv$ that strike a surface, per unit area and per unit time, is

$$\Phi_v dv = \frac{1}{4} v \underline{n}_v dv. \tag{11.37}$$

Here $\underline{n}_v dv$ is the number of molecules per unit volume with speeds between v and $v + dv$. Assuming that the molecules have a Maxwell-Boltzman speed distribution, we know that

$$\underline{n}_v dv = 4\pi \underline{n} \left(\frac{m}{2\pi kT}\right)^{3/2} v^2 e^{-mv^2/2kT} dv. \tag{11.38}$$

Substituting this in Equation (11.37), we have

$$\Phi_v dv = \pi \underline{n} \left(\frac{m}{2\pi kT}\right)^{3/2} v^3 e^{-mv^2/2kT} dv. \tag{11.39}$$

This is the number of molecules that escape through the hole in the container wall that have speeds from v to $v + dv$, per unit area and per unit time.

We wish to determine the mean speed $\bar{v}_e$ of escaping molecules. Following the usual procedure, we multiply v by $\Phi_v dv$, integrate over all possible values of v, and divide by the integral of $\Phi_v dv$. The last step is necessary to ensure that the weighting function used in calculating the mean value is a true probability density function. Thus

$$\bar{v}_e = \frac{\displaystyle\int_0^\infty v\Phi_v dv}{\displaystyle\int_0^\infty \Phi_v dv} = \frac{\displaystyle\int_0^\infty v^4 e^{-mv^2/2kT} dv}{\displaystyle\int_0^\infty v^3 e^{-mv^2/2kT} dv}.$$

Values of the integrals are given in Appendix D. The result is

$$\bar{v}_e = \frac{3}{4}\left(\frac{2\pi kT}{m}\right)^{1/2}. \tag{11.40}$$

Comparing this with the mean speed $\bar{v}$ of molecules within the container (Equation (11.29)), we note that $\bar{v}_e$ is slightly higher than $\bar{v}$:

$$\bar{v}_e = \frac{3\pi}{8}\bar{v} = 1.18\bar{v}. \tag{11.41}$$

This is because the faster molecules have a greater probability of "finding" the hole in a given time.

Effusion has found an important application in experimental physics. Molecules allowed to escape from a small hole in a container into surroundings maintained at low pressure can be collimated to form a well-defined molecular beam. (Standard methods involving electric or magnetic fields that are used to guide charged particles are not applicable to neutral molecules.) The effusion technique makes it possible to study individual molecules whose interactions with other molecules are negligibly small. Molecular beams have also been used to verify the Maxwell-Boltzman velocity distribution.

Another use of effusion is in the separation of isotopes. The principle is based on the fact that the rate of effusion of a molecule is dependent on its mass. The rate per unit area is the molecular flux $\Phi = n\bar{v}/4$; for an ideal gas, $PV = NkT$, so $n = N/V = P/kT$ and $\bar{v} = (8kT/\pi m)^{1/2}$. Combining these expressions, we see that

$$\Phi = \frac{P}{\sqrt{2\pi mkT}}. \tag{11.42}$$

Thus light molecules escape more rapidly than heavier ones.

A gas mixture of two isotopes is permitted to effuse through tiny holes in the wall of a container. After a period of time, the relative concentration of the heavier isotope will exceed that of the lighter isotope in the container. In the gas that has effused, the reverse will be the case. It is interesting to note that the effusion method was one of the methods used to separate the isotopes U^{235} and U^{238} in the making of the atomic bomb.

11.9 TRANSPORT PROCESSES

Transport processes involve the reaction of a gas to small departures from thermal equilibrium due to gradients in the medium. From a knowledge of the average speed and the mean free path of the molecules, it is possible to give an elementary derivation of three common properties of a dilute gas: its viscosity, thermal conductivity, and coefficient of diffusion.

In a transport process, each molecule conveys momentum, energy, and mass (the molecule itself) from one collision to the next. At each collision it acquires the average properties of the molecules in the neighborhood of the

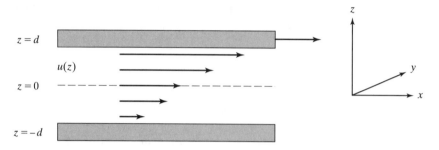

Figure 11.11   Flow of a viscous fluid between a moving upper plate and a stationary lower plate.

collision. Hence, as it travels along the next free path, it carries a sample of the properties of the gas in the vicinity of its previous collision. If there are gradients present, the properties in the region of its next collision may differ from those carried by the molecule. During the next collision the molecule will give up this difference to the gas in the new region. In their motion from one collision to the next, the molecules will couple the properties of the gas at one point with those at another. The ultimate result is an equalization of the properties throughout the gas. This is the essence of transport processes.

Consider a fluid such as a gas between two large parallel plates. The upper plate is moved horizontally while the lower plate remains at rest (Figure 11.11). Since real fluids are viscous, the fluid is pulled parallel with the moving plate. The magnitude of the fluid velocity u in the horizontal direction changes from the speed of the upper plate to zero at the stationary plate. In other words, a velocity gradient du/dz exists in the z direction, perpendicular to the plates.

Owing to the gradient, the molecules that are continually moving from positive values of z are passing through the $z = 0$ plane, transferring their higher momenta in the x-direction to the layer below, thereby increasing the average momentum of the layer. Similarly, the molecules moving from negative values of z through the plane are carrying with them the lower momenta corresponding to the region where they last underwent a collision. The net momentum transfer per unit time through unit area gives the tangential stress P_{zx} (the viscous force per unit area), which a layer below the xy plane exerts on the fluid above it due to the velocity gradient.

For small gradients, we expect a relation of the form

$$P_{zx} = \eta \frac{du}{dz}, \tag{11.43}$$

where the proportionality factor η is the coefficient of viscosity of the fluid. Viscosity has traditionally been measured in units of dyne $\cdot$ s cm^{-2}, called a

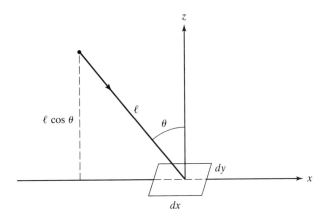

Figure 11.12   Molecule crossing the $z = 0$ plane after its last collision at a distance $\ell \cos\theta$ above the plane.

"poise," after Poiseuille*, who studied viscous flow. The SI unit is Pa · s, which is equal to 10 poise.

The coefficient of viscosity for a dilute gas can be calculated using kinetic theory and Newton's second law. We must first determine the average distance $\bar{z}$ above the xy plane at which a molecule makes its last collision before crossing the plane. From Figure 11.12, $z = \ell \cos\theta$. Then $\bar{z}$ can be found by multiplying this quantity by the incremental flux $d\Phi_\theta$ in the θ direction, integrating over θ, and dividing by the total flux Φ. Using Equations (11.7) and (11.8), we obtain

$$\bar{z} = \frac{\displaystyle\int \ell\cos\theta \, d\Phi_\theta}{\Phi} = \frac{\dfrac{1}{2}n\bar{v}\ell \displaystyle\int_0^{\pi/2} \sin\theta \, \cos^2\theta \, d\theta}{\dfrac{1}{4}n\bar{v}} = \frac{2}{3}\ell. \qquad (11.44)$$

Thus, on the average, a molecule crossing the xy plane makes its last collision before crossing at a distance equal to two-thirds of a mean free path.

If $u(0)$ is the forward velocity of the fluid at $z = 0$, then at a distance $2\ell/3$ above the plane, the fluid velocity is

$$u\left(\frac{2}{3}\ell\right) = u(0) + \frac{2}{3}\ell\frac{du}{dz}. \qquad (11.45)$$

---

*Jean Louis Marie Poiseuille, a French anatomist, was interested in the physics of the circulation of the blood. His work on the flow of liquids through capillaries was published in 1842.

Here we assume that the gradient does not vary appreciably over the distance of a mean free path. The momentum of a molecule with this velocity in the x-direction is $mu(2\ell/3)$ and the total momentum transferred across the plane per unit time and per unit area by all of the molecules crossing from above is

$$\frac{1}{4}nm\bar{v}\left[u(0) + \frac{2}{3}\ell\frac{du}{dz}\right].$$

Similarly, the total momentum transported per unit time and per unit area across the plane by the molecules crossing from below is

$$\frac{1}{4}nm\bar{v}\left[u(0) - \frac{2}{3}\ell\frac{du}{dz}\right].$$

The net rate of momentum transport per unit area is the difference between these two quantities, or

$$\frac{1}{3}nm\bar{v}\ell\frac{du}{dz}.$$

From Newton's second law, this is equal to the viscous force per unit area given by Equation (11.43). It follows that

$$\eta = \frac{1}{3}nm\bar{v}\ell. \tag{11.46}$$

Recalling that $\ell = (\sqrt{8/\pi}n\sigma)^{-1}$, we see that the coefficient of viscosity can be written

$$\eta = \frac{1}{6}\sqrt{\frac{\pi}{2}}\frac{m\bar{v}}{\sigma}. \tag{11.47}$$

The mean kinetic energy of a molecule is proportional to the temperature. Thus $\bar{v} \propto (T/m)^{1/2}$ and

$$\eta = \frac{(const)T^{1/2}}{\sigma}. \tag{11.48}$$

This result may seem surprising. Intuitively, we might expect the viscosity to decrease with increasing temperature and to increase with increasing collision cross section. But higher temperatures mean higher molecular speeds, and a small cross section means a large mean free path. Both factors contribute to a more effective transfer of momentum from one fluid layer to the next and hence to a higher viscosity. This behavior differs markedly from that of the viscosity of liquids, which decreases rapidly with increasing temperature. That is because in a liquid the molecules are closer together and momentum transport across a surface involves interaction forces between molecules on the two sides of the surface as well as molecular motion across the surface.

An especially useful expression for calculating the viscosity is obtained by solving for n in Equation (11.9) and substituting the result in Equation (11.46). Then, assuming that $\overline{v^2} \approx (\overline{v})^2$, we get

$$\eta \approx \frac{P\ell}{\overline{v}}. \tag{11.49}$$

For oxygen under standard conditions ($P = 1.01 \times 10^5$ Pa, $T = 273$K),

$$\eta = \frac{(1.01 \times 10^5 \text{Pa})(5.7 \times 10^{-8}\text{ m})}{(420 \text{ms}^{-1})} \approx 14 \times 10^{-6} \text{Pa} \cdot \text{s}.$$

This compares with the measured value of close to 19×10^{-6} Pa · s. The discrepancy is due to the approximate nature of kinetic theory, which treats the molecules of a gas as hard spheres. Also, the factor $1/3$ in Equation (11.46) is not to be taken too seriously, because it depends on how various quantities are averaged.

The calculation of the thermal conductivity proceeds in a similar way. The only difference is that it is the thermal energy that is transported rather than the momentum. The upper and lower plates in Figure 11.11 are both assumed to be stationary but are held at different temperatures so that there is a temperature gradient rather than a velocity gradient in the gas. The thermal conductivity results from the net flux of molecular kinetic energy across a surface. The average kinetic energy of a molecule is $c_v T / N_A$, where c_v is the specific heat capacity and N_A is Avogadro's number. The thermal conductivity λ is found to be

$$\lambda = \frac{1}{3} n \overline{v} \ell \left(\frac{c_v}{N_A}\right) = \frac{1}{6}\sqrt{\frac{\pi}{2}} \frac{c_v \overline{v}}{N_A \sigma}. \tag{11.50}$$

It is interesting to note that the ratio of the thermal conductivity to the viscosity is

$$\frac{\lambda}{\eta} = \frac{c_v}{m N_A} = \frac{c_v}{M},$$

so that

$$\frac{\lambda M}{\eta c_v} = 1,$$

where M is the molecular weight of the gas. Measured values are closer to 2, but kinetic theory gives the right order of magnitude.

Finally, the transport of mass across a surface as a result of random molecular motion gives rise to diffusion if a concentration gradient exists. The treatment of diffusion is complicated when there is more than one type of molecule present, but its main features can be understood by considering the motion of molecules in a single constituent gas, a phenomenon known as *self-diffusion*. The coefficient of self-diffusion D can be easily calculated by mean free path arguments like those used in the discussion of viscosity and heat conductivity. The result is

$$D = \frac{1}{3}\bar{v}\ell = \frac{1}{6}\sqrt{\frac{\pi}{2}}\frac{\bar{v}}{\underline{\underline{n}}\sigma}. \tag{11.51}$$

Since $\underline{\underline{n}} = 3P/m\overline{v^2}$ and both $\bar{v}$ and $(\overline{v^2})^{1/2}$ are proportional to $T^{1/2}$,

$$D = (const)T^{3/2}$$

at a fixed pressure and

$$D = \frac{(const)}{P}$$

at a fixed temperature. The ratio of the coefficients of viscosity and self-diffusion is

$$\frac{D}{\eta} = \frac{1}{\underline{\underline{n}}m} = \frac{1}{\rho},$$

where ρ is the mass density of the gas. Thus

$$\frac{D\rho}{\eta} = 1.$$

Experimentally, $D\rho/\eta$ lies in the range 1.3 to 1.5. The extent of the agreement between theory and experiment must be considered satisfactory in view of the approximations made in our physical model.

PROBLEMS:

11.1 The distribution of particle speeds of a certain hypothetical gas is given by

$$N(v)dv = Ave^{-v/v_0}dv,$$

where A and v_0 are constants.

(a) Determine A so that $f(v) \equiv N(v)/N$ is a true probability density function; i.e., $\int_0^\infty f(v)dv = 1$. Sketch $f(v)$ versus v.

(b) Find $\bar{v}$ and v_{rms} in terms of v_0.

(c) Differentiate $f(v)$ with respect to v and set the result equal to zero to find the most probable speed v_m.

(d) The standard deviation of the speeds from the mean is defined as

$$\sigma \equiv \left[\overline{(v - \bar{v})^2}\right]^{1/2},$$

where the bar denotes the mean value. Show that

$$\sigma = [\bar{v^2} - (\bar{v})^2]^{1/2}$$

in general. What is σ for this problem?

11-2 At standard temperature and pressure the mean speed of hydrogen molecules is $1.70 \times 10^3\,\text{ms}^{-1}$. What is the particle flux?

11-3 What is the number density of molecules at a temperature of 77 K in an ultra-high vacuum at 10^{-10} torr?

11-4 Compute the *rms* speed of helium atoms at 2 K; of nitrogen molecules at 27°C; and of mercury atoms at 100°C.

11-5 Compute the mean energy in electron volts and the *rms* speed in ms^{-1} of an electron at 1000 K. At 10,000 K, what fraction of the speed of light is the rms speed?

11-6 At what values of the speed does the Maxwell speed distribution have half its maximum value? Give your answers as a constant time $(kT/m)^{1/2}$.

11-7 (a) Show that the Maxwell speed distribution can be written as

$$N(x)dx = \frac{4N}{\sqrt{\pi}}\,x^2 e^{-x^2}\,dx,$$

where $x \equiv v/v_m$.

(b) Show that the number of molecules with speeds less than some specified speed v_0 is given by

$$N_{0 \to x_0} = N\left[erf(x_0) - \frac{2}{\sqrt{\pi}}x_0 e^{-x_0^2}\right],$$

where $x_0 \equiv v_0/v_m$. Here $erf(x_0)$ is the error function, defined by

$$erf(x) \equiv \frac{2}{\sqrt{\pi}}\int_0^x e^{-u^2}du.$$

(c) Find the fractional number of molecules with speeds less than v_m.

11-8 Compute v_m, $\bar{v}$, and v_{rms} for an oxygen molecule at 300 K. What are the corresponding values at 10,000 K?

11-9 (a) Show that the mean speed of a nitrogen molecule at 300 K is 476 ms^{-1}.

(b) What is the ratio of the probability of finding a nitrogen molecule whose speed is 476 ms^{-1} at 300 K to the probability of finding a nitrogen molecule with the same speed at 600 K?

11-10 Find the *rms* free path in terms of the mean free path. What is the most probable free path? Use $\ell^{-1}e^{-x/\ell}$ as the probability density function.

11-11 For carbon dioxide (molecular weight 44, molecular diameter 4.6×10^{-10} m) at 1 atm and 300 K, find the mean free path ℓ using Equation (11.34). What is the ratio of ℓ to the molecular diameter? What is the collision frequency?

11-12 The number density of the gas, mainly hydrogen, that fills interstellar space is one molecule per cubic centimeter ($10^6\,\mathrm{m}^{-3}$). If the molecular diameter is 10^{-10} m, what is the mean free path in interstellar space? What is the collision frequency of a hydrogen molecule in collisions per century if the temperature of interstellar space is 10 K?

11-13 A beam of oxygen molecules start together. The pressure is 1.8×10^3 torr and the temperature is 300 K. The diameter of an oxygen molecule is approximately 3.6×10^{-10} m. How long will half the molecules remain unscattered? (Assume that all the particles have a speed equal to the mean speed.)

11-14 A thin-walled vessel containing 1 liter of carbon dioxide is kept at a temperature of 0°C. The gas slowly leaks out through a circular hole of diameter 100 μm. The outside pressure is low enough that leakage back into the vessel is negligible. (The diameter of a CO_2 molecule is about 4.6×10^{-10} m and its molecular weight is 44 kg kilomole^{-1}.)
   (a) Estimate the upper limit of the pressure in the container for effusion to occur through the hole.
   (b) Estimate the time, starting at this pressure, for the pressure to drop to one-half its initial value.

11-15 A vessel is divided into two parts of equal volume by means of a plane partition, in the middle of which is a very small hole. Initially, both parts of the vessel contain ideal gas at a temperature of 300 K and a low pressure P. The temperature of one-half of the vessel is then raised to 600 K while the temperature of the other half remains at 300 K. Determine the pressure difference in terms of P between the two parts of the vessel when steady conditions are achieved.

11-16 A vessel has porous walls containing many tiny holes. Gas molecules can pass through these holes by effusion and then be pumped off to some collecting chamber. The vessel is filled with a dilute gas consisting of two types of molecules that have different masses m_1 and m_2 by virtue of the fact that they contain two different isotopes of the same atom. The concentrations of these molecules are c_1 and c_2 respectively. (The concentration c_i is the ratio of the number of molecules of type i to the total number of molecules.) The concentrations can be kept constant in the vessel by providing a steady flow of fresh gas through it so as to replenish any gas that has effused. If c_1' and c_2' denote the concentrations of the two types of molecules in the collecting chamber, what is the ratio c_2'/c_1'?

11-17 Diffusion can be regarded as a random walk problem in which the successive displacements of a gas molecule are statistically independent. In this model, the distance L traveled by a molecule after N displacements is related to the mean free path ℓ by the expression $L^2 = N\ell^2$, and the time t required to move a distance L is given by $t = N\ell/\bar{v}$. Using these relations, estimate how long it would

take a molecule in a room of macroscopically "still" air with uniform temperature and pressure to move a distance of 5 meters.

11-18 The experimental value of the viscosity of argon gas is found to be
22.0×10^{-6} Pa $\cdot$ s at 15°C and atmospheric pressure. The atomic weight of
argon is 39.94. Estimate the diameter of an argon atom.

11-19 The radius of an air molecule is approximately 1.8×10^{-10} m, its mass is about
4.8×10^{-26} kg, and it has 5 degrees of freedom. The molecular weight of air is 29
and its density is 1.29 kg m^{-3} under standard conditions.
  (a) Estimate the values of the coefficient of viscosity η, the thermal conductiv-
      ity λ, and the diffusion coefficient D under these conditions.
  (b) Check that for air,

$$\frac{\lambda M}{\eta c_v} \approx 1 \quad \text{and} \quad \frac{D\rho}{\eta} \approx 1.$$

Also, check that these quantities are indeed dimensionless.

Chapter *12*

Statistical Thermodynamics

12.1 INTRODUCTION

The union of the energy and entropy concepts led to a science of thermodynamics that combines great generality with reliability of prediction. Classical thermodynamics, however, provides little insight into *why* thermal phenomena occur the way they do.

The object of statistical thermodynamics is to present a particle theory resulting in an interpretation of the equilibrium properties of macroscopic systems. The foundation upon which the theory rests is necessarily quantum mechanics. Fortunately, a satisfactory theory can be developed using only the quantum mechanical concepts of quantum states, energy levels, and intermolecular forces.

The central idea is the probability density function applied to a large collection of identical particles. A thermodynamic system is regarded as an assembly of submicroscopic entities in an enormous number of ever-changing quantum states. The basic postulate of statistical thermodynamics is that all possible *microstates* of an isolated assembly are *equally probable.*

A few new concepts are needed. We shall use the term *assembly* to denote a number N of identical entities, such as molecules, atoms, electrons, photons, oscillators, etc. The word *system* is used synonymously.

The *macrostate* of a system, or *configuration,* is specified by the number of particles in each of the *energy levels* of the system. Thus N_j is the number of particles that occupy the jth energy level. If there are n energy levels, then

$$\sum_{j=1}^{n} N_j = N.$$

The macrostate of statistical thermodynamics is another word for the thermodynamic state of the classical theory, specified by a pair of state variables.

A *microstate* is specified by the number of particles in each *energy state.* In general, there will be more than one energy state (i.e., quantum state) for each energy level, a situation called *degeneracy.* A microstate is the most specific description one can get. In general, there will be many, many different microstates corresponding to a given macrostate. We will be interested in the number of microstates but not in their detailed specification.

The number of microstates leading to a given macrostate is called the *thermodynamic probability.* It is the number of ways in which a given configuration can be achieved. This is an "unnormalized" probability, an integer between zero and infinity, rather than a number between zero and one. For the *k*th macrostate, the thermodynamic probability is taken to be w_k. A *true probability* p_k could be obtained by dividing w_k by the total number of microstates Ω available to the system. For much of what we shall do, the thermodynamic probability will be adequate.

12.2 COIN-TOSSING EXPERIMENT

By way of introducing the central features of statistical thermodynamics, we shall apply some elementary concepts of probability theory to a coin-tossing experiment.

We assume that we have N coins that we toss on the floor and then examine to determine the number of *heads* N_1 and the number of *tails* $N_2 = N - N_1$. We consider the case of $N = 4$. Using the terms we last introduced, we can prepare a table listing the possible outcomes of the experiment (Table 12.1).

Each macrostate is defined by the number of heads and the number of tails. There are five. For each macrostate there are one or more microstates. A microstate is specified by the state, heads or tails, of each coin; it is the most detailed description possible. We are interested in the *number* of microstates for each macrostate, the thermodynamic probability. The true probability is the number divided by the total number of microstates. Thus

$$p_k = \frac{w_k}{\Omega}, \tag{12.1}$$

where

$$\Omega = \sum_{k=1}^{5} w_k = 16. \tag{12.2}$$

We might also wish to calculate the average occupation numbers—in this case, the average number of heads and tails. Let $j = 1$ or 2, where N_1 is the number of heads and N_2 the number of tails. Let N_{jk} be the occupation number for the kth macrostate. Then the average occupation number is

$$\overline{N_j} = \frac{\sum_k N_{jk}w_k}{\sum_k w_k} = \frac{\sum_k N_{jk}w_k}{\Omega} = \sum_k N_{jk}p_k. \tag{12.3}$$

TABLE 12.1   Possible outcomes of a coin-tossing experiment using four coins. We introduce the language of statistical thermodynamics.

Macrostate Label	Macrostate Specification		Microstate				Thermo-dynamic Probability	True Probability
k	N_1	N_2	Coin 1	Coin 2	Coin 3	Coin 4	w_k	P_k
1	4	0	H	H	H	H	1	1/16
2	3	1	H	H	H	T	4	4/16
			H	H	T	H		
			H	T	H	H		
			T	H	H	H		
3	2	2	H	H	T	T	6	6/16
			T	T	H	H		
			H	T	H	T		
			T	H	T	H		
			H	T	T	H		
			T	H	H	T		
4	1	3	H	T	T	T	4	4/16
			T	H	T	T		
			T	T	H	T		
			T	T	T	H		
5	0	4	T	T	T	T	1	1/16

The average number of heads is therefore

$$\overline{N_1} = \frac{1}{16}[(4 \times 1) + (3 \times 4) + (2 \times 6) + (1 \times 4) + (0 \times 1)] = 2.$$

Similarly, $\overline{N_2} = 2$, as we would expect. Then $\overline{N_1} + \overline{N_2} = 4 = N$.

Figure 12.1 is a plot of the thermodynamic probability w versus the number of heads N_1. The curve is symmetric about $N_1 = 2$, the most probable configuration.

Suppose now that we want to perform the coin-tossing experiment with a larger number of coins. We need a way of computing the thermodynamic probability without tabulating the actual state of each coin. We assume that we have N distinguishable coins and ask: how many ways are there to select from the N candidates N_1 heads and $N - N_1$ tails? The answer is given by the binomial coefficient

$$w = \binom{N}{N_1} = \frac{N!}{N_1!(N - N_1)!}. \tag{12.4}$$

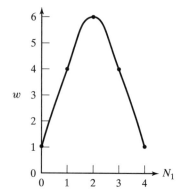

Figure 12.1 Thermodynamic probability versus the number of heads for a coin-tossing experiment with *four* coins.

This result can easily be seen by noting that there are N ways to pick the first coin, leaving $(N - 1)$ for the second, $(N - 2)$ for the third, etc. Thus the total number of ways of picking N_1 heads is

$$N(N - 1)(N - 2) \ldots (N - N_1 + 1) = \frac{N!}{(N - N_1)!}.$$

However, this counts the $N_1!$ permutations of the N_1 coins separately. The ordering of the coins is not important, so we divide by this factor, giving Equation (12.4).

If we increase the number of coins from four to eight, we obtain the plot of Figure 12.2. We see that the smallest value of w is again unity, but its maximum value has grown to 70; the peak has become considerably sharper.

The maximum value of the thermodynamic probability is of paramount interest to us. Its calculation is straightforward for $N = 8$. We know that the peak occurs at $N_1 = N/2$. Thus, Equation (12.4) gives

$$w_{\text{max}} = \frac{8!}{4!\,4!} = 70.$$

Suppose now that we perform a more ambitious experiment, with $N = 1000$. In this case,

$$w_{\text{max}} = \frac{1000!}{500!\,500!}, \tag{12.5}$$

and the computation becomes much more challenging. Fortunately, for such large numbers we can use Stirling's remarkable approximation (Appendix B) and write

$$\ln n! \approx n \ln n - n.$$

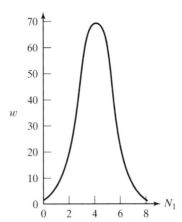

Figure 12.2   Thermodynamic probability versus the number of heads for a coin-tossing experiment with *eight* coins.

Taking the natural logarithm of both sides of Equation (12.5), we then obtain

$$\ln w_{\max} = \ln(1000!) - 2 \ln(500!) = 693.$$

But

$$\log_{10} x = (\log_{10} e)\ln x = 0.4343 \ln x,$$

so

$$\log_{10} w_{\max} = (0.4343)(693) = 300.$$

Thus

$$w_{\max} = 10^{300}.$$

(The error in using Stirling's approximation here is extremely small: even for n as small as 50, the error is less than 2 percent, and decreases rapidly as n gets larger.)

For $N = 1000$ we find that $w_{\max}$ is an astronomically large number. We conclude, then, from our coin-tossing experiments that the peak in our curve of thermodynamic probability versus the number of heads always occurs at $N/2$ but grows rapidly and becomes much sharper as N increases. Also, the tails of the curve always terminate at $w = 1$. The result is what we would expect: for "honest" coins, we get ever closer to an exact 50-50 split between heads and tails as the number of coins gets larger. In other words, the most probable configuration is that of total randomness.

Conceptually, this behavior for $N = 1000$ is shown in Figure 12.3. The most nearly random configuration (macrostate) $N_1 - N_2$ is the one that *almost always* occurs.

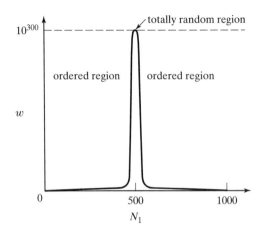

Figure 12.3 Thermodynamic probability versus the number of heads for a coin-tossing experiment with 1000 coins.

The "ordered regions" almost never occur; w is extremely small compared with w_{max}. We are led to a very important conclusion—namely, that the total number of microstates is very nearly equal to the maximum number:

$$\Omega = \sum w_k \approx w_{max}. \tag{12.6}$$

The figure also suggests a method for finding w_{max}. At the peak the slope of the curve is zero. Thus we need to find $w(N_1)$, take the derivative of the function with respect to N_1, and set the result equal to zero. This will give N_1 at $w = w_{max}$ and hence N_2. The outcomes will then be known for the most probable state.

For the thermodynamic problem, the "outcomes" are the occupation numbers for each of n energy levels. Our task is to find $w(N_1, N_2, \ldots N_j, \ldots N_n)$, possibly subject to constraints, and set its derivative equal to zero. In this way we will obtain the particular set $\{N_j\}$ of occupation numbers corresponding to the *most probable macrostate*. The most probable macrostate is the *equilibrium state* of the assembly. This is the fundamental problem of statistical thermodynamics—to determine the equilibrium state of the system.

Before proceeding, we need to extend our "counting formula" of Equation (12.4) from two "levels" (heads and tails) to n levels. We ask: in how many ways can N distinguishable objects be arranged if they are divided into n groups with N_1 objects in the first group, N_2 in the second, etc.? The answer is

$$w = \frac{N!}{N_1!N_2!\ldots N_n!} = \frac{N!}{\displaystyle\prod_{j=1}^{n} N_j!}. \tag{12.7}$$

Here the symbol $\prod$ represents the extended product, analogous to the summation sign $\sum$.

Equation (12.7) can be easily understood. Suppose we want to place N distinguishable objects in *three* boxes, with N_1 in the first box, N_2 in the second, and N_3 in the third. The number of ways of selecting N_1 (out of N) for box 1 is

$$\binom{N}{N_1} = \frac{N!}{N_1!(N - N_1)!}.$$

We are left with $N - N_1$ objects. We can choose N_2 of them for box 2 in

$$\binom{N - N_1}{N_2} = \frac{(N - N_1)!}{N_2!(N - N_1 - N_2)!} = \frac{(N - N_1)!}{N_2!N_3!}$$

ways. The number of ways of putting the remaining N_3 objects in box 3 is one. Thus the number of ways w of selecting objects to occupy the three boxes is the product of the three factors:

$$w = \frac{N!}{N_1!(N - N_1)!} \times \frac{(N - N_1)!}{N_2!N_3!} \times 1 = \frac{N!}{N_1!N_2!N_3!}.$$

Equation (12.7) follows by induction.

One additional important point: the coin-tossing model assumes that the coins are *distinguishable* (by a date, perhaps, or a mint mark). The thermodynamic analogy is with particles in a lattice, whose location distinguishes them. This would be the case for a crystalline solid, as an example. But for other cases, such as molecules in a gas, the particles are identical and indistinguishable and the number of microstates available to the assembly will be correspondingly fewer. We need to develop statistics for both situations.

12.3 ASSEMBLY OF DISTINGUISHABLE PARTICLES

The constituents of the system under study (a gas, liquid, or solid, for example) are considered to be a fixed number N of distinguishable particles, occupying a fixed volume V. We limit ourselves to isolated systems that do not exchange energy in any form with the surroundings. This implies that the internal energy U is also fixed. The macrostate in which the system finds itself is defined by (N, V, U), in correspondence with a classical thermodynamic state defined by (n, V, T), say, where n is the mass in kilomoles. Note that the temperature T is simply a measure of the internal energy for a system in thermal equilibrium.

We assume that the particles are *weakly interacting*. By this we mean that they interact sufficiently so that the system is in thermal equilibrium, but do not experience strong coupling through electromagnetic or nuclear forces.

We seek the distribution among energy levels $\{N_j\}$ for an equilibrium state of the system. That is, we want to determine the number of particles N_j with energy ε_j for all n energy levels of the system, subject to the restrictive conditions

$$\sum_{j=1}^{n} N_j = N \quad \text{(conservation of particles)}, \tag{12.8}$$

$$\sum_{j=1}^{n} N_j \varepsilon_j = U \quad \text{(conservation of energy)}, \tag{12.9}$$

where N and U are constants.

As with our coin-tossing experiments, we can clarify these ideas with a simple example. Consider three particles, labeled A, B, and C, distributed among four energy levels, $0, \varepsilon, 2\varepsilon$, and 3ε, such that the total energy is 3ε. Thus

$$\sum_{j=0}^{3} N_j = 3 \quad \text{and} \quad \sum_{j=0}^{3} N_j \varepsilon_j = 3\varepsilon.$$

The possible microstates and macrostates are listed in Table 12.2. The occupation numbers are: N_0 particles with energy 0, N_1 with energy ε, N_2 with

TABLE 12.2 Microstates and macrostates for $N = 3, U = 3\varepsilon$ with $\varepsilon_j = j\varepsilon, j = 0, 1, 2, 3$.

Macrostate Label	Macrostate Specification				Microstate Specification			Thermo-dynamic Probability	True Probability
k	N_0	N_1	N_2	N_3	A	B	C	w_k	p_k
1	2	0	0	1	0	0	3ε		
					0	3ε	0	3	0.3
					3ε	0	0		
2	1	1	1	0	0	ε	2ε		
					0	2ε	ε		
					ε	0	2ε	6	0.6
					ε	2ε	0		
					2ε	0	ε		
					2ε	ε	0		
3	0	3	0	0	ε	ε	ε	1	0.1

TABLE 12.3 Distribution of three particles among four energy levels with total energy $U = 3\varepsilon$.

3ε	1		
2ε		1	
ε		1	3
0	2	1	0
k	1	2	3
w_k	3	6	1

energy 2ε, and N_3 with energy 3ε. The total number of microstates is 10 and the number of possible macrostates is 3. For the most probable macrostate, the number of available microstates is 6. Another way of displaying the results is given in Table 12.3, where the occupation numbers are listed for each of the energy levels.

While the numbers here are small, it is evident that the most "disordered" macrostate is the state of highest probability. For the very large number of particles in a physically meaningful assembly, this state will be sharply defined and will be the observed equilibrium state of the system.

In classical thermodynamics we have seen that as a system proceeds toward a state of equilibrium the entropy increases, and *at* equilibrium attains its maximum value. Similarly, our statistical model suggests that systems tend to change spontaneously from states with low thermodynamic probability to states with high thermodynamic probability (large number of microstates). Thus the second law of thermodynamics is a consequence of the theory of probability: the world changes the way it does because it seeks a state of higher probability!

12.4 THERMODYNAMIC PROBABILITY AND ENTROPY

It was Boltzmann who made the connection between the classical concept of entropy and the thermodynamic probability. His argument is as follows. Assume that the entropy is some function of w, that is,

$$S = f(w). \tag{12.10}$$

Here S and w are properties of the state of the system (state variables). To be physically useful, $f(w)$ must be a single-valued, monotonically increasing function.

Consider two subsystems, A and B (Figure 12.4). The entropy is an *extensive* property; like the volume, it is doubled when the mass or number of

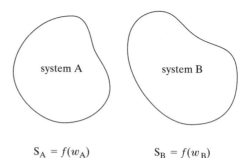

Figure 12.4 Two systems
with entropies S_A and S_B. $S_A = f(w_A)$ $S_B = f(w_B)$

particles is doubled. Thus the combined entropy of the two subsystems is simply the *sum* of the entropies of each subsystem:

$$S_{\text{total}} = S_A + S_B,$$

or

$$f(w_{\text{total}}) = f(w_A) + f(w_B). \tag{12.11}$$

However, one subsystem configuration can be combined with the other to give
the configuration of the total system. That is,

$$w_{\text{total}} = w_A w_B. \tag{12.12}$$

This follows from the fact that independent probabilities are multiplicative. To
return to our coin-tossing experiment, suppose that the two subsystems each
consist of two distinguishable coins. The possible configurations of the subsystems are listed in Table 12.4. The probability of two heads in subsystem A is
clearly 1/4, as it is in subsystem B. From Table 12.1, the probability of obtaining
all heads when *four* coins are tossed is 1/16, which is equal to $1/4 \times 1/4$. Thus
Equation (12.12) holds, and therefore

$$f(w_{\text{total}}) = f(w_A w_B). \tag{12.13}$$

TABLE 12.4 Possible configurations of tossed coins for subsystems
A and B, each consisting of two coins.

A	B
HH	HH
HT	HT
TH	TH
TT	TT

Figure 12.5 Ludwig Boltzmann's tombstone, showing his famous equation. (Institut fur Theoretische Physik der Universitat Wien/Central Library for Physics in Vienna.)

Combining Equations (12.11) and (12.13), we obtain

$$f(w_A) + f(w_B) = f(w_A w_B). \tag{12.14}$$

The only function for which this statement is true is the logarithm. Therefore

$$S = k \ln w, \tag{12.15}$$

where k is a constant with the units of entropy. It is, in fact, Boltzmann's constant:

$$k = 1.38 \times 10^{-23} \text{ J K}^{-1}.$$

Boltzmann's tombstone in Vienna bears this fundamental relation as its inscription (Figure 12.5).

12.5 QUANTUM STATES AND ENERGY LEVELS

We have tacitly assumed that the energy levels of our assembly of particles form a discrete set rather than a continuum. This is a consequence of the particles' confinement in a container of finite volume. Only in the case of completely

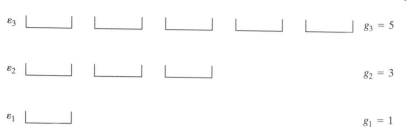

Figure 12.6   Energy levels and quantum states. The degeneracy g_j is the number of quantum states whose energy level is ε_j.

"free" particles can a continuous spectrum of energy exist. To understand how the discreteness arises, we must resort to quantum mechanics.

In quantum theory, to each energy level there corresponds one or more quantum states described by a wave function ψ. For so-called stationary states, ψ will be a function dependent on the position coordinates and the time.* When there are several quantum states that have the same energy, the states are said to be *degenerate*. The quantum state associated with the lowest energy level is called the *ground state* of the system; those that correspond to higher energies are called *excited states*.

The energy levels can be thought of as a set of shelves at different heights, while the quantum states correspond to a set of boxes on each shelf (Figure 12.6). For each energy level ε_j the number of quantum states is given by the degeneracy g_j.

As an instructive example, consider a basic problem in quantum mechanics, that of a particle of mass m in a one-dimensional box with infinitely high walls. (The particle is confined within the region $0 \leq x \leq L$.) Within the box the particle is free, subjected to no forces except those associated with the walls of the container (Figure 12.7).

The time-independent part of the wave function $\psi(x)$ is a measure of the probability of finding the particle at a position x in the box. The problem of finding $\psi(x)$ is analogous to the classical problem of transverse waves on a string with both ends fixed. The wave function satisfying the boundary conditions $\psi(0) = 0$ and $\psi(L) = 0$ is

$$\psi(x) = A \sin kx, \quad 0 \leq x \leq L, \tag{12.16}$$

with

$$k = n\frac{\pi}{L}, \quad n = 1,2,3.\ldots. \tag{12.17}$$

---

*Although Ψ is time-dependent, the observable quantity $\Psi^* \Psi$ is constant in time. Ψ^* is the complex conjugate of Ψ.

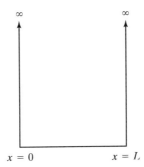

Figure 12.7   A one-dimensional potential well.

The boundary conditions simply state that the wave function must vanish outside the box. The probability of finding the particle there is zero.

Now, the de Broglie relation of wave mechanics is

$$p = \hbar k,$$

where p is the momentum of the particle, k is the wavenumber, and $\hbar$ is Planck's constant divided by 2π. The particle's kinetic energy is therefore

$$\varepsilon = \frac{1}{2}mv^2 = \frac{p^2}{2m} = \frac{\hbar^2 k^2}{2m}. \tag{12.18}$$

Substituting Equation (12.17) in Equation (12.18), we obtain

$$\varepsilon = \frac{\pi^2 \hbar^2}{2mL^2} n^2. \tag{12.19}$$

Here the integer n is the quantum number of the one-dimensional box. The important result is that the energy is proportional to the *square* of the quantum number.

We can easily extend this result to the case of a three-dimensional box with dimensions L_x, L_y, and L_z, for which the energy becomes

$$\frac{\pi^2 \hbar^2}{2m}\left(\frac{n_x^2}{L_x^2} + \frac{n_y^2}{L_y^2} + \frac{n_z^2}{L_z^2}\right).$$

In this case, any particular quantum state is designated by *three* quantum numbers n_x, n_y, and n_z. If $L_x = L_y = L_z = L$, then

$$\varepsilon_j \propto (n_x^2 + n_y^2 + n_z^2) = n_j^2, \tag{12.20}$$

where n_j is the total quantum number for states whose energy level is ε_j. The energy levels depend only on the values of n_j^2 and not on the individual values of the integers (n_x, n_y, n_z).

TABLE 12.5   First three states of a three-dimensional infinite potential well.

Level	Energy State	(n_x, n_y, n_z)	n_j^2	g_j
$j = 1$	Ground state	(1,1,1)	3	1
$j = 2$	First excited state	(1,1,2); (1,2,1); (2,1,1)	6	3
$j = 3$	Second excited state	(1,2,2); (2,1,2); (2,2,1)	9	3

The volume V of a cubical box equals L^3, so $L^2 = V^{2/3}$ and hence

$$\varepsilon_j = n_j^2 \frac{\pi^2 \hbar^2}{2m} V^{-2/3}. \tag{12.21}$$

This result applies to a container of any shape whose dimensions are large compared with the de Broglie wavelength $2\pi\hbar/p$. We observe that as the volume decreases, the value of the jth energy level increases.

The lowest energy level, $j = 1$, is that for which $n_x = n_y = n_z = 1$. Then $n_1^2 = 3$ and

$$\varepsilon_1 = 3 \frac{\pi^2 \hbar^2}{2m} V^{-2/3}.$$

This is the ground state. There is only one set of quantum numbers (n_x, n_y, n_z) that has this energy. Thus, the lowest level is nondegenerate and $g_1 = 1$. The degeneracy for the first three states is shown in Table 12.5. As a general rule, the degeneracy increases fairly rapidly with increasing j.

Now, each energy level is occupied by a number of particles N_j. Returning to our analogy of shelves with boxes on them, the degeneracy g_j of level j is the number of boxes on the jth shelf. The number of particles in any one box is the number in a particular quantum state. Those particles on any one shelf are in different states but have the same energy. Our interest is in the total number of particles in the boxes at the various levels, that is, in the occupation numbers N_j. The set of numbers $\{N_j\}$ then defines a macrostate of the system. A fictitious example is shown in Figure 12.8.

It is instructive to estimate the quantum number n_j for a one-liter volume of helium gas at room temperature. The mass of a helium atom is 6.65×10^{-27} kg, and 1 liter $= 10^{-3}$ m^3. Also, $\hbar = 1.054 \times 10^{-34}$ Js. Thus

$$\frac{\pi^2 \hbar^2}{2m} V^{-2/3} = \frac{\pi^2 (1.054 \times 10^{-34})^2}{2(6.65 \times 10^{-27})} (10^{-3})^{-2/3} = 8.24 \times 10^{-40} \text{ J} \approx 5 \times 10^{-21} \text{ eV}.$$

At room temperature the mean kinetic energy of a helium gas atom is

$$\varepsilon_j \approx kT = (1.38 \times 10^{-23})(293) = 4.04 \times 10^{-21} \text{ J} \approx 2.5 \times 10^{-2} \text{ eV}.$$

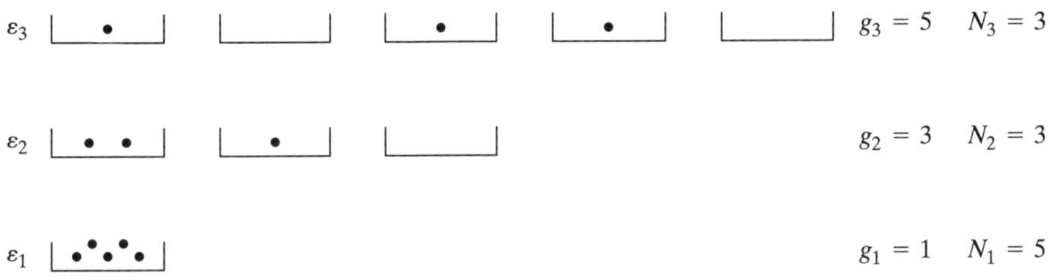

ε_3 ⌊ • ⌋ ⌊____⌋ ⌊ • ⌋ ⌊ • ⌋ ⌊____⌋ $g_3 = 5$ $N_3 = 3$

ε_2 ⌊ • • ⌋ ⌊ • ⌋ ⌊____⌋ $g_2 = 3$ $N_2 = 3$

ε_1 ⌊ • • • • ⌋ $g_1 = 1$ $N_1 = 5$

Figure 12.8   Distribution of eleven particles among three energy levels.

Then

$$n_j \approx \left(\frac{2.5 \times 10^{-2}}{5 \times 10^{-21}}\right)^{1/2} \approx 2 \times 10^9.$$

Thus, for the vast majority of the molecules of a gas at ordinary temperatures, the quantum numbers n_j are large indeed. In other words, most of the molecules occupy highly excited states. This means that the energy levels are very closely spaced and that the discrete spectrum may be treated as an energy continuum. That is the subject of the next section.

12.6 DENSITY OF QUANTUM STATES

Under the conditions that the quantum numbers are large and the energy levels are very close together, we can regard the n's and the ε's as forming a continuous function rather than a discrete set of values. We are interested in finding the density of states $g(\varepsilon)$.

Dropping the subscripts in Equation (12.21), using $\hbar = h/2\pi$, and solving for n^2, we obtain

$$n^2 = n_x^2 + n_y^2 + n_z^2 = \left(\frac{8mV^{2/3}}{h^2}\right)\varepsilon \equiv R^2. \tag{12.22}$$

The possible values of n_x, n_y, n_z — all positive integers — correspond to points in a cubic lattice in (n_x, n_y, n_z) space. According to Equation (12.22), for a given value of ε, the values of n_x, n_y, n_z that satisfy this equation lie on a sphere of radius R. A cross-section through the positive octant is shown in Figure 12.9.

We let $g(\varepsilon)d\varepsilon$ be the number of quantum states whose energy lies in the range ε to $\varepsilon + d\varepsilon$. This is the number of states whose quantum numbers (n_x, n_y, n_z) lie within the infinitesimally thin shell of the octant of a sphere with radius proportional to the square root of the energy (Figure 12.10). Evidently,

$$g(\varepsilon)d\varepsilon = n(\varepsilon + d\varepsilon) - n(\varepsilon) \approx \frac{dn(\varepsilon)}{d\varepsilon}d\varepsilon. \tag{12.23}$$

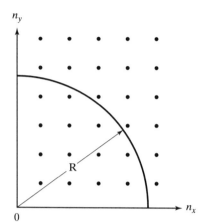

Figure 12.9   A cross-section
through the positive octant of
radius R showing lattice points
corresponding to the quantum
numbers n_x and n_y.

Here we have expanded the function $n(\varepsilon + d\varepsilon)$ in a Taylor series and
retained only the first two terms. Now $n(\varepsilon)$ is the number of states contained
within the octant of the sphere of radius R; that is,

$$n(\varepsilon) = \frac{1}{8} \cdot \frac{4}{3}\pi R^3 = \frac{\pi}{6}V\left(\frac{8m}{h^2}\right)^{3/2}\varepsilon^{3/2}, \tag{12.24}$$

so that

$$g(\varepsilon)d\varepsilon = \frac{dn(\varepsilon)}{d\varepsilon}d\varepsilon = \frac{4\sqrt{2}\pi V}{h^3}m^{3/2}\varepsilon^{1/2}d\varepsilon. \tag{12.25}$$

Equation (12.25) is not quite the result we seek. It takes into account the
translational motion only of a particle of the assembly. But quantum parti-
cles may have spin as well. The result for $g(\varepsilon)d\varepsilon$ is valid for bosons of zero
spin (bosons are particles with integer spin). For fermions of spin one-half
(fermions are particles with half-integer spin), there are *two* spin states for
each translational state. Thus we must multiply Equation (12.25) by a spin
factor γ_s:

$$g(\varepsilon)d\varepsilon = \gamma_s\frac{4\sqrt{2}\pi V}{h^3}m^{3/2}\varepsilon^{1/2}d\varepsilon, \tag{12.26}$$

where $\gamma_s = 1$ for spin zero bosons and $\gamma_s = 2$ for spin one-half fermions.
    For molecules, states associated with rotation and vibration may exist
in addition to translation and spin. We shall consider such cases in later
chapters.

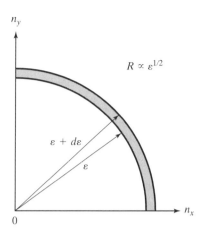

$$R \propto \varepsilon^{1/2}$$

$\varepsilon + d\varepsilon$

ε

Figure 12.10 Shell enclosing
quantum states with energies
between ε and $\varepsilon + d\varepsilon$.

PROBLEMS

12-1 Suppose that you flip 50 "honest" coins.
  (a) How many microstates are there? Give your answer as a factor of the order of unity times an integral power of 10.
  (b) How many microstates are there corresponding to the most probable macrostate?
  (c) What is the true probability of achieving the most probable macrostate?

  Note: Use a calculator that gives you $n!$ or a table of gamma functions ($\Gamma(n + 1) = n!$). Stirling's approximation will not give you sufficient accuracy.

12-2 Do Problem 12-1 for the case of 100 coins.

12-3 Using whatever computing means you have at your disposal, investigate the error in Stirling's approximation

$$\ln n! \approx n \ln n - n.$$

  (You might display the percentage error as a function of n for $n > 10$, say. Be creative.) For what value of n is the error less than 2 percent (approximately)?

12-4 Suppose you flip 1000 coins. We found that the thermodynamic probability of getting 500 heads and 500 tails is $w = 10^{300}$. Show that for the case of 600 heads and 400 tails w is smaller by a factor of 10^9.

12-5 For N distinguishable coins the thermodynamic probability is

$$w = \frac{N!}{N_1!(N - N_1)!},$$

  where N_1 is the number of heads and $N - N_1$ the number of tails.
  (a) Assume that N is large enough that Stirling's approximation is valid. Show that $\ln w$ is a maximum for $N_1 = N/2$.
  (b) Show that $w_{max} \approx e^{N\ln 2}$.

12-6 Consider a model thermodynamic assembly in which the allowed (nondegenerate) states have energies 0, ε, 2ε, 3ε. The assembly has *four distinguishable* (localized) particles and a total energy $U = 6\varepsilon$.

  (a) Tabulate the nine possible distributions of the four particles among the energy levels $n\varepsilon$, where $n = 0, 1, \ldots$.

  (b) Evaluate w_k for each of the macrostates and calculate $\Omega = \sum_k w_k$.

  (c) Calculate the *average* occupation numbers

$$\overline{N_j} = \sum_k N_{jk} w_k / \Omega$$

of the four particles in the energy states.

12-7 Six *distinguishable* particles are to be distributed among energy levels $0, \varepsilon, 2\varepsilon, 3\varepsilon$ $\ldots$, each with a degeneracy of *three*. The total energy is $U = 6\varepsilon$.

  (a) Tabulate the 11 possible macrostates of the assembly.

  (b) Calculate w_k (the number of microstates) for each of the 11 macrostates as a factor times 3^6.

  (c) What is the total number of available microstates Ω for the assembly?

  (d) Find the *average* occupation number for each energy level.

  (e) What is the *sum* of the average occupation numbers (it should be six)?

12-8 Two *distinguishable* particles are to be distributed among nondegenerate energy levels 0, ε, and 2ε such that the total energy $U = 2\varepsilon$.

  (a) What is the entropy of the assembly?

  (b) If a distinguishable particle with zero energy is added to the system, show that the entropy of the assembly is increased by a factor of 1.63.

12-9 In reference to Equation (12.14), prove that if

$$f(xy) = f(x) + f(y),$$

where x and y are independent variables, then $f(x) = C \ln x$, $C = $ constant. (Hint: Take partial derivatives of the equation.)

12-10 **(a)** Tabulate the values of the quantum numbers n_x, n_y, n_z for the 12 lowest energy levels of a particle in a container of volume $V = L^3$.

  (b) What is the degeneracy g of each level?

  (c) Find the energy of each level in units of $\pi^2 \hbar^2 / 2mL^2$.

  (d) Are the energy levels equally spaced?

12-11 Calculate the value of n_j in which an oxygen atom confined to a cubical box 1 cm on each side will have the same energy as the lowest energy available to a helium atom confined to a cubical box 2×10^{-10} m on each side.

12-12 Consider a gas consisting of one kilomole of helium atoms at standard temperature and pressure. Calculate the degeneracy $g(\varepsilon)$ for the energy level $\varepsilon = (3/2) kT$ (take $\gamma_s = 1$). What is the approximate ratio of $g(\varepsilon)$ to the number of atoms N?

Chapter 13

Classical and Quantum Statistics

13.1 BOLTZMANN STATISTICS

We wish to determine the equilibrium configuration for an assembly of N *distinguishable* noninteracting particles subject to the constraints (Equations (12.8) and (12.9))

$$\sum_{j=1}^{n} N_j = N, \tag{13.1}$$

$$\sum_{j=1}^{n} N_j \varepsilon_j = U. \tag{13.2}$$

Here N_j is the number of particles with single-particle energy ε_j; N and U are assumed to be fixed quantities. Our goal is to find the occupation number of each energy level when the thermodynamic probability is a maximum.

We need to allow for degeneracy. Consider the first energy level, $j = 1$. The number of ways of selecting N_1 particles from a total of N to be placed in the first level is

$$\binom{N}{N_1} = \frac{N!}{N_1!(N - N_1)!}.$$

We ask: in how many ways can these N_1 particles be arranged in the first level? There are g_1 quantum states in the first level, so for each particle there are g_1 choices. That is, there are $(g_1)^{N_1}$ possibilities in all. (Suppose that $g_1 = 2$ and $N_1 = 3$. For each particle there are two choices, so there are 2^3 choices total (Figure 13.1).)

Thus the number of ways to put N_1 particles into a level containing g_1 distinct options is

$$\frac{N! g_1^{N_1}}{N_1!(N - N_1)!}.$$

State a	State b
123	
	123
12	3
23	1
31	2
1	23
2	31
3	12

Figure 13.1 The eight arrangements of three distinguishable particles in two quantum states, a and b. The particles are labeled 1, 2 and 3.

For the second energy level, the situation is the same, except that there are only $(N - N_1)$ particles remaining to deal with:

$$\frac{(N - N_1)!g_2^{N_2}}{N_2!(N - N_1 - N_2)!}$$

Continuing the process, we obtain

$$w_B(N_1, N_2 \ldots N_n) = \frac{N!g_1^{N_1}}{N_1!(N - N_1)!} \times \frac{(N - N_1)!g_2^{N_2}}{N_2!(N - N_1 - N_2)!}$$

$$\times \frac{(N - N_1 - N_2)!g_3^{N_3}}{N_3!(N - N_1 - N_2 - N_3)!} \cdots$$

$$= N!\frac{g_1^{N_1}g_2^{N_2}g_3^{N_3}\cdots}{N_1!N_2!N_3!\cdots} = N!\prod_{j=1}^{n}\frac{g_j^{N_j}}{N_j!}. \tag{13.3}$$

Having developed the statistics w_B for the so-called Boltzmann distribution, our task is to maximize w, subject to the conditions of Equations (13.1) and (13.2).

13.2 THE METHOD OF LAGRANGE MULTIPLIERS

The so-called method of Lagrange multipliers is tailor-made for our problem. Suppose we wish to maximize a function $f(x, y)$ of two variables subject to the constraint $\phi(x, y) = $ constant. We must satisfy the condition

$$df = \frac{\partial f}{\partial x}dx + \frac{\partial f}{\partial y}dy = 0.$$

If dx and dy were independent, we would conclude that $\partial f/\partial x = \partial f/\partial y = 0$. However, they are not independent but are related by the constraint equation, which can be differentiated to give

$$d\phi = \frac{\partial \phi}{\partial x} dx + \frac{\partial \phi}{\partial y} dy = 0.$$

It follows that

$$\frac{\partial f/\partial x}{\partial \phi/\partial x} = \frac{\partial f/\partial y}{\partial \phi/\partial y}.$$

If we let the common ratio be the constant $-\alpha$, we then have

$$\frac{\partial f}{\partial x} + \alpha\frac{\partial \phi}{\partial x} = 0, \quad \frac{\partial f}{\partial y} + \alpha\frac{\partial \phi}{\partial y} = 0. \tag{13.4}$$

These are exactly the expressions we would get if we attempted to maximize the function $f + \alpha\phi$ without the constraint. The method adds a new unknown α called the Lagrange multiplier; it is the price we pay for the constraint.

The extension to a function of n variables is straightforward. Let

$$f = f(x_1, x_2, \ldots x_n) \tag{13.5}$$

be subject to the condition

$$\phi = \phi(x_1, x_2, \ldots x_n) = \text{constant.} \tag{13.6}$$

For a maximum, $df = 0$. Since ϕ is a constant, $d\phi = 0$. The method gives

$$\frac{\partial f}{\partial x_i} + \alpha\frac{\partial \phi}{\partial x_i} = 0, \quad i = 1, 2, 3 \ldots n. \tag{13.7}$$

These equations, together with Equation (13.6), constitute a set of $n + 1$ equations that can in principle be solved for the independent variables $x_1, x_2, \ldots x_n$ and the Lagrange multiplier α.

If there are *two* constraint relations, as in our case (Equations (13.1) and (13.2)),

$$\phi = \phi(x_1, x_2, \ldots x_n), \tag{13.8}$$

$$\psi = \psi(x_1, x_2, \ldots x_n), \tag{13.9}$$

we must introduce two arbitrary multipliers α and β. In this case, we have a set of $n + 2$ equations—the two constraint equations and the n equations of the form

$$\frac{\partial f}{\partial x_i} + \alpha \frac{\partial \phi}{\partial x_i} + \beta \frac{\partial \psi}{\partial x_i} = 0, \quad i = 1, 2, 3 \ldots n. \tag{13.10}$$

Thus the problem is fully specified. The multipliers are often called "undetermined," but they can be found.

13.3 THE BOLTZMANN DISTRIBUTION

Because $\ln w$ is a monotonically increasing function of w, maximizing $\ln w$ is equivalent to maximizing w. The logarithm is much easier to work with. From Equation (13.3), we have

$$\ln w = \ln N! + \sum_{i=1}^{n} N_i \ln g_i - \sum_{i=1}^{n} \ln N_i!. \tag{13.11}$$

Since we are concerned with very large numbers, we can safely use Stirling's approximation and write

$$\ln w = \ln N! + \sum_i N_i \ln g_i - \sum_i N_i \ln N_i + \sum_i N_i. \tag{13.12}$$

We needn't approximate $\ln N!$ because it is a constant that vanishes upon differentiation.

In applying the method of Lagrange multipliers, we take partial derivatives with respect to a particular value N_j of the set $(N_1, N_2, \ldots N_i \ldots N_n)$. Thus

$$\frac{\partial \ln w}{\partial N_j} + \alpha \frac{\partial \phi}{\partial N_j} + \beta \frac{\partial \psi}{\partial N_j} = 0, \tag{13.13}$$

where

$$\phi = \sum_i N_i \quad \text{and} \quad \psi = \sum_i N_i \varepsilon_i. \tag{13.14}$$

So

$$\frac{\partial}{\partial N_j} \left[\sum_i N_i \ln g_i - \sum_i N_i \ln N_i + \sum_i N_i \right] + \alpha \frac{\partial}{\partial N_j} \left(\sum_i N_i \right) + \beta \frac{\partial}{\partial N_j} \left(\sum_i N_i \varepsilon_i \right) = 0. \tag{13.15}$$

In working out the derivatives, we note that the only terms of the sums whose derivatives with respect to N_j are other than zero are those for which $i = j$. Therefore, Equation (13.15) reduces to

$$\ln g_j - \ln N_j \underbrace{- \frac{N_j}{N_j} + 1}_{=0} + \alpha + \beta\varepsilon_j = 0. \qquad (13.16)$$

Then

$$\ln\left(\frac{N_j}{g_j}\right) = \alpha + \beta\varepsilon_j,$$

or

$$\frac{N_j}{g_j} = e^{\alpha + \beta\varepsilon_j} \equiv f_j(\varepsilon_j). \qquad (13.17)$$

Equation (13.17) gives, for every energy level of the system, the number of particles per quantum state for the equilibrium configuration of the assembly. The result, obtained under the assumption of distinguishable particles, is called the Boltzmann distribution function.

The constants α and β are related to the physical properties of the assembly. We consider β first. If we multiply Equation (13.16) by N_j and sum over j, we obtain

$$\sum_j N_j \ln g_j - \sum_j N_j \ln N_j + \alpha \sum_j N_j + \beta \sum_j N_j \varepsilon_j = 0,$$

or

$$\sum_j N_j \ln g_j - \sum_j N_j \ln N_j = -\alpha N - \beta U.$$

The two terms on the left-hand side appear in the expression for $\ln w$ of Equation (13.12). Making the substitution, we get

$$\ln w = \ln N! + N - \alpha N - \beta U.$$

Simplifying, we have

$$\ln w = C - \beta U,$$

where C is a constant. Using the statistical definition of entropy,

$$S = k \ln w = S_0 - k\beta U, \tag{13.18}$$

where S_0 is the constant kC.

From classical theory,

$$dS = \frac{dU}{T} + \frac{PdV}{T} = \left(\frac{\partial S}{\partial U}\right)_V dU + \left(\frac{\partial S}{\partial V}\right)_U dV,$$

giving

$$\left(\frac{\partial S}{\partial U}\right)_V = \frac{1}{T}. \tag{13.19}$$

From Equation (13.18),

$$\left(\frac{\partial S}{\partial U}\right)_V = -k\beta. \tag{13.20}$$

It follows that

$$\beta = -\frac{1}{kT}. \tag{13.21}$$

The association of the classical expression with the statistical expression for the entropy suggests that the temperature is a Lagrange multiplier!* Note that the partial derivative in Equation (13.19) is taken with the volume held constant. The constancy of V is implied when we write Equation (13.2), since $\varepsilon_j \propto V^{-2/3}$, as we saw in Chapter 12.

Substituting Equation (13.21) in Equation (13.17), we obtain

$$N_j = g_j e^\alpha e^{-\varepsilon_j/kT}.$$

The value of α can easily be found from the first constraint equation (Equation (13.1)):

$$N = \sum_j N_j = e^\alpha \sum_j g_j e^{-\varepsilon_j/kT},$$

---

*Some texts give $\beta = +1/kT$. The positive sign comes from the choice of sign of the Lagrange multiplier, which is arbitrary.

so that

$$e^{\alpha} = \frac{N}{\sum\limits_{j} g_j e^{-\varepsilon_j/kT}},$$

and hence

$$f_j = \frac{N_j}{g_j} = \frac{N e^{-\varepsilon_j/kT}}{\sum\limits_{j} g_j e^{-\varepsilon_j/kT}} \quad \text{(Boltzmann distribution)}, \quad (13.22)$$

where f_j is the probability of occupation of a single state belonging to the jth energy level.

The sum in the denominator has a special significance. It is called the *partition function* or sum-over-states (*Zustandsumme* in German), and is represented by the symbol Z:

$$Z \equiv \sum_{j=1}^{n} g_j e^{-\varepsilon_j/kT} . \quad (13.23)$$

The partition function depends on the temperature and on the parameters that determine the energy levels and quantum states.

If the energy levels are crowded together very closely, as they are for a gaseous assembly, the degeneracy g_j may be replaced by $g(\varepsilon)d\varepsilon$, the number of states in the energy range from ε to $\varepsilon + d\varepsilon$. Correspondingly, N_j can be replaced by $N(\varepsilon)d\varepsilon$, the number of particles in the range ε to $\varepsilon + d\varepsilon$. We then obtain the continuous distribution function

$$f(\varepsilon) \equiv \frac{N(\varepsilon)}{g(\varepsilon)} = \frac{N e^{-\varepsilon/kT}}{\int g(\varepsilon) e^{-\varepsilon/kT} d\varepsilon}. \quad (13.24)$$

The equation is completely analogous to the expression for the discrete energy spectrum.

It is worth noting that the occupation numbers are fully determined by the temperature and the volume (which established the energy levels and the degeneracies). The set of occupation numbers, in turn, that maximize w specify the equilibrium macrostate. Thus two state variables define a thermodynamic state, exactly as in the classical theory.

The maximization of w does not absolutely require the use of Lagrange multipliers; an alternative method is given in Appendix C.

13.4 THE FERMI-DIRAC DISTRIBUTION

As we have discussed, Boltzmann statistics can be applied when the particles of the assembly are distinguishable—say, by their positions in a solid lattice. In this section we assume (1) that the particles are identical and *indistinguishable,* and (2) that they obey the *Pauli exclusion principle.* This means that no quantum state can accept more than one particle, taking spin into account. Such particles have half-integer spin and are given the generic name *fermion.* Examples of fermions are electrons, positrons, protons, neutrons, and muons. An important application of Fermi-Dirac statistics is to the behavior of free electrons in metals and semi-conductors.

Invoking once again our picture of energy level shelves with quantum state boxes resting on them, we emphasize that in the Fermi-Dirac case, one particle or no particle occupies a given state so that $N_j \le g_j$ for all j (Figure 13.2).

In determining the thermodynamic probability for a given macrostate we note that the group of g_j states is divisible into two subgroups: N_j of the states are to contain one particle and $(g_j - N_j)$ of the states must be unoccupied. The counting problem is precisely the same as that of the coin-tossing experiment, where we divided N particles into N_1 heads and $N - N_1$ tails and obtained

$$w = \frac{N!}{N_1!(N - N_1)!}.$$

Here

$$w_j = \frac{g_j!}{N_j!(g_j - N_j)!} \tag{13.25}$$

for the jth energy level. The total number of microstates corresponding to an allowable configuration is simply the product of the individual factors of the form of Equation (13.25) for all the levels:

$$w_{\mathrm{FD}}(N_1, N_2, \ldots N_n) = \prod_{j=1}^{n} \frac{g_j!}{N_j!(g_j - N_j)!}. \tag{13.26}$$

To maximize w_{FD} we follow exactly the same procedure as in the case of Boltzmann statistics. We use the method of Lagrange multipliers, assuming

jth
energy
level

$g_j = 7$

$N_j = 3$

Figure 13.2 Fermions in quantum states with energy ε_j.

conservation of particles and energy, to calculate the occupation numbers for the equilibrium macrostate. Taking the logarithm of w_{FD}, we have

$$\ln w_{FD} = \sum_i \ln g_i! - \sum_i \ln N_i! - \sum_i \ln(g_i - N_i)!. \qquad (13.27)$$

Using Stirling's approximation, this becomes

$$\ln w_{FD} = \sum_i [g_i \ln g_i - g_i - N_i \ln N_i + N_i - (g_i - N_i)\ln(g_i - N_i) + (g_i - N_i)]$$

$$= \sum_i [g_i \ln g_i - N_i \ln N_i - (g_i - N_i)\ln(g_i - N_i)]. \qquad (13.28)$$

Since $\sum_i N_i = N$ and $\sum_i N_i \varepsilon_i = U$, the equations

$$\frac{\partial(\ln w_{FD})}{\partial N_j} + \alpha \frac{\partial \phi}{\partial N_j} + \beta \frac{\partial \psi}{\partial N_j} = 0, \quad j = 1, 2, \ldots n.$$

lead to

$$-\frac{\partial}{\partial N_j}\left[\sum_i N_i \ln N_i + \sum_i (g_i - N_i)\ln(g_i - N_i)\right]$$

$$+\alpha \frac{\partial}{\partial N_j}\left(\sum_i N_i\right) + \beta \frac{\partial}{\partial N_j}\left(\sum_i N_i \varepsilon_i\right) = 0. \qquad (13.29)$$

Here we have taken into account that

$$-\frac{\partial}{\partial N_j}\left(\sum_i g_i \ln g_i\right) = 0,$$

since the g_i's are not functions of the N_j's. Working out the derivatives, we obtain

$$-\ln N_j - \underbrace{\frac{N_j}{N_j}}_{1} + \ln(g_j - N_j) - \underbrace{\frac{(g_j - N_j)}{(g_j - N_j)}(-1)}_{-1} = -\alpha - \beta\varepsilon_j,$$

or

$$\ln\left(\frac{g_j}{N_j} - 1\right) = -\alpha - \beta\varepsilon_j,$$

or

$$\frac{N_j}{g_j} = \frac{1}{e^{-\alpha - \beta \varepsilon_j} + 1}.$$ (13.30)

As before, we set $\beta = -1/kT$. However, the determination of the other Lagrange multiplier by summing over the N_j's doesn't work for this distribution. Provisionally, we associate α with the chemical potential μ divided by kT, and reserve for later the physical interpretation of this connection. Thus we set

$$\alpha = \frac{\mu}{kT},$$ (13.31)

so that (13.30) becomes

$$f_j \equiv \frac{N_j}{g_j} = \frac{1}{e^{(\varepsilon_j - \mu)/kT} + 1} \quad \text{(Fermi-Dirac distribution)}.$$ (13.32)

The corresponding result for the continuous energy spectrum is

$$f(\varepsilon) = \frac{1}{e^{(\varepsilon - \mu)/kT} + 1}.$$ (13.33)

13.5 THE BOSE-EINSTEIN DISTRIBUTION

Not all elementary particles obey the Pauli exclusion principle; photons are the most notable exception. Thus we need statistics for *indistinguishable* particles, any number of which can occupy a given quantum state. Such particles have zero or integer spin and are called *bosons*. The statistics developed for this case leads to the Bose-Einstein distribution, named after its inventors.

Counting microstates here is slightly more complicated than it is for Fermi-Dirac statistics. For the jth energy level there will be g_j quantum states containing a total of N_j identical particles with no restriction on the number of particles in each state. It is convenient to depict the arrangement of the N_j particles among the g_j states by $(g_j - 1)$ partitions or lines and N_j dots (Figure 13.3).

Figure 13.3   Bosons in quantum states with energy ε_j.

Now, we can obtain new microstates by shuffling the lines and dots while keeping the numbers g_j and N_j fixed. We ask: in how many ways can the $(N_j + g_j - 1)$ symbols (lines and dots) be arranged into $(g_j - 1)$ lines and N_j dots? Again we have the binomial problem of the coin-tossing experiment. Clearly the answer is

$$w_j = \frac{(N_j + g_j - 1)!}{N_j!(g_j - 1)!} \tag{13.34}$$

for the jth energy level. The number of microstates for a given macrostate is, then, the extended product

$$w_{BE}(N_1, N_2 \ldots N_n) = \prod_{j=1}^{n} \frac{(N_j + g_j - 1)!}{N_j!(g_j - 1)!}. \tag{13.35}$$

Taking the logarithm of Equation (13.35), we have

$$\ln w_{BE} = \sum_i \ln(N_i + g_i - 1)! - \sum_i \ln N_i! - \sum_i \ln(g_i - 1)!. \tag{13.36}$$

Invoking Stirling's approximation, we obtain

$$\ln w_{BE} = \sum_i [(N_i + g_i - 1)\ln(N_i + g_i - 1) - (N_i + g_i - 1) - N_i \ln N_i$$

$$+ N_i - (g_i - 1)\ln(g_i - 1) + (g_i - 1)]$$

$$= \sum_i [(N_i + g_i - 1)\ln(N_i + g_i - 1) - N_i \ln N_i - (g_i - 1)\ln(g_i - 1)]. \tag{13.37}$$

With the constraints taken into consideration, the method of Lagrange multipliers gives

$$\frac{\partial(\ln w_{BE})}{\partial N_j} + \alpha\frac{\partial\phi}{\partial N_j} + \beta\frac{\partial\psi}{\partial N_j} = 0, \quad j = 1, 2, 3, \ldots n,$$

which leads to

$$\frac{\partial}{\partial N_j}\left[\sum_i (N_i + g_i - 1)\ln(N_i + g_i - 1) - \sum_i N_i \ln N_i\right]$$

$$+ \alpha\frac{\partial}{\partial N_j}\left(\sum_i N_i\right) + \beta\frac{\partial}{\partial N_j}\left(\sum_i N_i \varepsilon_i\right) = 0, \tag{13.38}$$

or

$$\ln(N_j + g_j - 1) + \underbrace{\frac{N_j + g_j - 1}{N_j + g_j - 1}}_{1} - \ln N_j - \underbrace{\frac{N_j}{N_j}}_{1} = -\alpha - \beta\varepsilon_j.$$

This gives

$$\ln\left(\frac{N_j + g_j - 1}{N_j}\right) = -\alpha - \beta\varepsilon_j.$$

Neglecting unity compared with $N_j + g_j$, we obtain the result

$$\frac{N_j}{g_j} = \frac{1}{e^{-\alpha - \beta\varepsilon_j} - 1}. \tag{13.39}$$

With $\beta = -1/kT$ and $\alpha = \mu/kT$, we have

$$f_j \equiv \frac{N_j}{g_j} = \frac{1}{e^{(\varepsilon_j - \mu)/kT} - 1} \quad \text{(Bose-Einstein distribution)}. \tag{13.40}$$

For the continuous energy spectrum,

$$f(\varepsilon) = \frac{1}{e^{(\varepsilon - \mu)/kT} - 1}. \tag{13.41}$$

13.6 DILUTE GASES AND THE MAXWELL-BOLTZMANN DISTRIBUTION

Since elementary particles have half-integral or integral spins, the particles of a *gas* are either fermions or bosons; they are clearly one or the other. However, it is useful to consider another kind of statistics pertinent to a so-called *dilute gas*. The word "dilute" means that for all energy levels, the occupation numbers are very small compared with the available number of quantum states (most quantum states are empty). We assume that

$$N_j \ll g_j, \quad \text{for all } j. \tag{13.42}$$

This condition holds for real gases except at very low temperatures.

If in this region very few states are occupied at all, it is extremely unlikely that more than one particle will occupy a given state. Thus it is irrelevant whether or not the particles obey the Pauli exclusion principle, and we might therefore expect Fermi-Dirac and Bose-Einstein statistics to be approximately identical in the dilute gas limit. This is indeed the case, as can easily be seen. Recall Equation (13.26),

$$w_{FD} = \prod_j \frac{g_j!}{N_j!(g_j - N_j)!}. \tag{13.43}$$

Now

$$\frac{g_j!}{(g_j - N_j)!} = \frac{g_j(g_j - 1)(g_j - 2)\ldots(g_j - N_j + 1)(g_j - N_j)!}{(g_j - N_j)!} \approx g_j^{N_j},$$

so that

$$w_{FD} \approx \prod_j \frac{g_j^{N_j}}{N_j!}, \quad N_j \ll g_j. \tag{13.44}$$

The approximate value is slightly *greater* than the exact value since factors such as $(g_j - 1)$, $(g_j - 2)$, etc. are written as g_j.

Similarly, for Bose-Einstein statistics, Equation (13.35) is

$$w_{BE} = \prod_j \frac{(g_j + N_j - 1)!}{N_j!(g_j - 1)!}, \tag{13.45}$$

where

$$(g_j + N_j - 1)! = (g_j + N_j - 1)(g_j + N_j - 2)\ldots(g_j + N_j - N_j)(g_j - 1)!.$$

We see that in the numerator of Equation (13.45) there are N_j terms ahead of $(g_j - 1)!$ so that for $N_j \ll g_j$,

$$(g_j + N_j - 1)! \approx g_j^{N_j}(g_j - 1)!,$$

which is slightly *less* than the exact expression. Thus

$$w_{BE} \approx \prod_j \frac{g_j^{N_j}(g_j - 1)!}{N_j!(g_j - 1)!} \approx \prod_j \frac{g_j^{N_j}}{N_j!}, \quad N_j \ll g_j, \tag{13.46}$$

It is apparent that for a dilute gas, Fermi-Dirac and Bose-Einstein statistics give virtually the same thermodynamic probability. This "classical limit," called *Maxwell-Boltzmann statistics*, was investigated long before the development of quantum mechanics. Note the difference between Maxwell-Boltzmann statistics, Equation (13.46), and Boltzmann statistics, Equation (13.3):

$$w_B = N! w_{MB}, \quad w_{MB} = \prod_j \frac{g_j^{N_j}}{N_j!}. \tag{13.47}$$

The much larger Boltzmann probability includes the *permutation N!* of the N identifiable particles, giving rise to additional microstates.

The distribution of particles among energy levels can be found by the usual method of Lagrange multipliers. However, the result can be written down immediately by observing that w_{MB} and w_B differ only by a constant—the factor $N!$. Since maximizing the thermodynamic probability involves taking derivatives and the derivative of a constant is zero, we get precisely the Boltzmann distribution

$$f_j \equiv \frac{N_j}{g_j} = \frac{N e^{-\varepsilon_j/kT}}{Z} \quad \text{(Maxwell-Boltzmann distribution)}. \tag{13.48}$$

This should not come as a surprise: two functions differing only by a constant will have maximum values at the same point.

Because the two distributions are identical, Boltzmann statistics and Maxwell-Boltzmann statistics are frequently confused with each other. Boltzmann statistics assumes distinguishable (localizable) particles and therefore has limited application, largely to solids and some liquids. For gases, either Fermi-Dirac or Bose-Einstein statistics applies, depending on the spin of the particles. Maxwell-Boltzmann statistics is a very useful approximation for the special case of a dilute gas, which is a good model for a real gas under most conditions. It so happens that the corresponding distribution is the same as the Boltzmann distribution.

13.7 THE CONNECTION BETWEEN CLASSICAL AND STATISTICAL THERMODYNAMICS

The statistical expression for internal energy is

$$U = \sum_j N_j \varepsilon_j \,.$$

Taking the differential, we have

$$dU = \sum_j \varepsilon_j dN_j + \sum_j N_j d\varepsilon_j, \tag{13.49}$$

where ε_j is some function of an extensive property X such as the volume:

$$\varepsilon_j = \varepsilon_j(X).$$

Since $d\varepsilon_j = (d\varepsilon_j/dX)dX$, it follows that

$$\sum_j N_j d\varepsilon_j = \left[\sum_j N_j \frac{d\varepsilon_j}{dX} \right] dX.$$

Let

$$Y \equiv -\sum_j N_j \frac{d\varepsilon_j}{dX}.$$

Then

$$dU = \sum_j \varepsilon_j dN_j - Y dX.$$

For two states with X the same, $dX = 0$ and

$$(dU)_X = \sum_j \varepsilon_j dN_j \quad . \tag{13.50}$$

The classical analog to Equation (13.49) is

$$dU = T dS - Y dX,$$

so that

$$(dU)_X = T dS. \tag{13.51}$$

Comparing Equations (13.50) and (13.51), we see that

$$\sum_j \varepsilon_j dN_j = T dS.$$

Accordingly,

$$\sum N_j d\varepsilon_j = -Y dX.$$

Furthermore, since $đQ_r = T dS$ and $đW_r = Y dX$, it follows that

$$\sum \varepsilon_j dN_j = đQ_r \quad \text{and} \quad \sum N_j d\varepsilon_j = -đW_r. \qquad (13.52)$$

The first equation states that heat transfer is energy resulting in a net redistribution of particles among the available energy levels, involving no work. In Figure 13.4 heat added to a system shifts particles from lower to higher energy levels.

The second part of Equation (13.52) can be interpreted in terms of an adiabatic reversible process (no heat flow) in which work is done *on* the system (note the negative sign). An increase in the system's internal energy could therefore be brought about by a decrease in volume with an associated increase in the ε_j's. The energy levels are shifted to higher values with no redistribution of the particles among the levels (Figure 13.5).

The relationships of the two parts of Equation (13.52) provide a deeper understanding of the fundamental concepts of thermodynamics.

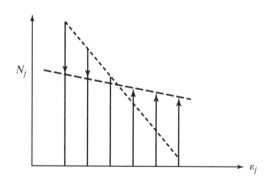

Figure 13.4   Heat added to a system moves particles from lower to higher energy levels.

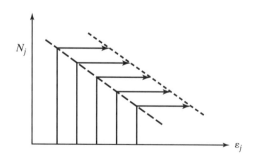

Figure 13.5   Work done on a system moves energy levels to higher values.

We turn our attention next to the chemical potential, the quantity appearing thus far in two of the distribution functions. In Chapter 9 we saw that when matter is added to or taken from an open system, the change in internal energy is

$$dU = T\,dS - P\,dV + \mu\,dN, \tag{13.53}$$

where μ is the chemical potential defined on a *per particle* basis.* Now the Helmholtz function is defined as $F = U - TS$ so that

$$dF = T\,dS - P\,dV + \mu\,dN - T\,dS - S\,dT,$$

or

$$dF = -S\,dT - P\,dV + \mu\,dN .$$

Therefore,

$$\mu = \left(\frac{\partial F}{\partial N}\right)_{T,V}. \tag{13.54}$$

Let us calculate first S and then F for Maxwell-Boltzmann statistics:

$$w_{\text{MB}} = \prod_j \frac{g_j^{N_j}}{N_j!}, \quad \text{with} \quad \frac{N_j}{g_j} = \frac{N}{Z}e^{-\varepsilon_j/kT}. \tag{13.55}$$

Then

$$S = k \ln w = k\left[\sum_j N_j \ln g_j - \sum_j \ln N_j!\right]$$

$$\approx k\left[\sum_j N_j \ln g_j - \sum_j N_j \ln N_j + \sum_j N_j\right]$$

$$= k\left[N - \sum_j N_j \ln\left(\frac{N_j}{g_j}\right)\right]. \tag{13.56}$$

Substituting the distribution function of Equation (13.55) in Equation (13.56), we obtain

$$S = k\left[N - \ln N \sum_j N_j + \ln Z \sum_j N_j + \frac{1}{kT}\sum_j N_j\varepsilon_j\right],$$

---

*Equation (9.1) is $dU = T\,dS - P\,dV + \mu^*dn$, where μ^* is the chemical potential defined as energy per kilomole. But $n = N/N_A$, where N_A is Avogadro's number. Thus, if we define μ as μ^*/N_A, then $\mu^*dn = \mu dN$; μ is the chemical potential energy *per particle*.

or

$$S = \frac{U}{T} + Nk(\ln Z - \ln N + 1). \tag{13.57}$$

Then

$$F = U - TS = -NkT(\ln Z - \ln N + 1). \tag{13.58}$$

Using Equation (13.54), we find that

$$\mu = -kT(\ln Z - \ln N + 1) + \frac{NkT}{N},$$

or

$$\mu = kT \ln\left(\frac{N}{Z}\right). \tag{13.59}$$

For gases, the partition function Z is proportional to the volume, as we shall see in Chapter 14. Thus μ is an increasing function of both the number of particles per unit volume and the temperature. This makes sense since particles flow from higher to lower chemical potential and from higher to lower concentrations. The difference in chemical potential is a true potential energy: for two systems it is the potential barrier that will bring the systems into diffusive equilibrium.

An example is the coexistence of two phases of the same substance—say, a liquid and its vapor at the same temperature. In Chapter 9, we observed that the chemical potentials for two phases of a given constituent must be equal for the system to be in diffusive equilibrium. This means that there is no net transfer of particles from one phase to the other; the total number of particles N is a constant. This is precisely the condition that allowed us to determine the Lagrange multiplier α. Thus the association of α with μ/kT is completely plausible.

Finally, we note that Equation (13.59) can be written

$$\frac{N}{Z} = e^{\mu/kT}. \tag{13.60}$$

Substituting this in the Maxwell-Boltzmann distribution gives

$$\frac{N_j}{g_j} = \frac{1}{e^{(\varepsilon_j - \mu)/kT}}. \tag{13.61}$$

We are now in a position to summarize and compare the distribution functions we have derived in the previous sections of this chapter.

13.8 COMPARISON OF THE DISTRIBUTIONS

The distribution functions for identical indistinguishable particles can be represented by the single equation

$$\frac{N_j}{g_j} = \frac{1}{e^{(\varepsilon_j - \mu)/kT} + a},\tag{13.62}$$

where

$$a = \begin{cases} +1 \text{ for FD statistics} \\ -1 \text{ for BE statistics} \\ 0 \text{ for MB statistics} \end{cases}.$$

For $N_j \ll g_j$, the denominator in Equation (13.62) is very large compared with unity, and the MB distribution is an approximation to the FD and BE distributions, as we have seen. Plots of $y \equiv N_j/g_j$ versus $x \equiv (\varepsilon_j - \mu)/kT$ are shown in Figure 13.6.

A word about the sign of μ is appropriate. The value of the chemical potential in some sense is arbitrary. When particles are being transferred from one system to another, it is important to establish a common zero-energy reference point from which all potential energies are measured. The problem is exactly analogous to defining a reference level for the gravitational potential in classical mechanics or for a potential well in quantum mechanics. Thus $\varepsilon_j - \mu$, the thermal energy of a particle, is the energy measured relative to the lowest point of the potential well. If a particle which itself has zero energy is inserted into a system, the ensuing interactions with its neighbors are most often attractive, in which case the chemical potential is negative. When

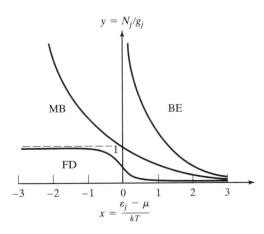

Figure 13.6 Comparison of the FD, BE and MB distributions.

systems are not exchanging particles, we can choose any convenient reference level, which is usually taken as zero.

Examining the curves of Figure 13.6, we note the following:

1. *BE curve:* $y = 1/(e^x - 1)$; for $x \to 0$, $y \to \infty$, and for large x, $y \approx e^{-x}$. The distribution is undefined for $x < 0$. Particles tend to condense (concentrate) in regions where ε_j is small, that is, in the lower energy states.

2. *FD curve:* $y = 1/(e^x + 1)$; for $x = 0$, $y = 1/2$, and for x large, $y \approx e^{-x}$. For $x \to -\infty$, $y \to 1$. At the lower levels with $\varepsilon_j - \mu$ negative, the quantum states are nearly uniformly populated with one particle per state.

3. *MB curve:* The curve $y = e^{-x}$ lies between the BE and FD curves and is only valid for $y \ll 1$. This is the dilute gas region: many, many states are unoccupied. Note that

$$\frac{N_j}{N_i} = \frac{g_j}{g_i} e^{-\frac{1}{kT}(\varepsilon_j - \varepsilon_i)}, \quad \varepsilon_i < \varepsilon_j.$$

At *high* temperatures, $\frac{1}{kT}(\varepsilon_j - \varepsilon_i) \ll 1$ and the exponential is approximately unity so that $N_j \approx g_j$. Since g_j tends to increase with increasing energy, so does the occupation number. At *low* temperatures, on the other hand, the population of the lower states is favored: for $g_i \approx g_j$, $N_i > N_j$, where $i < j$.

Statistical equilibrium is a balance between the randomizing forces of thermal agitation, tending to produce a uniform population of the energy levels, and the tendency of mechanical systems to sink to the states of lowest energy, resulting in highly ordered populations. For a negative chemical potential, Equation (13.60) gives

$$Z = Ne^{-\mu/kT} = Ne^{+|\mu|/kT}.$$

When many states are available to the system, Z is large and μ is highly negative. A particle seeks a position of lowest potential energy, choosing to be where the chemical potential has its largest negative value.

Having developed the statistics and equilibrium state distribution functions for all situations of physical interest, we are now in a position to apply the results to some specific problems in subsequent chapters.

13.9 ALTERNATIVE STATISTICAL MODELS

The statistical approach discussed in the previous two chapters describes an assembly of N particles with total energy U, where both N and U are fixed. The assembly is considered to be isolated in the sense that it does not exchange energy in any form with an external system. Such an assembly is known as a *microcanonical ensemble.*

In many experiments the system under investigation is not isolated. In chemical reactions, for example, the reaction vessel is often held at constant temperature by being kept in thermal contact with a heat reservoir such as a water bath that absorbs energy generated in the reaction. In this case it is convenient to consider an ensemble of a large number—say N_A—of identical assemblies in contact with a reservoir. The ensemble can be regarded as a single assembly in a heat reservoir provided by the remaining $N_A - 1$ assemblies. The reservoir fixes the temperature of the system rather than the energy. The individual members of this so-called *canonical ensemble* can exchange heat and work with the reservoir and hence the energy of the system can fluctuate. This is in contrast with the microcanonical ensemble in which there is no interaction with a reservoir at all.

In spite of this difference, the two ensembles give essentially the same results. The reason for this coincidence can be seen as follows. The assemblies of the canonical ensemble are separate macroscopic objects, like distinguishable particles, and can therefore be assumed to obey Boltzmann statistics, with the distribution

$$N_j = \frac{N}{Z} g_j e^{-\varepsilon_j/kT}, \quad Z = \sum_j g_j e^{-\varepsilon_j/kT}. \tag{13.63}$$

The corresponding distribution for the ensemble involves the energy states of a single sample assembly, which alone has a very large number of states with a given energy. The allowed values of the energy E are extremely close together, so that the degeneracy g_j can be replaced by a density of states $g(E)dE$, the number of one-assembly states (microstates) with energy between E and $E + dE$. Sums can be replaced by integrals and N_j/N becomes $N(E)/N_A = P(E)$, the probability that the sample assembly has an energy E. Thus

$$P(E) = \frac{g(E)e^{-E/kT}}{Z_A}, \quad Z_A = \int_0^\infty g(E)e^{-E/kT}dE. \tag{13.64}$$

The density of states $g(E)$ is a rapidly increasing function of E, whereas the exponential factor is a rapidly decreasing function. For definiteness, suppose that $g(E) \propto E^n$; then

$$P(E) \propto E^n e^{-E/kT}. \tag{13.65}$$

The maximum of $P(E)$ is found by setting its derivative equal to zero; the result gives $n = E/kT$ at the peak $E = U$. But $U = (3/2)NkT$ for an ideal monatomic gas, so $n = (3/2)N$, a very large number. Thus one factor in $P(E)$

is large and increasing at an astronomically high rate and the other is extremely small and decreasing. As a result, $P(E)$ has an exceptionally sharp peak.

We are interested in determining the probable fluctuation of E around the mean value U. To do this, it is convenient to expand the logarithm of $P(E)$ in a Taylor series about U:

$$\ln P(E) = \ln P(U) + \left(\frac{d\ln P(E)}{dE}\right)_{E=U} (E - U)$$

$$+ \frac{1}{2!}\left(\frac{d^2 \ln P(E)}{dE^2}\right)_{E=U} (E - U)^2 + \dots \quad . \qquad (13.66)$$

Here

$$\ln P(E) = n\ln E - E/kT,$$

$$\frac{d\ln P(E)}{dE} = \frac{n}{E} - \frac{1}{kT},$$

$$\frac{d^2 \ln P(E)}{dE^2} = -\frac{n}{E^2}.$$

The first derivative is zero at $E = U$ and the second derivative is $-2/3Nk^2T^2$. Using these values and neglecting terms higher than second order, we find that

$$P(E) = P(U)e^{-(E-U)^2/2\sigma_E^2}, \qquad (13.67)$$

where

$$\sigma_E^2 = \frac{3}{2}Nk^2T^2. \qquad (13.68)$$

This is the usual form of the Gaussian distribution in terms of the standard deviation σ_E, which is a measure of the peak width of $P(E)$. More useful for our purpose is the fractional width

$$\frac{\sigma_E}{U} = \frac{1}{\left(\frac{3}{2}N\right)^{1/2}}. \qquad (13.69)$$

Thus for a kilomole of an ideal gas, the deviation from the mean value U is of the order of one part in 10^{13}. It follows that predictions of the physical properties of a system using the canonical ensemble are virtually identical with those derived from the microcanonical ensemble in which the energy fluctuation is zero. When fluctuation effects are important, the canonical ensemble yields information not obtainable from the microcanonical ensemble. Otherwise the two are closely related and either one can be derived from the other.

In the canonical ensemble we remove the restriction that the energy U is fixed, but the assemblies are still assumed to have a fixed number N of particles. As a further generalization, we can consider assemblies with an indefinite number of particles. The resulting *grand canonical ensemble* is a system in contact with a reservoir with which it can exchange not only energy but particles as well. In this model only the mean number of particles is fixed. The grand canonical ensemble is useful when the constraint of a fixed number of particles is too restrictive, as in the problem of chemical equilibrium among a number of different species.

Summarizing the three approaches, the microcanonical ensemble treats a single material sample of volume V consisting of an assembly of N particles with fixed total energy U. The independent variables are V, N, and U. The central concept is the distribution N_j/N, the fractional number of particles occupying the jth energy level. The method is the least abstract of the three. It is mathematically the simplest, but is completely adequate for solving the problems usually discussed in undergraduate texts.

The canonical ensemble considers a collection of N_A identical assemblies, each of volume V. A single assembly is assumed to be in contact through a diathermal wall with a heat reservoir of the remaining $N_A - 1$ assemblies. The independent variables are V, N, and T, where T is the temperature of the reservoir. The averaging process involves the probability distribution function $P(E) = N(E)/N_A$. The approach is more mathematically abstract than the microcanonical ensemble, but one can be derived from the other.

The most general and most abstract of the three models is the grand canonical ensemble, which consists of open assemblies that can exchange both energies and particles with a reservoir. The independent variables describing the ensemble are V, T, and μ, where μ is the chemical potential. The grand canonical ensemble permits the treatment of more complicated problems than does either of the other methods and is correspondingly more mathematically demanding.

PROBLEMS

13-1 Use the method of Lagrange multipliers to solve the following problems:
  (a) A rectangle has a base x and a height y, where $x + y = 8$. Find the values of x and y that give the maximum area.
  (b) Find the area of the largest rectangle that can be inscribed in the ellipse

$$\frac{x^2}{a^2} + \frac{y^2}{b^2} = 1.$$

What percentage of the area of the ellipse does the rectangle occupy?

13-2 Five identical noninteracting particles occupy the kth energy level, which is ten-fold degenerate. How many possible microstates are there if
(a) the particles are bosons?
(b) the particles are fermions?

13-3 A model thermodynamic system in which the allowed nondegenerate states have energies $0, \varepsilon, 2\varepsilon, 3\varepsilon, \ldots$, consists of *four* particles with total energy $U = 6\varepsilon$.

Identify the possible distributions of particles, evaluate $\Omega = \sum_k w_k$, and work out the average occupation numbers for the various energy levels

(a) when the particles are gaseous bosons;
(b) when the particles are gaseous fermions.

13-4 Show that for a system of N particles obeying Maxwell-Boltzmann statistics, the occupation number for the jth energy level is given by

$$N_j = -NkT\left(\frac{\partial \ln Z}{\partial \varepsilon_j}\right)_T.$$

13-5 Show that it is possible to write the thermodynamic probability in the general form

$$w = \prod_{j=1}^{n} \frac{g_j(g_j - a)(g_j - 2a)\ldots[g_j - (N_j - 1)a]}{N_j!},$$

where

$$a = \begin{cases} + \text{ for FD statistics} \\ -1 \text{ for BE statistics} \\ 0 \text{ for MB statistics} \end{cases}.$$

13-6 Substitute the Maxwell-Boltzmann distribution function into the equation

$$\sum \varepsilon_j dN_j = T \, dS$$

to obtain

$$S = -k \sum N_j \ln\left(\frac{N_j}{g_j}\right).$$

Hint: note that $\sum dN_j = 0$ and $dN_j \ln\left(\dfrac{N_j}{g_j}\right) = d\left[N_j \ln\left(\dfrac{N_j}{g_j}\right) - N_j\right]$.

13-7 For the Fermi-Dirac distribution, sketch N_j/g_j versus ε_j for $T = 0$ and for T slightly greater than zero.

13-8 Show that for a system of a large number N of bosons at very low temperature (such that they are all in the nondegenerate lowest energy state $\varepsilon = 0$), the chemical potential varies with temperature according to

$$\mu \underset{T \to 0}{\to} -\frac{kT}{N}.$$

13-9 (a) At room temperature (300 K), what is the equilibrium number of bosons per state (N_j/g_j) for an energy ε_j
(1) 0.001 eV above the chemical potential μ?
(2) 0.1 eV above the chemical potential μ?

(b) At room temperature (300 K), what is the equilibrium number of fermions per state (N_j/g_j) for an energy ε_j
(1) 0.1 eV below the chemical potential μ?
(2) 0.1 eV above the chemical potential μ?

13-10 (a) Using results from Chapter 9, show that

$$\mu = -T\left(\frac{\partial S}{\partial N}\right)_{U,V}.$$

(b) It follows from the statistical definition of the entropy that

$$\Delta \ln w \approx -\frac{\mu}{kT}\Delta N.$$

Consider a system with a chemical potential $\mu = -0.3$ eV. By what factor is the number of possible microstates of the system increased when a single particle is added to it at room temperature?

13-11 A system has energy levels at $\varepsilon = 0$, $\varepsilon = 300k$, and $\varepsilon = 600k$, where k is Boltzmann's constant. The degeneracies of the levels are 1, 3, and 5 respectively.

(a) Calculate the partition function, the relative populations of the energy levels, and the average energy, all at a temperature of 300 K.

(b) At what temperature is the population of the energy level at $600k$ equal to the population of the energy level at $300k$?

13-12 A system with two nondegenerate energy levels ε_0 and ε_1 $(\varepsilon_1 > \varepsilon_0 > 0)$ is populated by N distinguishable particles at temperature T.

(a) Show that the average energy per particle is given by

$$u \equiv \frac{U}{N} = \frac{\varepsilon_0 + \varepsilon_1 e^{-\beta\Delta\varepsilon}}{1 + e^{-\beta\Delta\varepsilon}}, \quad \Delta\varepsilon \equiv \varepsilon_1 - \varepsilon_0, \quad \beta \equiv 1/kT.$$

(b) Show that when $T \to 0$,

$$u \approx \varepsilon_0 + \Delta\varepsilon e^{-\beta\Delta\varepsilon},$$

and when $T \to \infty$,

$$u \approx \frac{1}{2}(\varepsilon_0 + \varepsilon_1) - \frac{1}{4}\beta(\Delta\varepsilon)^2.$$

(c) Show that the specific heat at constant volume is

$$c_v = k\left(\frac{\Delta\varepsilon}{kT}\right)^2 \frac{e^{-\Delta\varepsilon/kT}}{(1 + e^{-\Delta\varepsilon/kT})^2}.$$

(d) Compute c_v in the limits $T \to 0$ and $T \to \infty$ and make a careful sketch of c_v versus $\Delta\varepsilon/kT$.

Chapter 14

The Classical Statistical Treatment of an Ideal Gas

14.1 THERMODYNAMIC PROPERTIES FROM THE PARTITION FUNCTION

The power of statistical thermodynamics is beautifully demonstrated in the application of the theory to an ideal gas. We shall begin by showing that all the thermodynamic properties can be expressed in terms of the logarithm of the partition function and its derivatives. Thereafter we need only to evaluate the partition function for a gas to obtain its thermodynamic properties.

In Section 13.7 we obtained the following relations for the Maxwell-Boltzmann distribution, valid for a dilute gas:

$$S = \frac{U}{T} + Nk(\ln Z - \ln N + 1), \tag{14.1}$$

$$F = -NkT(\ln Z - \ln N + 1), \tag{14.2}$$

$$\mu = -kT(\ln Z - \ln N). \tag{14.3}$$

To these we can add the other thermodynamic potentials and the pressure.

1. *Internal energy*
   Using the Maxwell-Boltzmann distribution function, we can write

$$U = \sum_j N_j \varepsilon_j = \frac{N}{Z} \sum_j g_j \varepsilon_j e^{-\varepsilon_j/kT}, \tag{14.4}$$

with

$$Z = \sum_j g_j e^{-\varepsilon_j/kT}. \tag{14.5}$$

Differentiating Equation (14.5) with respect to T, we have

$$\left(\frac{\partial Z}{\partial T}\right)_V = -\sum_j g_j \varepsilon_j e^{-\varepsilon_j/kT} \frac{d}{dT}\left(\frac{1}{kT}\right) = \frac{1}{kT^2} \sum_j g_j \varepsilon_j e^{-\varepsilon_j/kT}. \tag{14.6}$$

Since $\varepsilon_j = \varepsilon_j(V)$, keeping V constant means treating ε_j as a constant in taking the derivative. Solving for the sum in Equation (14.6) and substituting it in Equation (14.4), we obtain

$$U = \frac{N}{Z}(kT^2)\left(\frac{\partial Z}{\partial T}\right)_V,$$

or

$$U = NkT^2\left(\frac{\partial \ln Z}{\partial T}\right)_V. \tag{14.7}$$

2. *Gibbs function*
The calculation of G is trivially simple; in Chapter 9 we saw that $G = \mu N$ for a single component system. Using Equation (14.3), we have

$$G = -NkT(\ln Z - \ln N). \tag{14.8}$$

3. *Enthalpy*
The enthalpy is defined as $H \equiv U + PV$ and the Gibbs function is $G \equiv U - TS + PV$, so that $G = H - TS$, or

$$H = G + TS.$$

Substituting Equations (14.1), (14.7), and (14.8) in this equation yields $H = U + NkT$, or

$$H = NkT\left[1 + T\left(\frac{\partial \ln Z}{\partial T}\right)_V\right]. \tag{14.9}$$

4. *Pressure*
A reciprocity relation from Section 8.5 gives

$$P = -\left(\frac{\partial F}{\partial V}\right)_T.$$

Taking the derivative of Equation (14.2) with respect to V with T held constant gives

$$P = NkT\left(\frac{\partial \ln Z}{\partial V}\right)_T. \tag{14.10}$$

14.2 PARTITION FUNCTION FOR A GAS

The definition of the partition function is

$$Z \equiv \sum_j g_j e^{-\varepsilon_j/kT}.$$

Suppose that $g_j = 1$ for all j. For convenience, we assume temporarily that the sum runs from 0 to $n - 1$ and we take $\varepsilon_0 = 0$. Then

$$Z = 1 + e^{-\varepsilon_1/kT} + e^{-\varepsilon_2/kT} + \dots.$$

Suppose that the energy levels are far apart. What does Z look like? For low temperatures, with the ε_j's differing significantly from one another, the exponentials in the series get successively smaller quite rapidly, whereas at higher temperatures the falloff is more gradual (Figure 14.1). The partition function is a measure of the states *available* to the system and the distribution gives the occupation numbers, which are proportional to the factors $e^{-\varepsilon_j/kT}$. We see that for low T, essentially all the particles are in the ground state; for high T, the populations of the excited states are significant.

For a sample of gas in a container of macroscopic size, the energy levels are very closely spaced and many states are available to the system (Figure 14.2). It follows that the energy levels can be regarded as a continuum and we can use the result for the density of states derived in Chapter 12. In Equation (12.26) we use a spin factor $\gamma_s = 1$ since the gas is composed of molecules rather than

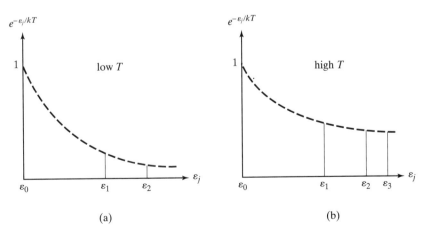

(a)   (b)

Figure 14.1   Successive terms of the partition function sum for (a) low temperature; and (b) high temperature. The energy level spacings are comparatively large for a small volume.

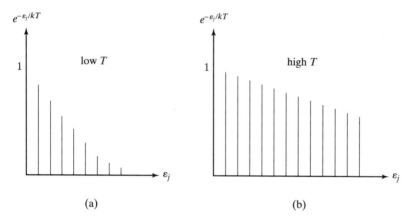

Figure 14.2 Successive terms of the partition function sum for (a) low temperature; and (b) high temperature. Here the energy levels are very closely spaced when the volume is large.

spin 1/2 particles; that is, the quantization associated with the spin does not apply. Thus

$$g(\varepsilon)d\varepsilon = \frac{4\sqrt{2}\pi V}{h^3}m^{3/2}\varepsilon^{1/2}d\varepsilon. \tag{14.11}$$

Then

$$Z = \int_0^\infty g(\varepsilon)e^{-\varepsilon/kT}d\varepsilon = \frac{4\sqrt{2}\pi V}{h^3}m^{3/2}\int_0^\infty \varepsilon^{1/2}e^{-\varepsilon/kT}d\varepsilon.$$

The integral can be found in tables and is

$$\int_0^\infty \varepsilon^{1/2}e^{-\varepsilon/kT}d\varepsilon = \frac{kT}{2}\sqrt{\pi kT}.$$

This gives the partition function for a gas under the assumption that the energy levels are so closely spaced that they form a continuum:

$$Z = V\left(\frac{2\pi mkT}{h^2}\right)^{3/2}. \tag{14.12}$$

14.3 PROPERTIES OF A MONATOMIC IDEAL GAS

The partition function depends on both the volume V and the temperature T. We need $\ln Z$ and its partial derivatives:

$$\ln Z = \frac{3}{2} \ln T + \ln V + \frac{3}{2} \ln\left(\frac{2\pi mk}{h^2}\right),$$

$$\left(\frac{\partial \ln Z}{\partial V}\right)_T = \frac{1}{V},$$

$$\left(\frac{\partial \ln Z}{\partial T}\right)_V = \frac{3}{2} \cdot \frac{1}{T}.$$

Substituting these expressions in the equations of Section 14.1 gives

$$PV = NkT \quad \text{and} \quad U = \frac{3}{2} NkT. \tag{14.13}$$

However familiar these equations are, their appearance here is an exhilarating development. It represents the capability of statistical thermodynamics to *produce* relations that occur in classical thermodynamics as empirical generalizations.

The calculation of the entropy is even more dramatic. It shows that the statistical theory can give results not at all obtainable in classical thermodynamics. Using Equation (14.12) and $U = (3/2)NkT$ in Equation (14.1), we obtain

$$S = Nk\left\{\frac{5}{2} + \ln\left[\frac{V(2\pi mkT)^{3/2}}{Nh^3}\right]\right\}. \tag{14.14}$$

This formula is known as the Sackur-Tetrode equation for the entropy of a monatomic gas. A rearrangement of terms yields

$$S = Nk\left(\frac{3}{2} \ln T + \ln \frac{V}{N}\right) + S_0, \tag{14.15}$$

where

$$S_0 = Nk\left\{\frac{5}{2} + \ln\left[\frac{(2\pi mk)^{3/2}}{h^3}\right]\right\}.$$

Recalling that $s = S/n$ and $Nk/n = R$, we obtain

$$s = c_v \ln T + R \ln v + s_0,$$

where $c_v = (3/2)R$. This is the equation for the specific entropy we obtained earlier. We note that s_0 is an undetermined constant in the classical theory.

The Sackur-Tetrode equation shows us how to calculate the number S_0, which could only have been previously estimated on empirical grounds. We therefore now have two different ways of evaluating the entropy of an ideal monatomic gas. From measurements of the heat capacity at various temperatures we can make an *empirical* determination and we can also make a theoretical calculation by using Equation (14.15). Excellent agreement has been found of the results obtained by the two methods. We note, however, that Equation (14.15) is obviously not valid down to absolute zero since S does not approach zero as $T \to 0$.

14.4 APPLICABILITY OF THE MAXWELL-BOLTZMANN DISTRIBUTION

Maxwell-Boltzmann statistics is valid under the dilute gas assumption $N_j \ll g_j$. How good is this approximation for gases under average conditions? We can write Equation (14.12) in the form

$$Z = n_Q V,$$

where

$$n_Q = \left(\frac{2\pi m k T}{h^2} \right)^{3/2}$$

has the dimension of the reciprocal of the volume (Z is a pure number), and is called the *quantum concentration* of the gas. Thus we can express the Maxwell-Boltzmann distribution as

$$\frac{N_j}{g_j} = \frac{N}{Z} e^{-\varepsilon_j/kT} = \left(\frac{N}{V} \right) \frac{1}{n_Q} e^{-\varepsilon_j/kT}.$$

Consider helium gas under standard (STP) conditions. Then $m = 6.65 \times 10^{-27}$ kg and $T = 273$ K, so that

$$n_Q \approx 7 \times 10^{30} \text{m}^{-3}.$$

The quantity N/V, the number of particles per unit volume, can be found from Avogadro's law:

$$\frac{N}{V} = \frac{6.02 \times 10^{26}}{22.4 \text{ m}^3} \approx 3 \times 10^{25} \text{m}^{-3}.$$

Since ε/kT is of the order of unity and so therefore is $e^{-\varepsilon_j/kT}$,

$$\frac{N_j}{g_j} \approx \frac{3 \times 10^{25}}{7 \times 10^{30}} \approx 4 \times 10^{-6}.$$

The result indicates that on the average only a few states in a million are occupied. Helium gas is thus highly dilute under normal conditions. When $N_j \ll g_j$, we say that the gas is in the *classical regime*. An ideal gas is defined as a gas of noninteracting molecules in the classical regime.

14.5 DISTRIBUTION OF MOLECULAR SPEEDS

The Maxwell-Boltzmann distribution of particles among energy levels leads directly to the speed distributions of molecules of an ideal gas. For a continuum of energy levels,

$$N(\varepsilon)d\varepsilon = f(\varepsilon)g(\varepsilon)d\varepsilon, \qquad (14.16)$$

where $N(\varepsilon)d\varepsilon$ is the number of particles having energies between ε and $\varepsilon + d\varepsilon$. For the Maxwell-Boltzmann distribution

$$f(\varepsilon) = \frac{N}{Z}e^{-\varepsilon/kT}, \qquad (14.17)$$

where, for an ideal gas,

$$Z = V\left(\frac{2\pi mkT}{h^2}\right)^{3/2}. \qquad (14.18)$$

The density of states is given by

$$g(\varepsilon) = \frac{4\sqrt{2}\pi V}{h^3}m^{3/2}\varepsilon^{1/2} \qquad (14.19)$$

for molecules. Combining these equations we obtain

$$N(\varepsilon)d\varepsilon = \frac{2\pi N}{(\pi kT)^{3/2}}\varepsilon^{1/2}e^{-\varepsilon/kT}d\varepsilon. \qquad (14.20)$$

Here ε is the single particle kinetic energy. Thus

$$\varepsilon = \frac{1}{2}mv^2, \quad d\varepsilon = mv\,dv,$$

and

$$\varepsilon^{1/2}d\varepsilon = \frac{1}{\sqrt{2}}m^{3/2}v^2\,dv.$$

Converting from kinetic energy to speed, we have $N(\varepsilon)d\varepsilon \to N(v)dv$, where $N(v)dv$ is the number of particles whose speeds are in the range v to $v + dv$. The computation gives the Maxwell speed distribution

$$N(v)dv = 4\pi N\left(\frac{m}{2\pi kT}\right)^{3/2}v^2 e^{-mv^2/2kT}\,dv. \tag{14.21}$$

This is precisely the result derived in Section 11.6 by using a more heuristic argument. The various averages discussed there need not be repeated here.

14.6 EQUIPARTITION OF ENERGY

In the kinetic theory of gases, an important result is that

$$\bar{\varepsilon} = \frac{f}{2}kT, \tag{14.22}$$

where $\bar{\varepsilon}$ is the average energy of a molecule and f is the number of degrees of freedom of its motion. Thus, for a monatomic gas there are three degrees of freedom, one for each direction of the molecule's translational motion, and $U = N\bar{\varepsilon} = (3/2)NkT$. It should be possible to obtain Equation (14.22) from statistical thermodynamics, and, indeed, that is the case.

Let z be a parameter such that $\varepsilon = \varepsilon(z)$. Each z is to be associated with a degree of freedom. For example, if v_x is the x-component of the molecular velocity, then $z = v_x$ and $\varepsilon = \varepsilon(v_x)$. From Equation (11.33) it is evident that we can write

$$N(z)dz = Ae^{-\varepsilon(z)/kT}\,dz, \tag{14.23}$$

where $N(z)dz$ is the number of molecules with parameter z in the range z to $z + dz$ and A is independent of z. From the definition of an average, we have

$$\bar{\varepsilon}(z) = \frac{\displaystyle\int \varepsilon(z) N(z)\, dz}{\displaystyle\int N(z)\, dz}, \tag{14.24}$$

where the denominator insures that $N(z)$ is a probability density function. We note in passing that $\bar{\varepsilon}(z) = U/N$.

We assume that $\varepsilon(z)$ is a *quadratic function* of z, that is, $\varepsilon(z) = az^2$, where a is a constant. This is certainly the case when $\varepsilon \propto v_x^2$. Therefore Equation (14.24) gives

$$\bar{\varepsilon}(z) = \frac{A \displaystyle\int_0^\infty az^2 e^{-az^2/kT}\, dz}{A \displaystyle\int_0^\infty e^{-az^2/kT}\, dz} = \frac{\dfrac{a}{4\alpha}\left(\dfrac{\pi}{\alpha}\right)^{1/2}}{\dfrac{1}{2}\left(\dfrac{\pi}{\alpha}\right)^{1/2}} = \frac{a}{2\alpha},$$

where $\alpha = a/kT$. Thus

$$\bar{\varepsilon}(z) = \frac{1}{2}kT. \tag{14.25}$$

The principle of the equipartition of energy states that for every degree of freedom for which the *energy is a quadratic function*, the mean energy per particle of a system in equilibrium at temperature T is $(1/2)kT$.

14.7 ENTROPY CHANGE OF MIXING REVISITED

In Section 9.5 we used the results of classical thermodynamics to determine the change in entropy due to the mixing of two *different* gases. We considered a volume V divided into two equal parts by a partition. On one side of the partition there are $n_1 = x_1 n$ kilomoles of an ideal gas and on the other side there are $n_2 = x_2 n$ kilomoles of a different gas, both gases being at the same temperature and pressure. The partition is removed, each gas diffuses into the other, and a new equilibrium state is obtained in which both gases occupy the total volume. The change in entropy as a result of the mixing process was found to be

$$\Delta S = -nR(x_1 \ln x_1 + x_2 \ln x_2), \tag{14.26}$$

where x_1 and x_2 are the kilomole fractions of the constituents.

Let's examine the problem from the point of view of statistical thermo-dynamics. First, we observe that $n = Nk/R$, so that

$$x_{1,2} = \frac{n_{1,2}}{n} = \frac{N_{1,2}}{N}.$$

That is, x_1 is the fractional number of molecules of gas 1 and x_2 is the corre-sponding quantity of gas 2, with $x_1 + x_2 = 1$.

The effect of the mixing is to *increase* the number of configurations available to the system by a factor of

$$\Delta w = \frac{N!}{N_1!N_2!} = \frac{N!}{(x_1N)!(x_2N)!}, \tag{14.27}$$

the number of ways of arranging the N molecules with N_1 in one container and N_2 in the other. The corresponding entropy change due to mixing is

$$\Delta S = k \ln \Delta w = k \ln N! - k \ln(x_1N)! - k \ln(x_2N)!.$$

Using Stirling's approximation, we obtain

$$\frac{\Delta S}{k} = N \ln N - N - x_1N \ln(x_1N) + x_1N - x_2N \ln(x_2N) + x_2N$$

$$= N \ln N - N - \underbrace{(x_1 + x_2)}_{1}N \ln N + \underbrace{(x_1 + x_2)}_{1}N - N(x_1 \ln x_1 + x_2 \ln x_2),$$

or

$$\Delta S = -Nk(x_1 \ln x_1 + x_2 \ln x_2). \tag{14.28}$$

Noting that $Nk = nR$, we see that this is precisely the classical result.

We ask: For what proportion of the two constituent molecules is the entropy of mixing a maximum? Let $x_1 = x$ and $x_2 = 1 - x$, so that

$$\sigma \equiv -\frac{\Delta S}{Nk} = x \ln x + (1 - x)\ln(1 - x).$$

To maximize σ, we take

$$\frac{d\sigma}{dx} = 0 = \ln x + 1 - \frac{1}{1 - x} - \ln(1 - x) + \frac{x}{1 - x} = \ln\left(\frac{x}{1 - x}\right).$$

Thus $x/(1 - x) = 1$ and $x = 1/2$. The entropy change due to mixing is there-fore a maximum when the two gases are present in equal proportions. This is in no way a surprise since the mixing problem is exactly equivalent to the coin-tossing experiment.

The statistical analysis of mixing sheds light on the Gibbs paradox described in Section 9.5. Clearly, if the gases are identical, there is no increase in the number of microstates when the partition is removed and the occupation numbers remain the same. As we have seen, the sharp distinction between distinguishable and nondistinguishable particles is fundamental in quantum mechanics and in statistical thermodynamics. The Gibbs paradox was viewed as a serious problem in the purely classical context in which it first arose.

14.8 MAXWELL'S DEMON

James Clerk Maxwell viewed the second law of thermodynamics as statistical even though it was left to Boltzmann to develop its theory. Maxwell, however, illustrated the statistical nature of the law with his famous "demon." In 1867 he suggested a conceivable way in which if two things are in contact, the hotter "*could* take heat from the colder without external agency."[*]

Consider a gas in a container divided by an adiabatic diaphragm into two sections, A and B (Figure 14.3). The gas in A is assumed to be hotter than the

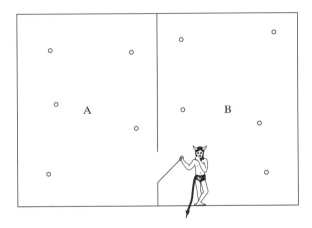

Figure 14.3 Maxwell's demon allows a molecule to pass from A to B if its kinetic energy is less than the average molecular kinetic energy in B. Passage from B to A is allowed only for molecules whose kinetic energies exceed the average kinetic energy per molecule in A. The demon is thumbing his nose at the second law of thermodynamics.

[*] J.C. Maxwell to P.G. Tait, 11 December 1867. Reprinted in C.G. Knott, *Life and Scientific Work of Peter Guthrie Tait*, Cambridge University Press, 1911.

gas in B. From the molecular point of view, the higher temperature means a higher average value of the kinetic energy of the molecules in A compared with those in B. Maxwell wrote, "Now conceive a finite being who knows the paths and velocities of all the molecules by simple inspection but who can do no work except open and close a hole in the diaphragm by means of a slide without mass." Maxwell's being is assigned to open the hole in the diaphragm to let through an approaching molecule in A when that molecule has a velocity less than the *rms* velocity of the molecules in B. At the same time, a molecule from B is allowed to pass through the hole into A only when its velocity exceeds the *rms* velocity of the molecules in A. These two procedures are carried out in such a way that the number of molecules in A and B do not change. As a result, "the energy in A is increased and that in B is diminished; that is, the hot system has got hotter and the cold colder and yet no work has been done; only the intelligence of a very observant and neat-fingered being has been employed." In this way the being violates the second law. "Only *we*

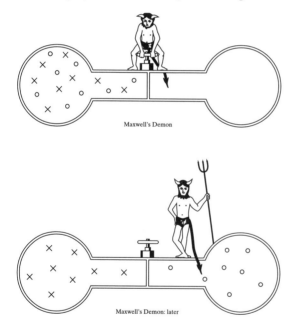

Maxwell's Demon

Maxwell's Demon: later

Figure 14.4   Maxwell's demon in action. In this version the demon operates a valve, allowing one species of a two-component gas (hot or cold) through a partition separating the gas from an initially evacuated chamber. Only fast molecules are allowed through, resulting in a cold gas in one chamber and a hot gas in the other. (Reproduced from Gasser, R.P.H. and Richards, W.G., *Entropy and Energy Levels,* Clarendon Press, Oxford, 1974, by permission of Oxford University Press.)

can't," added Maxwell, "not being clever enough." Kelvin immediately gave the being the name "demon" by which it has been known ever since.

Maxwell's demon has attracted an enormous amount of attention since it appeared well over a century ago. It has been said that the "demon is catlike; several times it has jumped up and taken on a new character after being left for dead."* Because of its connection with information theory and computer science, the subject of the demon continues to generate interest. We shall return briefly to it in the last chapter.

PROBLEMS

14-1 (a) Calculate the entropy S and the Helmholtz function F for an assembly of *distinguishable* particles.

(b) Show that the total energy U and the pressure P are the same for distinguishable particles as for molecules of an ideal gas while S is different. Explain why this makes sense.

14-2 Show that for an assembly of N particles that obeys Maxwell-Boltzmann statistics, the occupation numbers for the most probable distribution are given by:

$$N_j = -NkT\left(\frac{\partial \ln Z}{\partial \varepsilon_j}\right)_T.$$

14-3 (a) Show that for an ideal gas of N molecules,

$$\frac{g_j}{N_j} = \frac{Z}{N}e^{\varepsilon_j/kT},$$

where

$$\frac{Z}{N} = \frac{(kT)^{5/2}}{P}\left(\frac{2\pi m}{h^2}\right)^{3/2}.$$

(b) For $\varepsilon_j = (3/2)kT$, $T = 300$ K, $P = 10^3$ Pa, and $m = 10^{-26}$ kg, calculate g_j/N_j.

14-4 Calculate the entropy of a kilomole of neon gas at standard temperature and pressure. The mass of a neon atom is 3.35×10^{-26} kg.

14-5 Calculate the chemical potential in electron volts for a kilomole of helium gas at standard temperature and pressure. The mass of a helium atom is 6.65×10^{-27} kg.

---

*See *Maxwell's Demon: Entropy, Information, Computing*, by H.S. Leff and A.F. Rex, Princeton University Press, 1990. This book contains a chronological bibliography and reprints of 25 papers on the subject.

14-6 A tank contains one kilomole of argon gas at 1 atm and 300 K. The mass of an argon atom is 6.63×10^{-26} kg.

   (a) What is the internal energy of the gas in joules? What is the average energy of a molecule in eV?

   (b) What is the partition function Z?

   (c) What is the chemical potential μ in eV?

   (d) What is N_j/g_j?

14-7 The partition function of a system that obeys Maxwell-Boltzmann statistics is given by $Z = aVT^4$, where a is a constant. Calculate U, P, and S.

14-8 An ideal monatomic gas consists of N atoms in a volume V. The gas is allowed to expand isothermally to fill a volume $2V$. Show that the entropy change is $Nk \ln 2$.

Chapter 15

The Heat Capacity of a Diatomic Gas

15.1 INTRODUCTION

We have seen how statistical thermodynamics provides deep insight into the classical description of a monatomic ideal gas. We might have reason to hope, therefore, that the statistical model can resolve a thorny problem we encountered in the application of classical thermodynamics to a diatomic gas. As discussed in Chapter 11, the principle of equipartition of energy fails to give the observed value of the specific heat capacity. The explanation of this discrepancy was considered by Maxwell to be the most important challenge facing the statistical theory. In this chapter we shall see how the problem is solved.

15.2 THE QUANTIZED LINEAR OSCILLATOR

Until now we have confined our attention to systems of particles that have translational degrees of freedom only. To deal with particles such as diatomic molecules, we need to investigate so-called *internal* degrees of freedom such as vibrations, rotations, and electronic excitations. We shall address vibrations first.

We consider an assembly of N one-dimensional harmonic oscillators. We assume that the oscillators are *loosely coupled* in that the energy exchange among them is small. This means that each particle can oscillate nearly independently of the others. We further assume that each oscillator is free to vibrate in one dimension only with some natural frequency ν.

From quantum mechanics, the single particle energy levels are given by

$$\varepsilon_j = \left(j + \frac{1}{2} \right) h\nu, \quad j = 0, 1, 2, \ldots \quad . \tag{15.1}$$

Note that the energies are equally spaced and that the ground state has "zero-point" energy equal to $h\nu/2$. The states are nondegenerate in that $g_j = 1$ for all j.

For the internal degrees of freedom, Boltzmann statistics applies. The assumption may seem questionable since Boltzmann statistics characterizes *distinguishable* particles (localized in a crystal lattice, for example) and would therefore appear to be inappropriate for the treatment of an assembly of diatomic molecules. However, in the dilute gas approximation, the number of *translational* quantum states is so much larger than the number of particles ($g_j \gg N_j$) that the great majority of states are unoccupied, a few are occupied by a single particle, and virtually none have a population greater than one. Thus, in our treatment of the internal degrees of freedom of diatomic molecules, we can regard the particles as differentiated by the translational quantum states that they occupy.

We begin by evaluating the partition function

$$Z = \sum_{j=0}^{\infty} g_j e^{-\varepsilon_j/kT} = \sum_{j=0}^{\infty} e^{-(j+1/2)h\nu/kT}. \tag{15.2}$$

The temperature at which $kT = h\nu$ is called the characteristic temperature θ:

$$\theta = \frac{h\nu}{k}.$$

Using this in Equation (15.2), we have

$$Z = \sum_{j=0}^{\infty} e^{-(j+1/2)\theta/T} = e^{-\theta/2T} \sum_{j=0}^{\infty} e^{-j\theta/T}$$

$$= e^{-\theta/2T}(1 + e^{-\theta/T} + e^{-2\theta/T} + \dots)$$

$$= e^{-\theta/2T}[1 + e^{-\theta/T} + (e^{-\theta/T})^2 + \dots].$$

The sum in this expression is just an infinite geometric series of the form

$$1 + y + y^2 + \dots,$$

which equals $(1 - y)^{-1}$. This can easily be seen by taking the product $(1 - y)$ times $(1 + y + y^2 + \dots)$. Here $y = \exp(-\theta/T)$, so

$$Z = \frac{e^{-\theta/2T}}{1 - e^{-\theta/T}}. \tag{15.3}$$

We shall also be concerned with the occupation numbers, or with N_j/N, the fraction of the total number of particles with energy ε_j. The Boltzmann distribution for $g_j = 1$ is

$$\frac{N_j}{N} = \frac{e^{-\varepsilon_j/kT}}{Z} = e^{-\varepsilon_j/kT}[e^{\theta/2T}(1 - e^{-\theta/T})]$$

$$= (1 - e^{-\theta/T})e^{-\varepsilon_j/kT + \theta/2T}. \tag{15.4}$$

The exponent of the term outside the parentheses can be written

$$\frac{-\varepsilon_j}{kT} + \frac{\theta}{2T} = -\left(j + \frac{1}{2}\right)\frac{h\nu}{kT} + \frac{h\nu}{2kT} = -j\frac{h\nu}{kT} = -j\frac{\theta}{T}. \tag{15.5}$$

Thus Equation (15.4) becomes

$$\frac{N_j}{N} = e^{-j\theta/T}(1 - e^{-\theta/T}). \tag{15.6}$$

A sketch of Equation (15.6) for two temperatures shows that the lower the temperature, the more rapidly the occupation numbers decrease with j (Figure 15.1). At higher temperatures, more particles populate the higher energy levels.

Next we compute the internal energy of the assembly of oscillators. The expression we obtained for U in terms of the derivative of $\ln w$ (Equation (14.7)) applies here because the Boltzmann distribution is the same as the Maxwell-Boltzmann distribution. We have

$$U = NkT^2\left(\frac{\partial \ln Z}{\partial T}\right)_V. \tag{15.7}$$

Substituting Equation (15.3) in this equation and carrying out the differentiation, we obtain

$$U = Nk\theta\left(\frac{1}{2} + \frac{1}{e^{\theta/T} - 1}\right). \tag{15.8}$$

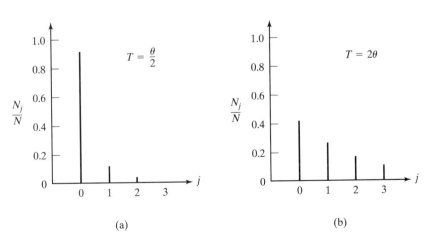

Figure 15.1 Fractional occupation numbers for quantized linear oscillators with (a) $T = \theta/2$, and (b) $T = 2\theta$.

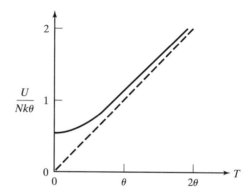

Figure 15.2 Variation with temperature of the internal energy of an assembly of linear oscillators.

Note that for $T \to 0$, $U = Nk\theta/2 = Nh\nu/2$, the zero-point energy. For high temperatures such that $T/\theta \gg 1$ or $\theta/T \ll 1$, we can expand in a series the denominator of the second term in Equation (15.8):

$$e^{\theta/T} - 1 = 1 + \left(\frac{\theta}{T}\right) + \frac{1}{2!}\left(\frac{\theta}{T}\right)^2 + \cdots - 1 \approx \frac{\theta}{T}.$$

Thus

$$U \approx Nk\theta\left(\frac{1}{2} + \frac{T}{\theta}\right) \approx NkT, \quad \frac{T}{\theta} \gg 1.$$

This is exactly what we would expect for a diatomic molecule with two vibrational degrees of freedom, one associated with the kinetic energy and the other with the potential energy. Note that in our model we have suppressed any kinetic energy of translation. The variation of U with T is sketched in Figure 15.2.

15.3 VIBRATIONAL MODES OF DIATOMIC MOLECULES

The most important application of these results is to the molecules of a diatomic gas. From classical thermodynamics,

$$C_V = \left(\frac{\partial U}{\partial T}\right)_V$$

for a reversible process. Using Equation (15.8), we have

$$C_V = Nk\theta\frac{\partial}{\partial T}(e^{\theta/T} - 1)^{-1},$$

or

$$C_V = Nk\left(\frac{\theta}{T}\right)^2 \frac{e^{\theta/T}}{(e^{\theta/T} - 1)^2}. \tag{15.9}$$

At very high temperatures $T/\theta \gg 1$ or $\theta/T \ll 1$ and

$$C_V \approx Nk\left(\frac{\theta}{T}\right)^2 \frac{1}{\left(\dfrac{\theta}{T}\right)^2} = Nk.$$

That is, the heat capacity approaches the constant Nk. In the low temperature limit, for $T/\theta \ll 1$ or $\theta/T \gg 1$, we have $e^{\theta/T} \gg 1$ and

$$\frac{e^{\theta/T}}{(e^{\theta/T} - 1)^2} \rightarrow \frac{1}{e^{\theta/T}} = e^{-\theta/T},$$

so that

$$C_V \rightarrow Nk\left(\frac{\theta}{T}\right)^2 e^{-\theta/T}.$$

The rate at which the exponential factor approaches zero as $T \rightarrow 0$ is greater than the rate of growth of $(\theta/T)^2$ so that

$$C_V \rightarrow 0 \quad \text{as} \quad T \rightarrow 0,$$

consistent with the third law. The variation of C_V with T is shown schematically in Figure 15.3.

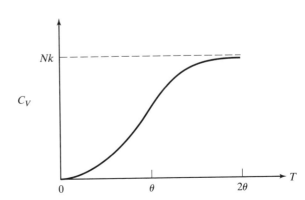

Figure 15.3 Variation with temperature of the heat capacity of an assembly of linear oscillators.

At an extremely low temperature $T \ll \theta$ or $kT \ll h\nu$; the oscillators (particles) are "frozen" into the ground state, in that virtually none are thermally excited at any one time. We have $U = Nh\nu/2$ and $C_V = 0$. This is the extreme "quantum limit." The discrete quantum nature of the states totally dominates the physical properties.

At the opposite extreme of temperature for which $T \gg \theta$ or $kT \gg h\nu$, we reach a "classical limit." Here the expressions for U and C_V do not involve $h\nu$. Planck's constant, the scale of the energy level separation, is irrelevant. Oscillators of any frequency have the same average energy kT. The dependence of U on kT is associated with the classical law of energy equipartition: each degree of freedom contributes $NkT/2$ to the internal energy. But the question left unanswered by the classical law is: when is a degree of freedom excited and when is it frozen out?

The total internal energy of a diatomic molecule is made up of four contributions that can be treated separately: (1) the kinetic energy associated with the translational motion of the center of the mass of the molecule (this is $(3/2)kT$, the same as that for a monatomic molecule); (2) the rotational energy due to the rotation of the two atoms about the center of mass of the molecule; (3) the vibrational motion of the two atoms along the axis joining them; and (4) the energy of excitation of the atomic electrons. The last three are internal modes of possessing energy. Because the four contributions can vary independently, it follows that the partition function is a product of the corresponding factors. This can be seen by noting that the internal energy is a function of the logarithm of Z.

Having considered the vibrational motion in Section 15.3, we turn our attention to the *rotational* contribution.

15.4 ROTATIONAL MODES OF DIATOMIC MOLECULES

The rotation of a diatomic molecule is modeled as the motion of a quantum mechanical rigid rotator. The rotation takes place about an axis through the center of mass of the molecule and perpendicular to the line joining the two atoms. We take I as the moment of inertia of the molecule about this axis; $I = \mu r_0^2$, where $\mu = m_1 m_2/(m_1 + m_2)$ is the reduced mass of the two atoms and r_0 is the equilibrium value of the distance between the nuclei.

Quantum mechanics states that the allowed values of the square of the angular momentum are $l (l + 1) \hbar^2$, where $\hbar$ is Planck's constant divided by 2π and $l = 0, 1, 2, 3. \ldots$. Recall from classical mechanics that the rotational energy is $(1/2)I\omega^2$, where ω is the angular velocity. The angular momentum L is $I\omega$ so the energy is $L^2/2I$. The quantized energy levels are therefore

$$\varepsilon_l = l(l + 1)\frac{\hbar^2}{2I}. \tag{15.10}$$

We define a characteristic temperature for rotation:

$$\theta_{rot} \equiv \frac{\hbar^2}{2Ik},\tag{15.11}$$

so that

$$\varepsilon_l = l(l+1)k\theta_{rot}.\tag{15.12}$$

Experimental values of θ_{rot} are given in Table 15.1 for several gases; they are found from infrared spectroscopy, in which the energies required to excite the molecules to higher rotational states are measured.

The energy levels of Equation (15.12) are degenerate; quantum mechanics gives

$$g_l = (2l+1)$$

states for level l corresponding to different possible directions of the angular momentum vector. Given these results, we can write down the partition function for rotation:

$$Z = \sum_l g_l e^{-\varepsilon_l/kT} = \sum_l (2l+1)e^{-l(l+1)\theta_{rot}/T}.\tag{15.13}$$

The important quantity in this expression is the argument of the exponential. For $T \ll \theta_{rot}$, virtually all the molecules are in the few lowest rotational states, and the series can be truncated with negligible error after the first two or three terms. It can be shown that in this case both the internal energy and the heat capacity at constant volume vanish as the temperature approaches zero (see Problem 15-4).

For all diatomic gases except hydrogen, the rotational characteristic temperature is of the order of 10 K or less. Since these gases have liquefied

---

TABLE 15.1 Characteristic temperature of rotation of diatomic molecules.

Substance	θ_{rot} (K)
H_2	85.4
O_2	2.1
N_2	2.9
HCl	15.2
CO	2.8
NO	2.4
Cl_2	0.36

(indeed, solidified) at such low temperatures, it is always true that $T \gg \theta_{rot}$ at ordinary temperatures. Hence a great many closely spaced energy states are excited and the sum in Equation (15.13) may be replaced by an integral. We write $x \equiv l(l + 1)$ and treat x as a continuous variable. Note that $dx = (2l + 1)dl$. In this approximation,

$$Z = \int_0^\infty [\exp(-\theta_{rot}/T)x] \, dx = \frac{T}{\theta_{rot}}, \quad T \gg \theta_{rot}. \tag{15.14}$$

This result is too large by a factor of 2 for homonuclear molecules such as H_2, O_2, and N_2. For two identical nuclei, two opposite positions of the molecular axis correspond to the same state of the molecule. This slight modification has no effect on the thermodynamic properties of the system such as the internal energy and the heat capacity.

Using Equation (15.14) in

$$U = NkT^2 \left(\frac{\partial \ln Z}{\partial T} \right)_V, \tag{15.15}$$

we find that for $T \gg \theta_{rot}$,

$$U_{rot} = NkT, \tag{15.16}$$

and

$$C_{V,rot} = Nk, \quad T \gg \theta_{rot}. \tag{15.17}$$

Thus, at significantly high temperatures, the rigid rotator exhibits the equipartition of energy between two degrees of freedom.

At very low temperatures such that $\theta_{rot} \gg T$, the partition function, Equation (15.13), can be expanded as

$$Z = 1 + 3e^{-2\theta_{rot}/T} + 5e^{-6\theta_{rot}/T} + \dots .$$

Retaining only the first two terms, we can write

$$\ln Z \approx \ln(1 + 3e^{-2\theta_{rot}/T})$$

Since the exponential term is small compared with unity, we can use the approximate relation

$$\ln(1 + \varepsilon) \approx \varepsilon, \quad \varepsilon \ll 1$$

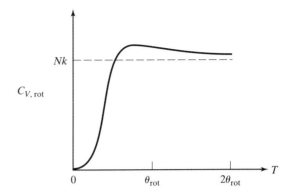

Figure 15.4 Variation with temperature of the heat capacity of an assembly of rigid rotators.

and obtain the result

$$\ln Z \approx 3e^{-2\theta_{\text{rot}}/T}.$$

Hence

$$U = NkT^2\left(\frac{\partial \ln Z}{\partial T}\right)_V = 6Nk\theta_{\text{rot}}e^{-2\theta_{\text{rot}}/T}, \quad \theta_{\text{rot}} \gg T \qquad (15.18)$$

and

$$C_{V,\text{rot}} = \left(\frac{\partial U}{\partial T}\right)_V = 3Nk\left(\frac{2\theta_{\text{rot}}}{T}\right)^2 e^{-2\theta_{\text{rot}}/T}. \quad \theta_{\text{rot}} \gg T \qquad (15.19)$$

Again, we see that C_V approaches zero as T approaches zero, as it must. Figure 15.4 shows schematically the contribution of rotation to the heat capacity as a function of temperature.

15.5 ELECTRONIC EXCITATION

The electronic partition function can be written

$$Z \approx g_0 + g_1 e^{-\theta_e/T}, \qquad (15.20)$$

where g_0 and g_1 are, respectively, the degeneracies of the ground state and the first excited state, and θ_e is the energy separation of the two lowest states divided by Boltzmann's constant k. For most gases, the higher electronic states are not excited ($\theta_e \sim 120,000$ K for hydrogen) and $Z \approx g_0$. Thus the partition function at practical temperatures makes no contribution to the external energy or heat capacity.

15.6 THE TOTAL HEAT CAPACITY

We are finally in a position to account fully for the behavior of C_V as a function of T for a diatomic gas at temperatures in the usual range of interest. Assuming that electronic excitations are negligible, we have

$$C_V = C_{V,\text{tr}} + C_{V,\text{rot}} + C_{V,\text{vib}}.$$

Thus, at ordinary temperatures,

$$C_V = \frac{3}{2}Nk + Nk + Nk\left(\frac{\theta_{\text{vib}}}{T}\right)^2 \frac{e^{\theta_{\text{vib}}/T}}{(e^{\theta_{\text{vib}}/T} - 1)^2},$$

or

$$C_V = Nk\left[\frac{5}{2} + \left(\frac{\theta_{\text{vib}}}{T}\right)^2 \frac{e^{\theta_{\text{vib}}/T}}{(e^{\theta_{\text{vib}}/T} - 1)^2}\right]. \tag{15.21}$$

Table 15.2 gives values for θ_{vib} for several diatomic gases.

At the lowest temperatures at which the system exists in the gaseous phase, the molecular motion is solely translational, contributing $(3/2)Nk = (3/2)nR$ to the heat capacity. As the temperature increases and approaches θ_{rot}, rotational quantum states are excited. Eventually, for $T \gg \theta_{\text{rot}}$, the rotational contribution becomes equal to nR, corresponding to two rotational degrees of freedom. Therefore, the heat capacity steps up to $(5/2)nR$ in the region $\theta_{\text{rot}} \ll T \ll \theta_{\text{vib}}$. Since room temperature lies within this range, this is the value usually given for the heat capacity of a diatomic gas.

TABLE 15.2 Characteristic temperatures of vibration of diatomic molecules.

Substance	θ_{vib} (K)
H_2	6140
O_2	2239
N_2	3352
HCl	4150
CO	3080
NO	2690
Cl_2	810

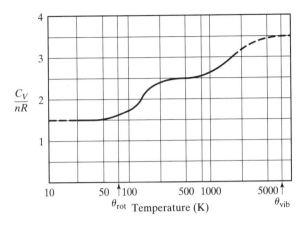

Figure 15.5 Values of C_V/nR for hydrogen (H_2) as a function of temperature. The temperature scale is logarithmic. (Adapted from *Thermodynamics, Kinetic Theory, and Statistical Thermodynamics*, 3rd edition, by F.W. Sears and G.L. Salinger, Addison-Wesley, 1975.)

At elevated temperatures, the higher vibrational states are excited and the heat capacity exhibits two additional degrees of freedom, rising asymptotically to a value of $(7/2)nR$ for $T \gg \theta_{vib}$. The characteristic temperature of vibration depends on the bond strength between the two atoms of the molecule and on their masses. In some cases, the diatomic molecules disassociate as the vibrational energy overcomes the bonding energy.

Experimental values of the heat capacity at constant volume are close to predicted values over a wide range of temperature. Values of C_V/nR are shown for hydrogen on a logarithmic scale in Figure 15.5. An understanding of the temperature variation of the heat capacity of diatomic gases is surely an outstanding triumph of quantum statistical theory.

PROBLEMS

15-1 (a) Calculate the fractional number N_j/N of oscillators in the three lowest quantum states ($j = 0, 1, 2$) for $T = \theta/4$ and for $T = \theta$.

(b) Sketch N_j/N versus j for the two temperatures.

15-2 (a) For a system of localized *distinguishable* oscillators, Boltzmann statistics applies. Show that the entropy S is given by

$$S = -k \sum_j N_j \ln\left(\frac{N_j}{N}\right).$$

(b) Substitute the Boltzmann distribution in the previous result to show that

$$S = \frac{U}{T} + Nk \ln Z.$$

(c) Using the expressions derived in the text for U and T, prove that

$$S = Nk\left[\frac{\theta/T}{e^{\theta/T} - 1} - \ln(1 - e^{-\theta/T})\right],$$

where $\theta = h\nu/k$. Examine the behavior of S as T approaches zero.

15-3 Consider 1000 diatomic molecules at a temperature $\theta_{vib}/2$.
  (a) Find the number in each of the three lowest vibrational states.
  (b) Find the vibrational energy of the system.

15-4 (a) In the low temperature approximation of Section 15-4, show that the Helmholtz function for rotation is

$$F_{rot} = -3NkTe^{-2\theta_{rot}/T}.$$

  (b) Use the reciprocal relation $S = -(\partial F/\partial T)_V$ to find the entropy S_{rot} in the same approximation. Note that $S \to 0$ as $T \to 0$, in agreement with the third law.

15-5 As an alternative evaluation of Z_{rot} to that given in the text, assume that for $T \gg \theta_{rot}$ the numbers l in the sum of Equation (15.13) are large compared with unity and replace the summation by integration with respect to l. Show that $Z_{rot} = T/\theta_{rot}$.

15-6 Use the data of Table 15.1 to determine r_0, the equilibrium distance between the nuclei, for
  (a) an H_2 molecule;
  (b) a CO molecule.

15-7 Consider a diatomic gas near room temperature. Show that the entropy is

$$S = Nk\left\{\frac{7}{2} + \ln\left[\frac{V}{N}\left(\frac{2\pi mk}{h^2}\right)^{3/2}\frac{T^{5/2}}{2\theta_{rot}}\right]\right\},$$

if the atoms of the diatomic molecule are identical.

15-8 For a kilomole of nitrogen (N_2) at standard temperature and pressure, compute (a) the internal energy U; (b) the Helmholtz function F; and (c) the entropy S.

15-9 Using the relation

$$P = NkT\left(\frac{\partial \ln Z}{\partial V}\right)_T,$$

show that the equation of state of a diatomic gas is the same as that of a monatomic gas.

15-10 Calculate the specific heat capacity at constant volume for hydrogen (H_2) at $T = 2000$ K.

Chapter 16

The Heat Capacity of a Solid

16.1 INTRODUCTION

The investigation of the heat capacity of solids is important in the study of condensed matter. Like the subject of Chapter 15, this topic is an example of the shortcomings of classical kinetic theory and the need for statistical mechanics and quantum theory to provide answers that agree with experimental results.

In 1819, Dulong and Petit observed that the specific heat capacity at constant volume of all elementary solids is approximately $2.49 \times 10^4 \, \text{J kilomole}^{-1} \, \text{K}^{-1}$ at temperatures near room temperature. This result can be explained by the principle of equipartition of energy by treating every atom of the solid as a linear oscillator with six degrees of freedom and then associating an energy of $(1/2)kT$ with each degree of freedom. Then $c_v = 6(R/2) = 3R$, in agreement with the observation of Dulong and Petit.

However, additional measurements showed that the specific heat of solids varies with temperature, decreasing to zero as the temperature approaches zero. This behavior cannot be explained by the "freezing" of degrees of freedom when the temperature is decreased since the specific heat varies gradually with temperature and does not exhibit abrupt jumps by any multiple of $R/2$ (in contrast to the specific heat of a diatomic gas). Even at room temperature the specific heat capacities of certain substances such as beryllium, boron, carbon, and silicon were found to be much smaller than $3R$. Quantum statistics is needed to explain these discoveries.

16.2 EINSTEIN'S THEORY OF THE HEAT CAPACITY OF A SOLID

It was Einstein who developed the first reasonably satisfactory theory of the specific heat capacity of a solid. He assumed that the crystal lattice structure of a solid comprising N atoms can be treated as an assembly of $3N$ distinguishable one-dimensional oscillators (three oscillators for each atom since the atoms of a solid are free to move in three dimensions). From Equation (15.8), the internal energy of a solid made up of N atoms is

$$U = 3Nk\theta_{\mathrm{E}}\left(\frac{1}{2} + \frac{1}{e^{\theta_{\mathrm{E}}/T} - 1}\right), \tag{16.1}$$

where θ_{E} is the Einstein temperature given by $\theta_{\mathrm{E}} \equiv h\nu/k$. The heat capacity at constant volume is

$$C_V = \left(\frac{\partial U}{\partial T}\right)_V = 3Nk\left(\frac{\theta_{\mathrm{E}}}{T}\right)^2 \frac{e^{\theta_{\mathrm{E}}/T}}{(e^{\theta_{\mathrm{E}}/T} - 1)^2}. \tag{16.2}$$

For temperatures very large compared with the Einstein temperature,

$$C_V \approx 3Nk = 3nR. \tag{16.3}$$

Thus the high temperature limit of Einstein's equation gives the value of Dulong and Petit. The failure of their law becomes evident when we examine the low temperature limit. For $\theta_{\mathrm{E}}/T \gg 1$,

$$C_V \approx 3Nk\left(\frac{\theta_{\mathrm{E}}}{T}\right)^2 e^{-\theta_{\mathrm{E}}/T}. \tag{16.4}$$

As T approaches zero, C_V also goes to zero, since the exponential decay overpowers the growth of $(\theta_{\mathrm{E}}/T)^2$.

Einstein's theory also explains the low heat capacities of some elements at moderately high temperatures. If an element has a large Einstein temperature, the ratio θ_{E}/T will be large even for temperatures well above absolute zero, and C_V will be small. For such an element $\theta_{\mathrm{E}} \equiv h\nu/k$ must be very large and, accordingly, ν must be large. For an oscillator with force constant κ and reduced mass μ, the oscillator frequency is

$$\nu = \frac{1}{2\pi}\sqrt{\frac{\kappa}{\mu}}. \tag{16.5}$$

A large frequency value suggests a small reduced mass or a large force constant, corresponding to lighter elements and elements that produce very hard crystals. The theory correctly predicts the failure of the law of Dulong and Petit for those elements. As an example, the heat capacity of diamond approaches $3Nk$ only at extremely high temperatures ($\theta_{\mathrm{E}} = 1450$ K for diamond).

The essential behavior of the specific heat capacity of solids is incorporated in the ratio θ_{E}/T. When this ratio is large, the partition function reduces to the zero-point term, implying that all the atoms are in the ground state and the vibrational degrees of freedom are not excited (Equation (15.6)). The specific heat remains close to zero since small temperature increases are not sufficient to excite a significant number of atoms to the first vibrational state. When θ_{E}/T is small, the difference between energies corresponding to various vibrational states is small compared with thermal energies. Thus the vibrational states can be approximated by an energy continuum and treated by classical theory. For values of θ_E/T between the two extremes, there is a transition region of partial excitations.

A consequence of the fact that C_V/N depends only on the ratio θ_E/T is that a single measurement of the heat capacity at one temperature determines its value at all other temperatures. In addition, different elements at different temperatures will possess the same specific heat capacity if the ratio θ_E/T is the same in each case. The elements are said to be in "corresponding states." The graph in Figure 16.1 clearly shows this behavior.

Careful measurements of heat capacities show that Einstein's model gives results that fall slightly below experimental values in the transition range of θ_E/T between the two limiting values. The discrepancy can be seen in Figure 16.2 in which the heat capacity of lead is shown for temperatures in the range 0–50 K.

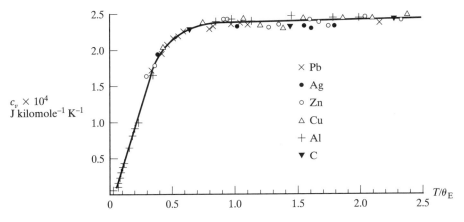

Figure 16.1   The specific heat capacity of various solids as a function of T/θ_E. Note that $c_v = (N_A/N)C_V$. (Adapted from *Elements of Statistical Thermodynamics*, 2nd edition, by L.K. Nash, Addison-Wesley, 1972.)

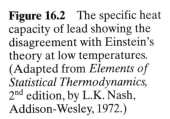

Figure 16.2   The specific heat capacity of lead showing the disagreement with Einstein's theory at low temperatures. (Adapted from *Elements of Statistical Thermodynamics*, 2nd edition, by L.K. Nash, Addison-Wesley, 1972.)

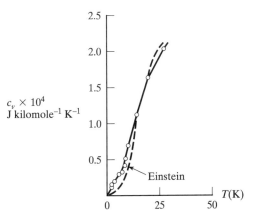

16.3 DEBYE'S THEORY OF THE HEAT CAPACITY OF A SOLID

The disagreement between Einstein's result and the experimental data is due to the fact that Einstein's assumptions about the atoms in a crystal do not strictly apply to real crystals. The main problem lies in the assumption that a single frequency of vibration characterizes all $3N$ oscillators. Debye improved on Einstein's theory by considering the vibrations of a body as a whole, regarding it as a continuous elastic solid. He associated the internal energy of the solid with stationary elastic sound waves. Each independent mode of vibration is treated as a degree of freedom.

In Debye's theory a solid is viewed as a *phonon gas*. Vibrational waves are matter waves, each with its own de Broglie wavelength and associated particle. The particle is called a phonon, with characteristics similar to those of a photon. We are interested in determining the number of possible wavelengths or frequencies within a given range.

For quantum waves in a one-dimensional box, we saw in Chapter 12 that the wave function is $\psi = A \sin kx$, where

$$k = \frac{2\pi}{\lambda} = \frac{n\pi}{L}, \quad n = 1, 2, 3, \dots \quad . \tag{16.6}$$

Here λ is the de Broglie wavelength and L is the dimension of the box. Using the fundamental equation of wave motion,

$$c = \lambda \nu,$$

where c is the wave velocity and ν the frequency, we obtain

$$n = \frac{2L}{c}\nu.$$

If we consider an elastic solid as a cube of volume $V = L^3$, we get

$$n = \frac{2V^{1/3}}{c}\nu, \tag{16.7}$$

where, in this case $n^2 = n_x^2 + n_y^2 + n_x^2$. The quantum numbers n_x, n_y, and n_z are positive integers. Thus the possible values that they can assume occupy the first octant of a sphere of radius

$$n = (n_x^2 + n_y^2 + n_z^2)^{1/2}$$

(Figure 16.3).

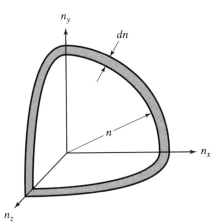

Figure 16.3 A shell of thickness dn of an octant of a sphere of radius n.

Let $g(\nu)d\nu$ be the number of possible frequencies in the range ν to $\nu + d\nu$. Since n is proportional to ν, $g(\nu)d\nu$ is the number of positive sets of integers in the interval n to $n + dn$—that is, within a shell of thickness dn of an octant of a sphere with radius n:

$$g(\nu)d\nu = \frac{1}{8}4\pi n^2 dn = \frac{\pi}{2}n^2 dn.$$

Substituting Equation (16.7) for n, we obtain

$$g(\nu)d\nu = \frac{4\pi V}{c^3}\nu^2 d\nu. \tag{16.8}$$

In a vibrating solid, there are three types of waves: one longitudinal with velocity c_l and two transverse with velocity c_t. All are propagated in the same direction. When all three waves are taken into account, Equation (16.8) becomes

$$g(\nu)d\nu = 4\pi V\left[\frac{1}{c_l^3} + \frac{2}{c_t^3}\right]\nu^2 d\nu. \tag{16.9}$$

Since each oscillator of the assembly vibrates with its own frequency, and we are considering an assembly of $3N$ linear oscillators, there must be an upper limit to the frequency spectrum. The maximum frequency ν_m is determined from the fact that there are only $3N$ phonons:

$$3N = \int_0^{\nu_m} g(\nu)d\nu = \frac{4\pi V}{3}\left[\frac{1}{c_l^3} + \frac{2}{c_t^3}\right]\nu_m^3. \tag{16.10}$$

Combining this result with Equation (16.9), we get

$$g(\nu)d\nu = \frac{9N\nu^2 d\nu}{\nu_m^3}. \tag{16.11}$$

Equation (16.10) provides us with some insight into the cutoff frequency and wavelength. Since $\nu_m \propto (N/V)^{1/3}$ and $\lambda_{min} \propto 1/\nu_m$, it follows that

$$\lambda_{min} \propto \left(\frac{V}{N}\right)^{1/3}.$$

The minimum possible wavelength is determined by the average interatomic spacing. Thus the structure of the crystal sets a lower limit to the wavelength; shorter wavelengths do not lead to new modes of atomic vibration.

The principal difference between Einstein's description and Debye's model is in the assumption about the frequency spectrum of the lattice vibrations. This is shown graphically in Figure 16.4. More rigorous calculations have led to more complex spectra.

Now, there is no restriction on the number of phonons per energy level $jh\nu$, where j is an integer. Thus phonons are bosons.* This means that the occupation numbers must be given by the Bose-Einstein distribution. For the continuum, Equation (13.41) applies:

$$\frac{N(\varepsilon)}{g(\varepsilon)} = \frac{1}{e^{(\varepsilon-\mu)/kT} - 1}.$$

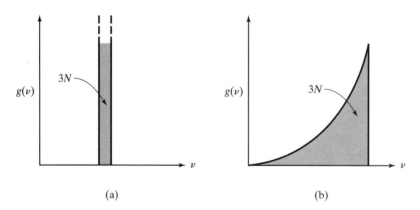

(a)   (b)

Figure 16.4 Frequency spectra of crystal vibrations: (a) Einstein model; (b) Debye model.

*Some physicists like to call them "phony bosons."

In this expression the chemical potential μ must be set equal to zero. This is because the total number N of phonons is not an independent variable but rather is determined by the volume and temperature of the particular crystal being considered. Specifically, N is the number of phonons that causes the Helmholtz function to be a minimum at equilibrium. Since $\mu = (\partial F/\partial N)_{T,V}$, it follows that $\mu = 0$. With $\varepsilon = h\nu$, we have

$$N(\nu)d\nu = \frac{g(\nu)d\nu}{e^{h\nu/kT} - 1},\tag{16.12}$$

where $N(\nu)d\nu$ is the number of phonons with frequencies in the range ν to $\nu + d\nu$. When we substitute Equation (16.11) in this result, we obtain

$$N(\nu)d\nu = \begin{cases} \dfrac{9N}{\nu_m^3} \cdot \dfrac{\nu^2 d\nu}{e^{h\nu/kT} - 1}, & \nu \le \nu_m \\ 0, & \nu > \nu_m \end{cases}.\tag{16.13}$$

The total energy of the phonons in the frequency range ν to $\nu + d\nu$ is $h\nu\, N(\nu)d\nu$. Hence the internal energy of the assembly is

$$U = \int_0^{\nu_m} h\nu N(\nu)d\nu$$

$$= \frac{9Nh}{\nu_m^3} \int_0^{\nu_m} \frac{\nu^3 d\nu}{e^{h\nu/kT} - 1}.\tag{16.14}$$

(We leave out the constant zero-point energy since this term has no effect on the heat capacity. Its value is calculated in Problem 16-2.)

The *Debye temperature* θ_D is defined as

$$\theta_D \equiv \frac{h\nu_m}{k}.\tag{16.15}$$

That is, the Debye temperature is proportional to the cutoff frequency ν_m. Some values are given in Table 16.1

To obtain the heat capacity, we need to differentiate Equation (16.14) with respect to the temperature. Now

$$\frac{d}{dT}\left(\frac{1}{e^{h\nu/kT} - 1}\right) = \frac{h\nu}{kT^2} \cdot \frac{e^{h\nu/kT}}{(e^{h\nu/kT} - 1)^2}.$$

TABLE 16.1 Debye temperatures of some materials.

Substance	θ_D (K)
Lead	88
Mercury	97
Sodium	172
Silver	215
Copper	315
Iron	453
Diamond	1860

Thus

$$C_V = \frac{dU}{dT} = \frac{9Nh^2}{v_m^3} \cdot \frac{1}{kT^2} \int_0^{v_m} \frac{v^4 e^{hv/kT}}{(e^{hv/kT} - 1)^2} dv. \tag{16.16}$$

We let $x = hv/kT$ and $x_m = hv_m/kT = \theta_D/T$. With the change of variable we have

$$C_V = 9Nk\left(\frac{T}{\theta_D}\right)^3 \int_0^{\theta_D/T} \frac{x^4 e^x}{(e^x - 1)^2} dx. \tag{16.17}$$

For high temperatures, $\theta_D/T \ll 1$ and $x \ll 1$. So $e^x - 1 \approx x$ in the denominator and $e^x \approx 1$ in the numerator, so the integral becomes

$$\int_0^{\theta_D/T} x^2 dx = \frac{1}{3}\left(\frac{\theta_D}{T}\right)^3.$$

Hence

$$C_V = 9Nk\left(\frac{T}{\theta_D}\right)^3 \cdot \frac{1}{3}\left(\frac{\theta_D}{T}\right)^3 = 3Nk.$$

This is the law of Dulong and Petit. For low temperatures, θ_D/T is large and we can let the upper limit of the integral be infinity. Then

$$\int_0^\infty \frac{x^4 e^x}{(e^x - 1)^2} dx = \frac{4\pi^4}{15}, \tag{16.18}$$

and

$$C_V = \frac{12\pi^4}{5} Nk \left(\frac{T}{\theta_D}\right)^3. \tag{16.19}$$

This equation is known as Debye's T^3 law. It is valid when the temperature is lower than about $0.1\theta_D$, which means for most substances about 10–20 K. The relation gives a better fit to experimental data at very low temperatures than the Einstein model, and is valid for all monatomic solids. When the temperature is above the Debye temperature, the heat capacity is very nearly equal to the classical value $3Nk$. For temperatures below the Debye temperature, quantum effects become important and C_V decreases to zero. Note that diamond, with a Debye temperature of 1860 K, is a "quantum solid" at room temperature.

Recent work has centered on the behavior of solids at low temperatures. Experiments suggest that amorphous materials do not follow the Debye T^3 law even at temperatures below $0.01\theta_D$. There is more yet to be learned.

PROBLEMS

16-1 The partition function of an Einstein solid is

$$Z = \frac{e^{-\theta_E/2T}}{1 - e^{-\theta_E/T}},$$

where θ_E is the Einstein temperature. Treat the crystalline lattice as an assembly of $3N$ distinguishable oscillators.
(a) Calculate the Helmholtz function F.
(b) Calculate the entropy S.
(c) Show that the entropy approaches zero as the temperature goes to absolute zero. Show that at high temperatures, $S \approx 3Nk[1 + \ln(T/\theta_E)]$. Sketch $S/3Nk$ as a function of T/θ_E.

16-2 Show that the inclusion of the zero-point energy in Equation (16.14) gives a term $(9/8)Nk\theta_D$ that is added to the integral. Since the term is a constant, it does not contribute to the heat capacity.

16-3 (a) Referring to Equation (16.14) and Problem 16-2, show that the internal energy of the assembly of phonons can be written

$$U = \frac{9}{8} Nk\theta_D + 3NkT D\left(\frac{\theta_D}{T}\right),$$

where $D(\theta_D/T)$ is one of several Debye functions. Its definition is

$$D\left(\frac{\theta_D}{T}\right) = \frac{3}{(\theta_D/T^3)} \int_0^{\theta_D/T} \frac{x^3 dx}{e^x - 1}.$$

(b) For very low temperatures, the upper limit of the integral can be taken to be infinity. Show that

$$I \equiv \int_0^\infty \frac{x^3 dx}{e^x - 1} = \frac{\pi^4}{15}$$

by expanding the integrand in powers of e^{-x} and integrating term by term. The result can be expressed as an infinite series whose sum is well known. Note that $D(\infty) = 0$.

(c) Show that at low temperatures

$$U = 3Nk\theta_D\left[\frac{3}{8} + \frac{\pi^4}{5}\left(\frac{T}{\theta_D}\right)^4\right].$$

16-4 (a) The heat capacity can be expressed in terms of the Debye function $D(\theta_D/T)$ by noting that

$$\int_0^{\theta_D/T} \frac{x^4 e^x}{(e^x - 1)^2} dx = \int_0^{\theta_D/T} x^4 d\left(\frac{-1}{e^x - 1}\right)$$

and integrating by parts. Show that

$$\frac{C_V}{3Nk} = 4D\left(\frac{\theta_D}{T}\right) - \frac{3(\theta_D/T)}{e^{\theta_D/T} - 1}.$$

(d) Calculate and plot $C_V/(3Nk)$ for $T/\theta_D = 0.2, 0.4, 0.6, 0.8,$ and 1.0. The corresponding values of the Debye function are:

T/θ_D	0.2	0.4	0.6	0.8	1.0
$D(\theta_D/T)$	0.117	0.354	0.496	0.608	0.674

16-5 The partition function of a Debye solid is

$$\ln Z = -9\left(\frac{T}{\theta_D}\right)^3 \int_0^{\theta_D/T} x^2 \ln(1 - e^{-x}) dx.$$

Show that the Helmholtz function can be written

$$F = 9NkT \ln(1 - e^{-\theta_D/T}) - 3NkTD(\theta_D/T).$$

16-6 The experimental value of c_v for diamond is 2.68×10^3 J kilomole^{-1} K^{-1} at a temperature of 207 K. For diamond the Einstein temperature is 1450 K and the Debye temperature is 1860 K. Calculate c_v at 207 K using the Einstein and Debye models and compare the results with the experimental value.

16-7 (a) The Debye temperature θ_D can be determined from measurements of the speed of sound in the solid. Use Equation (16.10) to show that

$$\theta_D = \frac{hc}{k}\left(\frac{3N}{4\pi V}\right)^{1/3},$$

where

$$\frac{1}{c^3} = \frac{1}{3}\left(\frac{1}{c_l^3} + \frac{2}{c_t^3}\right).$$

(b) Show that

$$\theta_D(K) = 2.51 \times 10^{-2}\left(\frac{\rho}{M}\right)^{1/3} c,$$

where ρ is the density in $kg\,m^{-3}$, M is the molecular weight, and c is the sound speed in ms^{-1}. For copper, $\rho = 8.95 \times 10^3\,m^{-3}$, $M = 63.6$, and $c = 2.24 \times 10^3\,ms^{-1}$. Calculate θ_D.

16-8 The equation of state for a monatomic solid is

$$Pv + f(v) = \Gamma u,$$

where v is the specific volume and u is the specific internal energy of the lattice vibrations of the crystal. The so-called Grüneisen constant Γ is important in solid state theory.

(a) Show that

$$\Gamma = \frac{Bv}{c_v \kappa}.$$

(b) Show that

$$\gamma = 1 + \Gamma B T,$$

where γ is the ratio of specific heats.

(c) The Grüneisen constant is defined as

$$\Gamma = -\frac{d\ln\theta_D}{d\ln v}.$$

Assume that θ_D in the Debye model is a function of v only and show that this definition leads to the result of (a). Hint: Use Maxwell's relation

$$\left(\frac{\partial s}{\partial v}\right)_T = \left(\frac{\partial P}{\partial T}\right)_v$$

and assume that the specific entropy s is a function of T/θ_D.

Chapter 17

The Thermodynamics of Magnetism

17.1 INTRODUCTION

Magnetism comprises physical phenomena involving magnetic fields and their effects on materials. All magnetic fields are due to electric charges in motion. On an atomic scale, individual atoms can give rise to magnetic fields when their electrons have a net magnetic moment as a result of their orbital and/or spin angular momentum.

Each electron of an atom contributes a magnetic dipole that experiences a torque when placed in an external magnetic field. The phenomenon of *paramagnetism* results from the tendency of unpaired dipole moments to align themselves *parallel* to the applied field. Paramagnetism in solids occurs in compounds containing transition metal ions that have either incomplete *d* shells (the iron, palladium and platinum groups) or incomplete *f* shells (the lanthanide and actinide groups). Both the net spin and orbital magnetic moments may contribute to paramagnetism, but for an ion with an incomplete *d* shell, the effective orbital moment may be "quenched" by electrostatic interactions with its neighboring ions, leading to a predominant spin moment. Paramagnetism is therefore due mainly to the spin angular momenta of the electrons.

In an external magnetic field, all the orbiting electrons of an atom speed up or slow down, depending on the direction of the field. In either case, a change in each electron's magnetic moment is produced *antiparallel* to the field. The effect is known as *diamagnetism*. It is present in all atoms, but is usually much weaker than paramagnetism.

In *ferromagnetic* materials, each atom has a comparatively large dipole moment caused primarily by uncompensated electron spins. Interatomic forces produce parallel alignments of the spins over regions containing large numbers of atoms. These regions, or *domains*, have a variety of shapes and sizes (with dimensions ranging from a micron to several centimeters), depending on the material sample and its magnetic history. The domain moments are generally randomly oriented, so the material as a whole has no magnetic moment. Upon application of an external field, however, those domains with moments in the direction of the applied field increase their size at the expense of their neighbors, and the internal field becomes much larger than the external field alone. When the external field is removed, a random domain alignment does not usually occur, and a residual dipole field remains in the sample. This effect is called *hysteresis*. The only elements that are ferromagnetic at room temperature are iron, nickel, and cobalt. At sufficiently high temperatures, a ferromagnet becomes paramagnetic.

The modern quantum mechanical theory of magnetism was to a large extent developed by J. H. Van Vleck. One of the triumphs of this theory is the remarkable agreement between the observed and theoretically calculated magnetic moments of paramagnetic salts. These are the magnetic systems of primary interest in statistical thermodynamics.

17.2 PARAMAGNETISM

Most experiments on magnetic materials are performed at constant pressure and involve insignificant volume changes. Thus, pressure and volume variations can be ignored, and equilibrium states can be represented by a relation involving three thermodynamic variables: the magnetic field B, the magnetization M, and the temperature T. The methods of statistical thermodynamics are appropriate since the alignment of large numbers of atomic magnets is a statistical process.

Paramagnetic effects are produced in a crystal when some of its atoms have a net magnetic dipole moment associated with the electron orbital angular momentum, the electron spin, or both. To understand the relationship between the magnetic moment and the angular momentum, it is instructive to consider first the hydrogen atom and then give the result for more complex atoms or ions. The basic ideas are the same for other systems as for hydrogen.

In the quantum-mechanical description, the magnitude of the orbital angular momentum vector of an isolated hydrogen atom is $\sqrt{l(l + 1)}\hbar$, where the allowed values of the quantum number l are $0, 1, 2 \ldots$. The z component of the orbital angular momentum is $m_l \hbar$ where $-l \le m_l \le l$. It follows that there are $2l + 1$ values of m_l for a given value of l. The magnetic dipole moment μ is proportional to the angular momentum; the constant of proportionality is $e/2m_e$, where e is the charge of the electron and m_e is its mass. Thus

$$|\mu| = \frac{e\hbar}{2m_e}\sqrt{l(l + 1)} = g\mu_B\sqrt{l(l + 1)},$$

and

$$\mu_z = g\mu_B m_l, \qquad -l \le m_l \le l.$$

Here $\mu_B = e\hbar/2m_e$ is the so-called Bohr magneton and g is a number of the order of unity referred to as the Landé g factor; g is difficult to determine exactly but is known to be 1 for orbital angular momentum.

The analogous expressions for the electron's intrinsic angular momentum or spin are

$$|\boldsymbol{\mu}| = g\mu_B\sqrt{s(s + 1)},$$

$$\mu_z = g\mu_B m_s, \qquad -s \le m_s \le s,$$

where g is very close to 2 for spin. For an electron the spin quantum number s is 1/2, so that m_s can assume only two values, $-1/2$ and 1/2. (Both m_l and m_s vary in integral steps.)

The magnitude of the total angular momentum, the vector sum of the orbital and spin parts, is $\sqrt{j(j + 1)}\hbar$, where the possible values of j are $|l - s|, |l - s| + 1, \dots |l + s|$. The z component of angular momentum has the value $m_j\hbar$, where the possible values of m_j are $-j, -j + 1, \dots j - 1, j$.

These relations are restricted to one orbital electron. For a many-electron atom, similar expressions apply, with capital letters instead of small letters to indicate that they pertain to the total number of particles. The generalization is

$$|\boldsymbol{\mu}| = g\mu_B\sqrt{J(J + 1)}, \tag{17.1}$$

$$\mu_z = g\mu_B m, \qquad -J \le m \le J, \tag{17.2}$$

where the possible values of J are $|L - S|, |L - S| + 1, \dots |L + S|$ and the allowed values of m are $-J, -J + 1, \dots J - 1, J$. Here L and S are the total orbital and total spin quantum numbers.* For simplicity, the so-called magnetic quantum number m is written without a subscript. Since the total angular momentum includes spin, J can be integral or half-integral, depending on whether the atom has an even or an odd number of electrons, respectively. To a good approximation, the Landé g factor is given by

$$g = 1 + \frac{J(J + 1) + S(S + 1) - L(L + 1)}{2J(J + 1)}. \tag{17.3}$$

---

*For paramagnetic ions in solids it is often convenient to use an effective spin S' to describe the magnetic behavior of the lowest group of energy levels. In this case, we may write

$$|\boldsymbol{\mu}| = g'\mu_B\sqrt{S'(S' + 1)},$$

where g', the effective value of g, is close to 2 for many ions with incomplete d shells. However, we shall continue to use S and g rather than S' and g'.

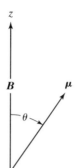

Figure 17.1 Magnetic moment
μ in a magnetic field B.

A dipole with magnetic moment μ placed in an external magnetic field B pointing in the z direction (Figure 17.1) will experience a torque N given by the vector product

$$N = \mu \times B. \tag{17.4}$$

The magnetic potential energy when the dipole is in the angular position θ (the angle between B and μ) is the work that must be done to rotate the dipole from its zero energy position $\theta = \pi/2$ to the given position θ:

$$\varepsilon = \int_{\pi/2}^{\theta} N d\theta = \mu B \int_{\pi/2}^{\theta} \sin\theta d\theta = -\mu B \cos\theta$$

$$= -\mu \cdot B = -\mu_z B. \tag{17.5}$$

By substituting Equation (17.2) in this equation, we get

$$\varepsilon_m = -g\mu_B Bm, \tag{17.6}$$

the magnetic energy of the atom in the state designated by the quantum number m.* As an example, for an electron with a single net electron spin, there would be only two possible energies, with $m = 1/2$ (spin parallel to B) and $m = -1/2$ (spin antiparallel to B). Note that the energy is lowest when the dipole moment is parallel to the field.

In a paramagnetic material, because the atoms occupy definite sites in the crystal lattice, they can be regarded as distinguishable particles and therefore obey Boltzmann statistics. Accordingly, the probability P_m that an atom is in an energy state ε_m is equal to the fractional occupation number corresponding to that energy level (Equation (13.22)). Since the degeneracy is removed

---

*We used the classical argument to derive Equation (17.5), but Equation (17.6) is the same as the result obtained from quantum theory.

by the applied magnetic field, the degeneracy factors are unity for all possible states. Thus

$$P_m = \frac{N_m}{N} = \frac{e^{-\varepsilon_m/kT}}{Z}, \tag{17.7}$$

where

$$Z = \sum_{m=-J}^{J} e^{-\varepsilon_m/kT}. \tag{17.8}$$

It follows that the mean z component of the magnetic moment of the atom is

$$\overline{\mu}_z = \sum_{m=-J}^{J} \mu_z P_m = \frac{1}{Z} \sum_{m=-J}^{J} g\mu_B m e^{g\mu_B Bm/kT}, \tag{17.9}$$

with

$$Z = \sum_{m=-J}^{J} e^{g\mu_B Bm/kT}. \tag{17.10}$$

We can write the expression for $\overline{\mu}_z$ more compactly as follows. Differentiating Equation (17.10) with respect to B, we obtain

$$\left(\frac{\partial Z}{\partial B} \right)_T = \frac{1}{kT} \sum_{m=-J}^{J} g\mu_B m e^{g\mu_B Bm/kT}.$$

Then

$$\overline{\mu}_z = \frac{kT}{Z} \left(\frac{\partial Z}{\partial B} \right)_T = kT \left(\frac{\partial \ln Z}{\partial B} \right)_T. \tag{17.11}$$

The next step is to evaluate the partition function Z. We introduce the abbreviation

$$\eta \equiv \frac{g\mu_B B}{kT}, \tag{17.12}$$

and write

$$Z = \sum_{m=-J}^{J} e^{\eta m}.$$

Setting $x = e^{\eta}$, we have

$$Z = \sum_{m=-J}^{J} x^m = x^{-J} + x^{-J+1} + \cdots + x^{J-1} + x^J$$
$$= x^{-J}(1 + x + \cdots + x^{2J-1} + x^{2J}).$$

The terms in parentheses constitute a finite geometric series whose sum is

$$\frac{1 - x^{2J+1}}{1 - x}.$$

Hence

$$Z = e^{-J\eta}\left(\frac{1 - e^{(2J+1)\eta}}{1 - e^{\eta}}\right) = \frac{e^{-J\eta} - e^{(J+1)\eta}}{1 - e^{\eta}}.$$

If we multiply both the numerator and the denominator by $e^{-\eta/2}$, we obtain the following result:

$$Z = \frac{e^{-(J+1/2)\eta} - e^{(J+1/2)\eta}}{e^{-\eta/2} - e^{\eta/2}} = \frac{\sinh(J + \frac{1}{2})\eta}{\sinh \eta/2}. \tag{17.13}$$

Finally, using Equation (17.11), we have

$$\bar{\mu}_z = kT\left(\frac{\partial \ln Z}{\partial B}\right)_T = kT\left(\frac{\partial \eta}{\partial B}\right)\left(\frac{\partial \ln Z}{\partial \eta}\right)_T = g\mu_B\left(\frac{\partial \ln Z}{\partial \eta}\right)_T$$
$$= g\mu_B\left[\frac{(J + \frac{1}{2})\cosh(J + \frac{1}{2})\eta}{\sinh(J + \frac{1}{2})\eta} - \frac{1}{2} \cdot \frac{\cosh(\eta/2)}{\sinh(\eta/2)}\right].$$

Therefore,

$$\bar{\mu}_z = g\mu_B J B_J(\eta), \tag{17.14}$$

where

$$B_J(\eta) \equiv \frac{1}{J}\left[\left(J + \frac{1}{2}\right)\coth\left(J + \frac{1}{2}\right)\eta - \frac{1}{2}\coth\frac{1}{2}\eta\right] \tag{17.15}$$

is called the *Brillouin function*. Its variation with η is sketched in Figure 17.2.

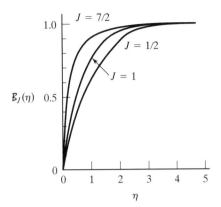

Figure 17.2   The Brillouin function for several values of J.

Let us examine the behavior of $\mathcal{B}_J(\eta)$ for large and small values of its argument. For $\eta \gg 1$, the hyperbolic cotangent approaches unity and

$$\mathcal{B}_J(\eta) \to \frac{1}{J}\left[\left(J + \frac{1}{2}\right) - \frac{1}{2}\right] = 1, \qquad \eta \gg 1, \tag{17.16}$$

corresponding to magnetic saturation. For small values of y, $\coth y \approx 1/y + y/3$; hence

$$
\begin{aligned}
\mathcal{B}_J(\eta) &\approx \frac{1}{J}\left\{\left(J + \frac{1}{2}\right)\left[\frac{1}{(J + \frac{1}{2})\eta} + \frac{1}{3}\left(J + \frac{1}{2}\right)\eta\right] - \frac{1}{2}\left[\frac{2}{\eta} + \frac{\eta}{6}\right]\right\} \\
&= \frac{1}{J}\left[\frac{1}{\eta} + \frac{1}{3}\left(J + \frac{1}{2}\right)^2\eta - \frac{1}{\eta} - \frac{\eta}{12}\right] \\
&= \frac{\eta}{3J}\left(J^2 + J + \frac{1}{4} - \frac{1}{4}\right) \\
&= \left(\frac{J + 1}{3}\right)\eta, \qquad \eta \ll 1.
\end{aligned}
\tag{17.17}
$$

From Equations (17.12), (17.14), (17.16), and (17.17), we obtain

$$\bar{\mu}_z = g\mu_B J, \qquad g\mu_B B \gg kT, \tag{17.18}$$

and

$$\bar{\mu}_z = \frac{g^2\mu_B^2 J(J + 1)B}{3kT}, \qquad g\mu_B B \ll kT. \tag{17.19}$$

Of particular interest is the second region, corresponding to high temperatures and a weak magnetic field. The magnetization M of a magnetic material is defined as the total mean dipole moment per unit volume. Thus, for a material sample of N atoms and volume V,

$$M = \frac{N\bar{\mu}_z}{V} = \frac{Ng^2\mu_B^2 J(J+1)B}{3VkT}, \qquad g\mu_B B \ll kT. \qquad (17.20)$$

This is the experimentally observed Curie law, discovered by Pierre Curie, which states that the magnetization of a paramagnetic solid is proportional to the ratio of the magnetic field to the temperature.

For $B = 1$ T,* $T = 300$ K and $g = 2$,

$$\frac{g\mu_B B}{kT} = \frac{(2)(5.79 \times 10^{-5}\,\text{eV/T})(1\,\text{T})}{(8.62 \times 10^{-5}\,\text{eV/K})(3 \times 10^2\,\text{K})} = 4.4 \times 10^{-3},$$

so the Curie law certainly applies under these conditions. Even at 3 K, with $g = 2$ and $J = 1/2$, a field of 1 T (a comparatively strong laboratory magnetic field) would produce less than 25 percent magnetic saturation.

The paramagnetic salts most widely used contain paramagnetic ions surrounded by a large number of nonmagnetic particles. A typical example is $Cr_2(SO_4)_3 \cdot K_2SO_4 \cdot 24\,H_2O$ (chromium potassium alum). Its magnetic properties are due solely to the chromium atoms that exist in the crystal as ions. The chromium ion Cr^{+++} has three unpaired electron spins and therefore a magnetic moment of $3\,\mu_B$. Besides the two chromium ions there are 4 sulfur atoms, 2 potassium atoms, 40 oxygen atoms, and 48 hydrogen atoms. Hence there are a total of 94 particles that are nonmagnetic. The magnetic ions are so widely separated in the molecule that the interaction between them is negligibly small. At the same time, the effect of the orbital motions of the valence electrons is quenched by the fields of neighboring ions. What remains is a net electron spin.

The configuration of the chromium atom $_{52}^{24}Cr$ is $(Ar)(4s)^1(3d)^5$. Argon has 18 electrons in closed shells, so chromium has an additional 6 electrons—one in the $4s$ subshell and 5 in the $3d$ subshell. In the chromium ion Cr^{+++} the $4s$ electron and two of the $3d$ electrons are missing. Thus the structure of Cr^{+++} is $(Ar)(3d)^3$. It is the three $3d$ electrons that produce paramagnetism.

Since the total orbital angular momentum is essentially zero, $J = 3/2$ (1/2 for each of the unpaired electrons). With $g = 2$, the mean dipole moment in the z direction is

$$\bar{\mu}_z = 3\mu_B B_{3/2}(\eta), \qquad \eta = \frac{2\mu_B B}{kT}. \qquad (17.21)$$

---

*The SI unit of the magnetic field B is the tesla (T); in terms of fundamental units, $1\,\text{T} = 1\,\text{kg/A} \cdot \text{s}^2$. Also, $1\,\text{T} = 10^4$ gauss.

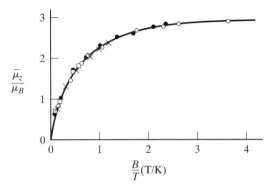

Figure 17.3 The mean magnetic moment in units of the Bohr magneton for the chromium ion Cr^{+++} in chromium potassium alum. The solid curve is the theoretical result. The experimental values are those reported by W. E. Henry, *Phys. Rev.* 88, 561 (1952).

In Figure 17.3 $\bar{\mu}_z/\mu_B$ is plotted against B/T for chromium potassium alum. Note the excellent agreement between the theoretical result and the experimental data.

17.3 PROPERTIES OF A SPIN-1/2 PARAMAGNET

The simplest paramagnetic system is an atom or ion with a spin-1/2 ground state and zero orbital angular momentum. Then $J = 1/2, g = 2$, and there are just two energy levels. The thermodynamic properties of a two-level system exhibit the same features as those of more complex systems and are easier to calculate.

For this case, the partition function of Equation (17.13) takes the form

$$Z = \frac{\sinh \eta}{\sinh \dfrac{\eta}{2}}, \qquad \eta = \frac{2\mu_B B}{kT}. \tag{17.22}$$

This can be simplified by using the identity

$$\sinh \eta \equiv 2 \sinh \frac{\eta}{2} \cosh \frac{\eta}{2}$$

in the numerator and canceling one of the factors with the denominator, giving

$$Z = 2 \cosh \frac{\eta}{2} = 2 \cosh\left(\frac{\varepsilon}{kT}\right), \tag{17.23}$$

where $\varepsilon \equiv \mu_B B$. The total magnetic energy U_B can then be found from

$$U_B = NkT^2\left(\frac{\partial \ln Z}{\partial T}\right)_B. \tag{17.24}$$

This is Equation (14.7), calculated for the Maxwell-Boltzmann distribution and also valid here. (Boltzmann statistics and Maxwell-Boltzmann statistics differ by a constant factor, but the distributions are the same.) Then

$$\left(\frac{\partial \ln Z}{\partial T}\right)_B = -\frac{\varepsilon}{kT^2} \cdot \frac{\sinh\left(\frac{\varepsilon}{kT}\right)}{\cosh\left(\frac{\varepsilon}{kT}\right)},$$

so that

$$U_B = -N\varepsilon \tanh\left(\frac{\varepsilon}{kT}\right). \tag{17.25}$$

Since $\tanh(0) = 0$ and $\tanh(\infty) = 1$, the energy is evidently $-N\varepsilon$ at $T = 0$ and approaches zero asymptotically at high temperatures. This behavior, shown in Figure 17.4, is markedly different from that of the corresponding energy of an assembly of harmonic oscillators treated in Section 15.2. The latter increases without limit as the temperature increases. Here higher temperatures produce increased randomization of the dipole moments and the magnetic potential energy goes to zero.

The magnetic contribution to the heat capacity also has distinctive characteristics, given by

$$C_B = \left(\frac{\partial U_B}{\partial T}\right)_{B,N} = Nk\left(\frac{\varepsilon}{kT}\right)^2 \operatorname{sech}^2\left(\frac{\varepsilon}{kT}\right). \tag{17.26}$$

This expression can be put in an alternative form from which the limiting behavior is more readily seen (Problem 17-8):

$$C_B = Nk\left(\frac{2\varepsilon}{kT}\right)^2 \frac{e^{2\varepsilon/kT}}{(e^{2\varepsilon/kT} + 1)^2}. \tag{17.27}$$

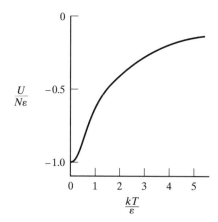

Figure 17.4   Internal energy of a spin-1/2 paramagnet.

Then it is clear that

$$C_B \approx \left(\frac{2\varepsilon}{kT}\right)^2 e^{-2\varepsilon/kT}, \qquad kT/\varepsilon \ll 1,$$

and the exponential factor reduces C_B rapidly to zero as the temperature decreases, in accordance with the third law. At the other limit,

$$C_B \to 0, \qquad kT/\varepsilon \gg 1.$$

For ε/kT in the vicinity of unity, the heat capacity has a fairly sharp peak known as a *Schottky anomaly* (Figure 17.5). The anomaly is useful for determining energy level splittings of ions in rare-earth and transition-group metals. Like the energy, the heat capacity is qualitatively different from other systems in its variation with temperature. The most important thermodynamic property of two-level systems is the entropy. For Boltzmann statistics,

$$S = \frac{U}{T} + Nk \ln Z. \tag{17.28}$$

Substituting Equations (17.23) and (17.25) in this expression, we get

$$S = Nk\left\{\ln\left[2\cosh\left(\frac{\varepsilon}{kT}\right)\right] - \left(\frac{\varepsilon}{kT}\right)\tanh\left(\frac{\varepsilon}{kT}\right)\right\}. \tag{17.29}$$

Figure 17.6 is a plot of S/Nk versus kT/ε.
    At low temperatures,

$$\ln\left[2\cosh\left(\frac{\varepsilon}{kT}\right)\right] \to \ln(e^{\varepsilon/kT}) = \frac{\varepsilon}{kT},$$

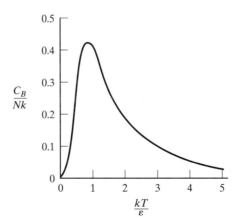

Figure 17.5   Heat capacity of a spin-1/2 paramagnet.

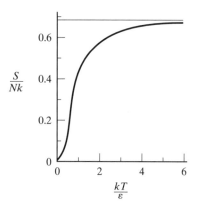

Figure 17.6   Entropy of a
spin-1/2 paramagnet.

and

$$\left(\frac{\varepsilon}{kT}\right)\tanh\left(\frac{\varepsilon}{kT}\right) \to \left(\frac{\varepsilon}{kT}\right),$$

so $S \to 0$.

At high temperatures, the second term in curly brackets approaches zero, $\cosh(\varepsilon/kT) \to 1$, and $S \to Nk \ln 2$. This is exactly what we would expect. At the upper temperature limit, $w = 2^N$, the number of equally probable microstates, and $S = k \ln 2^N = Nk \ln 2$. This corresponds to a pattern of random dipole orientations, involving equal numbers of parallel and antiparallel magnets in any chosen direction. In this disordered state the entropy is a maximum. As $T \to 0$ all the dipoles are in the lower energy state pointing in a direction parallel to the applied magnetic field. There is only one possible microstate so the thermodynamic probability w is 1 and $S = k \ln 1 = 0$.

The foregoing discussion is significant because many systems can be treated as two-level systems.

17.4 ADIABATIC DEMAGNETIZATION

Paramagnetic salts have been used to attain very low temperatures with a process known as *adiabatic demagnetization,* alluded to in Chapter 10. The method rests on the fact that the entropy is a monotonically increasing function of kT/ε—that is, of T/B.

When the magnetic field is increased, the degree of alignment of the magnetic moments increases and the disorder of the spin system decreases, thereby lowering the entropy. The entropy is also lowered if the temperature decreases, again because the moments tend to line up. But if the field is

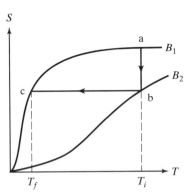

Figure 17.7 Entropy versus temperature for two values of the magnetic field B_1 and B_2, with $B_2 > B_1$. The spin system goes from state a to state b at a constant temperature T_i as the magnetic field is increased. When the magnetic field is reduced to its original value along the adiabatic path from b to c, the temperature is reduced to its final value T_f.

reduced adiabatically, without changing the entropy, the temperature must drop if the same degree of disorder is maintained.

This method is illustrated in Figure 17.7. A paramagnetic salt is first cooled to a temperature of 1 K or less by contact with liquid helium. A magnetic field acting on the salt is then increased from B_1 to B_2 under isothermal conditions at temperature T_i. The spins are in a lower energy state so heat is evolved and the entropy is lowered. At this point the salt is thermally isolated so that no heat is exchanged with the surroundings and the field is gradually reduced from B_2 to B_1. Since the entropy is constant,

$$\frac{T_i}{B_2} = \frac{T_f}{B_1} \quad \text{or} \quad T_f = \left(\frac{B_1}{B_2}\right) T_i.$$

At very low temperatures the heat capacity of the spin system is much larger than that of the crystal lattice, so the final temperature of the paramagnetic salt is only slightly greater than T_f. Temperatures of less than 1 mK have been achieved with this technique. The reason a final temperature of absolute zero cannot be reached is discussed in Chapter 10.

Demagnetization cooling can be understood by combining results from classical thermodynamics with the statistically derived Curie law. The work done by the paramagnetic salt when an applied magnetic field changes by an amount dB is $\bar{d}W = \mathcal{M}dB$ where $\mathcal{M}$ is the total magnetization $(= MV)$. With $dV = 0$, the Helmholtz function takes the form

$$dF = -SdT - \mathcal{M}dB.$$

If F is considered a function of T and B,

$$dF = \left(\frac{\partial F}{\partial T}\right)_B dT + \left(\frac{\partial F}{\partial B}\right)_T dB.$$

Since F is a state variable, dF is a perfect differential, so that

$$\frac{\partial^2 F}{\partial B\, \partial T} = \frac{\partial^2 F}{\partial T\, \partial B}.$$

From this we immediately obtain the Maxwell relation

$$\left(\frac{\partial S}{\partial B}\right)_T = \left(\frac{\partial \mathcal{M}}{\partial T}\right)_B. \qquad (17.30)$$

The magnetization $\mathcal{M}$ is given approximately by the Curie law

$$\mathcal{M} = \frac{aB}{T}, \qquad (17.31)$$

where a is a positive constant. Taking the derivative with respect to T while keeping B constant, we have

$$\left(\frac{\partial S}{\partial B}\right)_T = -\frac{aB}{T^2}. \qquad (17.32)$$

Hence the entropy always decreases with an isothermal increase in the magnetic field.

For the adiabatic leg in Figure 17.7, we can use the fundamental relation

$$dU = T\,dS - \mathcal{M}\,dB.$$

Here it is convenient to let $S = S(T, B)$. Then

$$T\,dS = T\left(\frac{\partial S}{\partial T}\right)_B dT + T\left(\frac{\partial S}{\partial B}\right)_T dB,$$

or, for a reversible process,

$$T\,dS = C_B dT + T\left(\frac{\partial S}{\partial B}\right)_T dB,$$

where C_B is the heat capacity of the system in a constant magnetic field. Setting $dS = 0$ and using Equation (17.30), we find that

$$\frac{dT}{T} = -\frac{1}{C_B}\left(\frac{\partial \mathcal{M}}{\partial T}\right)_B dB. \qquad (17.33)$$

Because thermal motions tend to disalign the magnetic dipoles of the paramagnetic material, the magnetization decreases with increasing temperature. Therefore $(\partial M/\partial T)_B$ is negative and consequently the temperature drops as the magnetic field is lowered, according to Equation (17.33). This is known as the *magnetocaloric effect.*

17.5 NEGATIVE TEMPERATURE

The two-level case can be used to discuss an extension of the notion of temperature. As noted, the energy of the lower level ($\boldsymbol{\mu}$ parallel to $\boldsymbol{B}$) is $\varepsilon_0 = -\mu_B B$ whereas that of the upper level ($\boldsymbol{\mu}$ antiparallel to $\boldsymbol{B}$) is $\varepsilon_1 = \mu_B B$. We set $\varepsilon = \mu_B B$ as before, so that the separation of levels is 2ε, the excitation gap. The number of particles in the lower level $-\varepsilon$ is N_0 and the number in the upper level ε is N_1, with

$$N_0 + N_1 = N, \tag{17.34}$$

and

$$N_0 = \frac{N}{Z}e^{\varepsilon/kT}, \quad N_1 = \frac{N}{Z}e^{-\varepsilon/kT}. \tag{17.35}$$

The total magnetic energy of the system is

$$U = N_0\varepsilon_0 + N_1\varepsilon_1 = (N_1 - N_0)\varepsilon. \tag{17.36}$$

The ratio N_1/N_0 is

$$\frac{N_1}{N_0} = e^{-2\varepsilon/kT},$$

which leads to the following expression for the temperature:

$$T = \frac{1}{k}\left(\frac{2\varepsilon}{\ln N_0 - \ln N_1}\right). \tag{17.37}$$

In a state of stable equilibrium the occupation number N_0 of the lower energy level is greater than the occupation number N_1 of the higher energy level, so the absolute temperature N_1 is positive in this equation.

Imagine now that the direction of the applied magnetic field is suddenly reversed. The elementary magnets that were parallel to the original field and in the lower energy state are now antiparallel to the field and in the higher energy state. Conversely, the magnetic moments originally in the higher energy state are now in the lower energy state. A *population inversion* has taken place. In Equation (17.37) the sign changes and the temperature T becomes negative!

To be sure, after some relaxation time has passed, the moments in the higher energy state will flop over into the new low-energy state and a new state of equilibrium will be established. But immediately after the field is reversed, the temperature of the spin system will be negative.

The situation can be made clearer by examining how the entropy and temperature vary with the total energy. The entropy of the two-level system is $S = k \ln w$, where

$$w = \frac{N!}{N_0! N_1!}. \tag{17.38}$$

Using Equations (17.34) and (17.36), we can write

$$N_0 = \frac{N}{2}\left(1 - \frac{U}{N\varepsilon}\right), \qquad N_1 = \frac{N}{2}\left(1 + \frac{U}{N\varepsilon}\right). \tag{17.39}$$

Substituting these relations in Equation (17.38), we have

$$w = \frac{N!}{\left[\frac{N}{2}\left(1 - \frac{U}{N\varepsilon}\right)\right]! \left[\frac{N}{2}\left(1 + \frac{U}{N\varepsilon}\right)\right]!}. \tag{17.40}$$

The use of Stirling's approximation leads to an expression for the entropy in terms of the dimensionless quantity $x = U/N\varepsilon$ (Problem 17-10):

$$\frac{S}{Nk} = \ln 2 - \frac{1}{2}[(1 - x)\ln(1 - x) + (1 + x)\ln(1 + x)]. \tag{17.41}$$

Evidently x varies from -1 to $+1$; in this interval S/Nk is an even function of x with a maximum at $x = 0$ (Figure 17.8).

Referring to Equation (17.39), we see that when $U/N\varepsilon = -1$, $N_0 = N$, and $N_1 = 0$, all of the particles occupy the lower energy level. Similarly, when $U/N\varepsilon = 1$, $N_0 = 0$, and $N_1 = N$, the upper level is fully occupied. In both cases $T = 0$.

When the particles are equally distributed between the two levels, the entropy reaches its maximum value of $Nk \ln 2$. Here the temperature suddenly jumps from $+\infty$ to $-\infty$ as the direction of the field is reversed.

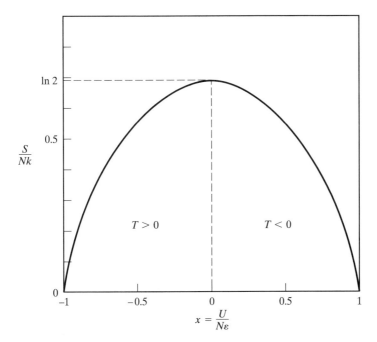

Figure 17.8   Entropy as a function of energy for a two-level system.

To obtain a more detailed picture of the relationship between the temperature and the energy, we can calculate T as a function of x by using a reciprocity formula (see Chapter 8):

$$T = \left(\frac{\partial U}{\partial S}\right)_N.$$

Then

$$\frac{1}{T} = \left(\frac{\partial S}{\partial U}\right)_N = \frac{1}{N\varepsilon}\left(\frac{\partial S}{\partial x}\right)_N = \frac{k}{2\varepsilon}\ln\left(\frac{1-x}{1+x}\right). \qquad (17.42)$$

In Figure 17.9, $kT/2\varepsilon$ is plotted versus $x = U/N\varepsilon$. Clearly shown is the abrupt shift from $+\infty$ to $-\infty$ when slightly more particles occupy the upper energy level.

Three conditions must be satisfied if a system is to have a negative temperature: (1) the system must be in thermal equilibrium, (2) the energy spectrum of the system must have a finite upper bound, and (3) the system must be energetically isolated from states that are at a positive temperature. The second condition rules out systems such as a harmonic oscillator, which has no upper limit to its possible energy.

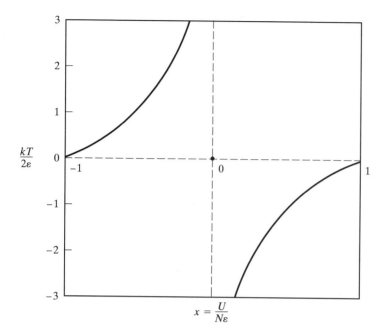

Figure 17.9 Temperature as a function of energy for a two-level system. Note the jump in temperature from $+\infty$ to $-\infty$ when more particles are in the higher energy state.

In paramagnetic materials the interaction between the magnetic ions and the crystal lattice, though weak, is sufficiently large that the substance cannot exist in a state of population inversion for any appreciable length of time. However, in a series of elegant experiments, Purcell and Pound in 1951 used nuclear magnetic resonance techniques to show that a negative temperature could exist in a *nuclear* spin subsystem. In a lithium fluoride crystal, nuclear spins were aligned by a strong magnetic field. At a positive temperature, lithium nuclei absorbed energy supplied by a radio-frequency oscillator with angular frequency equal to the excitation frequency $2\varepsilon/\hbar$. When the magnetic field was reversed, the investigators detected resonant emission of electromagnetic energy instead of resonant absorption, indicating a population inversion and a resultant negative temperature. The relaxation time necessary to reestablish thermal equilibrium between the spins and the crystal lattice was 2–5 minutes, long enough to demonstrate the existence of the inverted population. In the experiments only the spin subsystem was affected by the field reversal; the whole crystal remained at laboratory temperature.

In radio astronomy, negative temperature systems have been used to amplify very weak radio-frequency signals. Also, the action of a laser depends on a population inversion, which is sustained by some external influence. In the case of a ruby laser, the method is optical pumping using light radiation.

17.6 FERROMAGNETISM

Pierre Weiss was the first to understand that underlying the domain structure of ferromagnetism is its fundamental atomic nature. The atoms of a ferromagnet have a net magnetic moment and these couple together to form the domains. Much of the previous discussion of this chapter is applicable to ferromagnetism, with one important modification.

A ferromagnet has a magnetic moment even when it is not in an external field, a property called *spontaneous magnetization*. This observation led Weiss to postulate the existence of an internal field proportional to the magnetization, and to replace B with the sum of the external and internal fields:

$$B \rightarrow B + \lambda M, \tag{17.43}$$

where λ is an undetermined constant. From Equation (17.14) we know that the magnetization is

$$M = \frac{N\bar{\mu}_z}{V} = \frac{N}{V} g\mu_B J B_J(\eta). \tag{17.44}$$

For the case $J = 1/2$ and $g = 2$, this becomes

$$M = \frac{N}{V} \mu_B B_{1/2}(\eta), \qquad \eta = \frac{2\mu_B B}{kT}, \tag{17.45}$$

where the Brillouin function is

$$B_{1/2}(\eta) = 2 \coth \eta - \coth \frac{\eta}{2}. \tag{17.46}$$

This can be simplified at once by using the identities

$$\coth \eta \equiv \frac{\cosh \eta}{\sinh \eta}, \quad \coth \frac{\eta}{2} \equiv \frac{\cosh \eta + 1}{\sinh \eta}, \quad \tanh \frac{\eta}{2} \equiv \frac{\cosh \eta - 1}{\sinh \eta}.$$

Then

$$B_{1/2}(\eta) = \tanh \frac{\eta}{2}$$

and

$$M = \frac{N\mu_B}{V} \tanh\left(\frac{\mu_B B}{kT}\right). \tag{17.47}$$

With the replacement indicated in Equation (17.43), we find that

$$M = \frac{N\mu_B}{V} \tanh\left[\frac{\mu_B(B + \lambda M)}{kT}\right]. \tag{17.48}$$

From this equation we can learn a great deal about ferromagnetic behavior. To investigate spontaneous magnetization we set B equal to zero in Equation (17.48) and obtain a transcendental equation for M:

$$M = \frac{N\mu_B}{V} \tanh\left(\frac{\lambda\mu_B M}{kT}\right). \tag{17.49}$$

If we set

$$\xi \equiv \frac{MV}{N\mu_B} \quad \text{and} \quad \xi_0 \equiv \left(\frac{VkT}{\lambda N\mu_B^2}\right), \tag{17.50}$$

Equation (17.49) becomes

$$\xi = \tanh\left(\frac{\xi}{\xi_0}\right). \tag{17.51}$$

Figure 17.10 is a graphical solution of this equation. Nonzero solutions evidently exist for ξ_0 in the range $0 < \xi_0 < 1$. The corresponding temperature range is $0 < T < T_c$, where

$$T_c \equiv \frac{\lambda N\mu_B^2}{kV}. \tag{17.52}$$

This critical temperature is called the Curie temperature or Curie point.

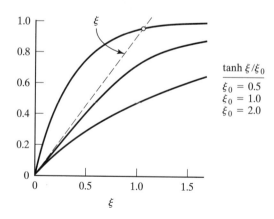

Figure 17.10 Graphical solution of Equation (17.51). The dashed curve is the left-hand side, a straight curve with unit slope. The solid curves represent the right-hand side for the indicated values of ξ_0. Non-zero intersections occur only for $\xi_0 < 1$.

$\tanh \xi/\xi_0$

$\xi_0 = 0.5$
$\xi_0 = 1.0$
$\xi_0 = 2.0$

At $T = 0$ the magnetization takes on its saturation value $N\mu_B/V$. As T increases, M decreases, until at the Curie point an abrupt change in the material properties takes place. This can be seen by solving for the temperature in Equation (17.49):

$$kT = \frac{\lambda\mu_B M}{\tanh^{-1}(MV/N\mu_B)}. \tag{17.53}$$

Assuming that $MV/N\mu_B$ is small, we can approximate the inverse hyperbolic function using

$$\tanh^{-1}x \approx x + \frac{x^3}{3}, \qquad x \ll 1,$$

and write

$$kT = \frac{\lambda\mu_B M}{(MV/N\mu_B) + \frac{1}{3}(MV/N\mu_B)^3}. \tag{17.54}$$

With a little algebra this becomes

$$M = \sqrt{3}\frac{N\mu_B}{V}\left(\frac{T_c}{T} - 1\right)^{1/2}. \tag{17.55}$$

We note that as $T \to T_c$, the spontaneous magnetization vanishes abruptly: $M \to 0$ with infinite slope. For $T \to T_c$, the ferromagnetic material becomes paramagnetic. This is an example of a second-order phase transition.* A plot of the reduced magnetization as a function of T/T_c is shown in Figure 17.11, along with experimental values for nickel.

We have seen how the spontaneous magnetization varies with temperature in the absence of an applied field. In the paramagnetic region, $T > T_c$, we would like to know how M varies with T in an external field B. At temperatures above the Curie point, the magnetization is small compared with its saturation value. Therefore the hyperbolic tangent is small in Equation (17.48) and can be set equal to its argument. The resultant approximate expression for the magnetization is

$$M = \frac{N\mu_B^2 B}{Vk(T - T_c)}, \qquad T > T_c. \tag{17.56}$$

---

*A second-order phase transition occurs when a system passes through a critical region corresponding to incipient instability. Here the Curie temperature is the ferromagnetic critical point and the phase transition consists of an order-disorder transition of the electron pairs.

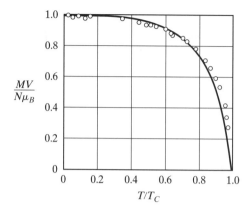

Figure 17.11 Magnetization of nickel as a function of temperature. The experimental values were reported by P. Weiss and R. Forrer. The solid curve is the theoretical result for $J = 1/2$.

This relation is the Weiss-Curie law, which is a modification of Curie's law discussed in Section 17.2.

Equation (17.56) suggests a method for determining the Curie temperature. For a given material, measurements are made of the magnetic susceptibility defined as $\chi \equiv (B/\mu_0 M - 1)^{-1}$. For temperatures well above the Curie point, a plot of experimental values of $1 + \chi^{-1}$ versus T gives a straight line whose intercept with the temperature axis is proportional to T_c. Except for minor departures at very high temperatures, linear behavior is observed and extrapolation is straightforward. For many paramagnetic substances T_c is less than a kelvin so that Curie's law gives an adequate representation at all but the lowest temperatures.

A more extensive treatment of ferromagnetism leads to refinements of these results and attempts to incorporate a detailed description of domains and hysteresis in the physical model.

PROBLEMS

17-1 Show that the Bohr magneton is given by $\mu_B = e\hbar/2m_e$. Assume that the electron in a hydrogen atom moves in a circular orbit of radius a about the proton and that its angular momentum is $\hbar$. The magnetic moment μ is the product of the electron current and the area swept out by the orbiting electron.

17-2 (a) For the hydrogen atom of Problem 17-1 show that the orbital magnetic dipole moment is $ea^2\omega/2$, where ω is the electron's angular velocity.

(b) Show that the torque produced by a magnetic field parallel to the plane of the orbit is $ea^2\omega B/2$.

(c) By equating the Coulomb force and the mass times the centripetal acceleration, show that

$$\omega = \left(\frac{4\pi\varepsilon_0 ma^3}{e^2}\right)^{-1/2}.$$

(d) Find values for the angular velocity, torque, and orbital magnetic moment for a hydrogen atom, where $a = 5.29 \times 10^{-11}$ m; let $B = 1$ T.

17-2 Suppose that the hydrogen atom of Problems 17-1 and 17-2 is subjected to a magnetic field B perpendicular to the plane of the orbit. It can be shown that the forces caused by B result in a decrease in the orbital magnetic moment given by $\Delta\mu = e^2 a^2 B / 4m_e$. For the relatively large magnetic field of 1 T, what is $\Delta\mu/\mu$ in parts per million? (This is the mechanism responsible for diamagnetism.)

17-3 The paramagnetic salt iron ammonium alum has the magnetic ion Fe^{+++}. The spin system has $S = 5/2$ and the orbital angular momentum is quenched $(L = 0)$. Thus $J = 5/2$ and $g = 2$. Find the mean dipole moment $\bar{\mu}_z$ of the paramagnetic salt in a magnetic field of 1 tesla at a temperature of 2 K. What is the saturation value of $\bar{\mu}_z$?

17-5 In a paramagnetic solid containing N particles, the total magnetic moment μ_z^{TOTAL} is $N\bar{\mu}_z$. The Curie law is often written in the form

$$\mu_z^{\text{TOTAL}} = \frac{C_c B}{T},$$

where C_c is the so-called Curie constant, given by

$$C_c = \frac{Ng^2\mu_B^2 J(J + 1)}{3k}.$$

Consider a kilomole of a paramagnetic material made up of molecules whose electron cloud has zero orbital angular momentum, spin $1/2$, and total angular momentum $J = 3/2$. Compute C_c.

17-6 A sample of paramagnetic material containing 10^{25} atoms is placed in an external magnetic field of 1 tesla at room temperature. Assume that the electrons of each atom have zero orbital angular momentum and spin angular momentum of $1/2$. Thus $J = 1/2$ and $g = 2$.
(a) Find the total magnetic moment of the sample.
(b) What would be the total magnetic moment at a temperature of 0.1 K?

17-7 Show that the magnetic moment M of a paramagnetic material is a state variable by proving that dM is an exact differential. (Use the Curie law $M \propto B/T$.)

17-8 Consider a two-level system with an energy 2ε separating the upper and lower states. Assume that the energy splitting is the result of an external magnetic field B. Given that the total magnetic energy is

$$U_B = -N\varepsilon \tanh\left(\frac{\varepsilon}{kT}\right),$$

show that the associated heat capacity is

$$C_B = Nk\left(\frac{2\varepsilon}{kT}\right)^2 \frac{e^{2\varepsilon/kT}}{(e^{2\varepsilon/kT} + 1)^2},$$

thereby verifying Equation (17.27).

17-9 The energy levels of a localized particle are 0, ε, and 2ε. The middle level is doubly degenerate and the other levels are nondegenerate.
  (a) Write and simplify the partition function.
  (b) Find the total energy, the heat capacity, and the entropy of a system of these particles. Sketch these properties as a function of the temperature and compare them with the corresponding properties of the spin-1/2 system.

17-10 Prove Equation (17.41) starting with Equation (17.40) and using Stirling's approximation. Show that the entropy is a maximum when the total energy U is zero, corresponding to equal populations of the two energy levels of the system.

17-11 Verify Equation (17.42), referring to Equation (17.41).

17-12 (a) For iron find the saturation value of the magnetization. (The atomic weight of iron is 55.9 and its density is 7880 kg m^{-3}.)
  (b) Estimate the magnitude of the internal field λM for iron ($T_c = 1042$ K).

17-13 What is the magnetization of iron in an external magnetic field of 1 tesla at a temperature twice the Curie temperature?

Chapter 18

Bose-Einstein Gases

18.1 BLACKBODY RADIATION

Statistical thermodynamics is applicable to radiant energy as well as material particles. It is a familiar observation that a hot body loses heat by radiation. The energy loss is attributable to the emission of electromagnetic waves from the body. The distribution of the energy flux over the wavelength spectrum does not depend on the nature of the body but does depend on its temperature.

Here we are concerned with the thermodynamic properties of electromagnetic radiation in thermal equilibrium. The radiation can be regarded as a photon gas. We consider an enclosure or cavity of volume V at a constant temperature T. The walls of the cavity are thermally insulated and perfectly reflecting. Since the system is isolated, it has a fixed energy U. However, the photons emitted by one energy level may be absorbed at another, so the number of photons is not constant. This means that the restriction $\sum N_j = N$ does not apply. Correspondingly, the Lagrange multiplier α that was determined by this condition is zero and $e^{-\alpha} = 1$.

Photons are bosons of spin 1 and hence obey Bose-Einstein statistics. The number of photons per quantum state is therefore given by Equation (13.40) with μ set equal to zero (recall that $\alpha = \mu/kT$):

$$f_j = \frac{N_j}{g_j} = \frac{1}{e^{\varepsilon_j/kT} - 1}.$$

For a continuous spectrum of energies,

$$f(\varepsilon) = \frac{N(\varepsilon)}{g(\varepsilon)} = \frac{1}{e^{\varepsilon/kT} - 1}. \tag{18.1}$$

The energy of a photon is $h\nu$, so this equation can be written

$$f(\nu) = \frac{N(\nu)}{g(\nu)} = \frac{1}{e^{h\nu/kT} - 1}. \tag{18.2}$$

Here $g(\nu)d\nu$ is the number of quantum states with frequencies in the range ν to $\nu + d\nu$. This number was obtained in Chapter 16 for a phonon gas; however, it must be doubled in this application because in a photon gas there are two states of polarization corresponding to the two independent directions of polarization of an electromagnetic wave. Each photon may be in either polarization state. With this modification, Equation (16.8) becomes

$$g(\nu)d\nu = \frac{8\pi V}{c^3}\nu^2 d\nu, \tag{18.3}$$

where c in this case is the speed of light.

The energy $u(\nu)d\nu$ in the range ν to $\nu + d\nu$ is the number of photons in this range times the energy $h\nu$ of each:

$$u(\nu)d\nu = N(\nu)d\nu \times h\nu. \tag{18.4}$$

But

$$N(\nu) = g(\nu)f(\nu).$$

Hence

$$u(\nu)d\nu = \frac{8\pi hV}{c^3}\left[\frac{\nu^3 d\nu}{e^{h\nu/kT} - 1}\right]. \tag{18.5}$$

This is the Planck radiation formula. It gives the spectral distribution of the radiant energy inside the enclosure, i.e., the energy per unit frequency.

The blackbody spectrum is often expressed in terms of the wavelength. Then

$$u(\nu)d\nu \propto u(\lambda)d\lambda$$

and

$$d\nu = d\left(\frac{c}{\lambda}\right) = -\frac{c}{\lambda^2}d\lambda,$$

or

$$|d\nu| = \frac{c}{\lambda^2}|d\lambda|.$$

The wavelength spectrum is therefore

$$u(\lambda)d\lambda = 8\pi hcV\left[\frac{d\lambda}{\lambda^5(e^{hc/\lambda kT} - 1)}\right], \tag{18.6}$$

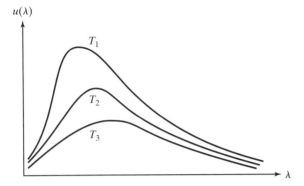

Figure 18.1   The wavelength spectrum of blackbody
radiation energy for three temperatures: $T_1 > T_2 > T_3$.

(Figure 18.1). Here $u(\lambda)$ is the energy per unit wavelength.

Prior to Planck's analysis, various empirical formulas existed. All of them can be found from Equations (18.5) or (18.6). The Stefan-Boltzmann law states that the total radiation energy is proportional to T^4 (the area under the curves of Figure 18.1).

The total energy density (energy per unit volume) is

$$\frac{U}{V} = 8\pi hc \int_0^\infty \frac{d\lambda}{\lambda^5(e^{hc/\lambda kT} - 1)}. \tag{18.7}$$

Setting $x = hc/\lambda kT$, we have

$$\frac{U}{V} = \frac{8\pi h}{h^3 c^3}(kT)^4 \int_0^\infty \frac{x^3 dx}{e^x - 1}.$$

The integral has the value $\pi^4/15$ (see Appendix D). Thus

$$\frac{U}{V} = aT^4, \tag{18.8}$$

where

$$a = \frac{8\pi^5 k^4}{15h^3 c^3} = 7.55 \times 10^{-16} \text{ J m}^{-3} \text{ K}^{-4}.$$

Since radiation inside the cavity is continually absorbed and emitted by the inner surface, we can relate the energy per unit volume, which moves at the speed of light c, to the energy emitted per unit area of the surface per unit time. The latter is the power per unit area or energy flux. In Chapter 11 we found the particle flux to be $(\bar{v}/4)n$, where $\bar{v}$ is the mean speed and n is the number of particles per unit volume. In a similar way, the energy flux e is $(c/4)(U/V)$. Therefore,

$$e = \sigma T^4, \tag{18.9}$$

where

$$\sigma = \frac{ca}{4} = \frac{2\pi^5 k^4}{15h^3c^2} = 5.67 \times 10^{-8}\,\mathrm{Wm^{-2}\,K^{-4}}. \tag{18.10}$$

Equation (18.9) is known as the Stefan-Boltzmann law, and σ is called the Stefan-Boltzmann constant. The theoretical and experimental values of σ agree to within at least three significant figures.

The wavelength $\lambda_{\max}$ at which $u(\lambda)$ is a maximum satisfies a relation known as Wien's displacement law. It can be found by setting the derivative of $u(\lambda)$ equal to zero or, equivalently, by minimizing the denominator in Equation (18.6):

$$\frac{d}{d\lambda}\{\lambda^5[e^{hc/\lambda kT} - 1]\} = 0.$$

With $x = hc/\lambda kT$, this reduces to

$$\frac{x}{5} = 1 - e^{-x}.$$

This is a transcendental equation whose numerical solution is 4.96. Hence

$$\lambda_{\max} T = \frac{hc}{4.96k} = 2.90 \times 10^{-3}\,\mathrm{m\,K}. \tag{18.11}$$

Equation (18.11) is Wien's displacement law. The experimental values of the constant on the right-hand side and of the Stefan-Boltzmann constant can be used to determine the values of h and k, assuming that c is known.

For long wavelengths, $hc/\lambda kT \ll 1$ in Equation (18.6), and the exponential can be approximated by the first two terms of its Taylor series expansion:

$$e^{hc/\lambda kT} \approx 1 + \frac{hc}{\lambda kT}.$$

Then

$$u(\lambda)d\lambda \approx V\frac{8\pi kT}{\lambda^4}d\lambda. \qquad (18.12)$$

This is the so-called Rayleigh-Jeans formula, which exhibits an "ultraviolet catastrophe" as the wavelength approaches zero ($u(\lambda)$ becomes infinite).

For short wavelengths,

$$e^{hc/\lambda kT} \gg 1$$

and

$$u(\lambda)d\lambda \approx V\left(\frac{8\pi hc}{\lambda^5}\right)e^{-hc/\lambda hT}d\lambda. \qquad (18.13)$$

This is Wien's law, valid in the short wavelength region (Figure 18.2).

To summarize, the total blackbody radiant energy per unit volume increases with the fourth power of the temperature, and the wavelength of the peak of the radiation curve $u(\lambda)$ is inversely proportional to T.

The temperature of the Sun's surface is approximately 6000 K, and λ_{max} is 483 nm, a wavelength in the visible range of the electromagnetic spectrum. At the surface temperature of the Earth, roughly 300 K, λ_{max} is about 10μm, which is in the infrared region.

The cosmic background microwave radiation that permeates all of space was discovered by accident in 1964 by Arno Penzias and Robert W. Wilson while making measurements of radio signals at a wavelength of 7.35 centimeters (4 GHz). They observed a nonzero value for the radiation in every direction in which they pointed their radio telescope. In 1990 the Cosmic Background Explorer (COBE) satellite measured the background radiation

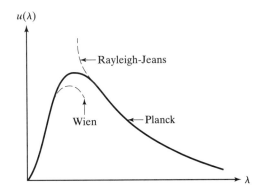

Figure 18.2 Sketch of Planck's law, Wien's law and the Rayleigh-Jeans law.

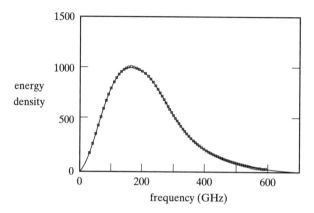

Figure 18.3   Cosmic background measurements by the COBE satellite. The smooth curve is the theoretical blackbody spectrum at 2.735 K. The energy density is in eV m^{-3} per gigahertz. (Adapted from *The Inflationary Universe* by Alan H. Guth, Addison-Wesley, Reading, Massachusetts, 1997.)

at a large number of wavelengths. The data fit the theoretical Planck curve with almost unbelievable precision (Figure 18.3). The cosmic background radiation is evidently a blackbody with a temperature of 2.735 $\pm$ 0.06 K.

18.2 PROPERTIES OF A PHOTON GAS

It is instructive to ask: what is the average value of the ratio $h\nu/kT$ in a volume V at temperature T? The number of photons having frequencies between ν and $\nu + d\nu$ is found by combining Equations (18.2) and (18.3):

$$N(\nu)d\nu = \frac{8\pi V}{c^3}\left(\frac{\nu^2 d\nu}{e^{h\nu/kT} - 1}\right). \tag{18.14}$$

The total number of photons in the cavity is determined by integrating this expression over the infinite range of frequencies. The result is

$$N = 8\pi V\left(\frac{kT}{hc}\right)^3 \int_0^\infty \frac{x^2 dx}{e^x - 1}, \tag{18.15}$$

where the substitution $x = h\nu/kT$ has been made. The integral has the numerical value 2.404. Hence

$$N = 2.02 \times 10^7 \, T^3 \, V, \tag{18.16}$$

where T is in kelvins and V is in m^3. To find the mean energy of the photons in the cavity, we divide the total energy $U = 7.55 \times 10^{-16} T^4 V$ J (Equation (18.8)) by N. The result is

$$3.74 \times 10^{-23} T \approx 2.7 \, kT.$$

Thus, the average value of $h\nu/kT$ is of the order of unity.

The heat capacity is easily calculated using Equation (18.8):

$$C_V = \left(\frac{\partial U}{\partial T}\right)_V = 4aT^3 V = \frac{32\pi^5}{15} k \left(\frac{kT}{hc}\right)^3 V. \tag{18.17}$$

Since this relation holds down to a temperature of absolute zero, it can be used to determine the absolute entropy:

$$S = \int_0^T \frac{C_V}{T'} dT' = 4aV \int_0^T (T')^2 dT' = \frac{4}{3} aT^3 V = \frac{32\pi^5}{45} k \left(\frac{kT}{hc}\right)^3 V. \tag{18.18}$$

Thus both the heat capacity and the entropy increase with the third power of the temperature.

For an open system, we know that

$$dU = TdS - PdV + \mu dN.$$

Substituting the differential of the Helmholtz function $F = U - TS$ in this equation gives

$$dF = -SdT - PdV + \mu dN. \tag{18.19}$$

Therefore,

$$\mu = \left(\frac{\partial F}{\partial N}\right)_{T,V}. \tag{18.20}$$

For the photon gas,

$$F = U - TS = aT^4 V - \frac{4}{3} aT^4 V = -\frac{1}{3} aT^4 V. \tag{18.21}$$

Thus, as we discussed in the previous section, the chemical potential μ is zero, because F does not depend explicitly on N.

Finally, Equation (18.19) also yields

$$P = -\left(\frac{\partial F}{\partial V}\right)_{T,N}.$$ (18.22)

The pressure of the photon gas on the walls of the cavity is therefore

$$P = \frac{1}{3}aT^4 = \frac{1}{3}\left(\frac{U}{V}\right).$$ (18.23)

Note that this differs from the relation $P = 2U/3V$ that obtains for an ideal gas of weakly interacting particles (see Chapter 11). Equation (18.23) can also be obtained from electromagnetic theory if it is assumed that the radiant energy in the enclosure is isotropic.

18.3 BOSE-EINSTEIN CONDENSATION

In this section we shall be concerned with a gas of noninteracting particles (atoms or molecules) of comparatively large mass such that quantum effects only become important at very low temperatures. The particles are assumed to comprise an ideal Bose-Einstein gas. The discussion is relevant to ^{4}He, which undergoes a remarkable phase transition known as *Bose-Einstein condensation*. This phenomenon is intimately related to the superfluidity of liquid helium at low temperatures.

As we noted previously, bosons are particles of integral spin that obey Bose-Einstein statistics. There is no limit to the number of bosons that can occupy any single particle state. We consider an ideal boson gas consisting of N bosons in a container of volume V held at absolute temperature T. The Bose-Einstein continuum distribution is

$$f(\varepsilon) = \frac{N(\varepsilon)}{g(\varepsilon)} = \frac{1}{e^{(\varepsilon-\mu)/kT} - 1}.$$ (18.24)

Our initial concern is determining how the chemical potential μ varies with the temperature. We shall adopt the convention of choosing the ground state energy to be zero. At $T = 0$ all N bosons will be in the ground state since there is no restriction on the number of bosons in a given state. Setting $\varepsilon = 0$ in Equation (18.24), we see that if $f(\varepsilon)$ is to make sense, μ must be intrinsically negative. Furthermore, μ must be zero at a temperature of absolute zero and only slightly less than zero at nonzero low temperatures, assuming N to be a large number.

At high temperatures, in the classical limit of a dilute gas, the Maxwell-Boltzmann distribution applies:

$$f(\varepsilon) = e^{-(\varepsilon-\mu)/kT}. \tag{18.25}$$

From Chapter 14,

$$\mu = -kT \ln\left(\frac{Z}{N}\right), \tag{18.26}$$

where

$$Z = \left(\frac{2\pi mkT}{h^2}\right)^{3/2} V. \tag{18.27}$$

Thus

$$\frac{\mu}{kT} = -\ln\left[\left(\frac{2\pi mkT}{h^2}\right)^{3/2} \frac{V}{N}\right]. \tag{18.28}$$

As an example, for one kilomole of a boson gas comprising ^{4}He atoms at standard temperature and pressure, we have

$$\frac{\mu}{kT} = -\ln\left\{\left[\frac{2\pi(6.65 \times 10^{-27})(1.38 \times 10^{-23})(273)}{(6.63 \times 10^{-34})^2}\right]^{3/2}\left(\frac{22.4}{6.02 \times 10^{26}}\right)\right\}$$

$$= -12.43.$$

The chemical potential *per se* is -0.29 eV. In comparison, the average energy of an ideal monatomic gas atom is $\varepsilon = (3/2)kT = 0.035$ eV at 273 K. Also, $(\varepsilon - \mu)/kT = 1.5 + 12.4 = 13.9$; substituting this value in Equation (18.25) gives $f(\varepsilon) = 9.2 \times 10^{-7}$, confirming the validity of the dilute gas assumption. In this classical limit,

$$e^{-\mu/kT} = \left(\frac{2\pi mkT}{h^2}\right)^{3/2}\frac{V}{N} \tag{18.29}$$

is a positive number that increases with temperature and decreases with the particle density N/V.

The obvious way to determine the temperature dependence of μ is to use the conservation of particles condition that led to the statistical definition of the chemical potential in the first place. For the continuum approximation, this is

$$\int_0^\infty N(\varepsilon)d\varepsilon = \int_0^\infty f(\varepsilon)g(\varepsilon)d\varepsilon = N. \tag{18.30}$$

For $g(\varepsilon)$, we have the result of Equation (12.26) with the spin factor γ_s equal to unity:

$$g(\varepsilon)d\varepsilon = \frac{4\sqrt{2}\pi V}{h^3}m^{3/2}\varepsilon^{1/2}d\varepsilon. \tag{18.31}$$

Thus

$$N = 2\pi V\left(\frac{2m}{h^2}\right)^{3/2}\int_0^\infty \frac{\varepsilon^{1/2}d\varepsilon}{e^{(\varepsilon-\mu)/kT} - 1}. \tag{18.32}$$

There is a significant flaw in this formulation. In using the integral approximation of Equation (18.30) rather than the sum, the ground state $\varepsilon = 0$ is left out. This term, in the sum of particle states at low temperature, is the largest term of all. It is omitted here because the density of states function $g(\varepsilon)$ depends on $\varepsilon^{1/2}$, which is zero for the ground state. Under ordinary circumstances this doesn't matter, since, for ε small compared with kT, the omission of the one term makes a negligible difference in the result. But at low temperatures, bosons condense into this lowest state and its occupation becomes much greater than for any other state.

We can surmount the difficulty in the following way. The total number of bosons consists of N_0 in the ground state and N_{ex} in the excited states. Hence

$$N = N_0 + N_{ex}. \tag{18.33}$$

Since the ground state is excluded from Equation (18.32), the integration only gives the number of bosons in excited states. Therefore

$$N_{ex} = 2\pi V\left(\frac{2m}{h^2}\right)^{3/2}\int_0^\infty \frac{\varepsilon^{1/2}d\varepsilon}{e^{(\varepsilon-\mu)/kT} - 1}. \tag{18.34}$$

The integral can be solved numerically to obtain a relationship between μ, the temperature, and the particle density. It is more instructive, however, to note that at temperatures very close to zero, $N_0 \approx N$. We can therefore put $\varepsilon = 0$ in Equation (18.24) and write

$$N \approx \frac{1}{e^{-\mu/kT} - 1}.$$

Then

$$-\mu/kT = \ln\left(1 + \frac{1}{N}\right) \approx \frac{1}{N}$$

for N large. (Even for a small macroscopic system N will be a huge number.) For low temperatures, then, we can safely put $\exp(-\mu/kT)$ equal to unity in Equation (18.34). With the change of variable $x = \varepsilon/kT$, we get

$$N_{ex} = V\frac{2}{\sqrt{\pi}}\left(\frac{2\pi mkT}{h^2}\right)^{3/2}\int_0^\infty \frac{x^{1/2}dx}{e^x - 1}. \tag{18.35}$$

The value of the integral is $2.612\sqrt{\pi}/2$. Thus

$$N_{ex} = 2.612V\left(\frac{2\pi mkT}{h^2}\right)^{3/2}. \tag{18.36}$$

The so-called Bose temperature T_B is the temperature above which *all* the bosons should be in excited states. Thus we set $N_{ex} = N$ and $T = T_B$ in Equation (18.36) and obtain

$$N = 2.612V\left(\frac{2\pi mkT_B}{h^2}\right)^{3/2}. \tag{18.37}$$

Solving for T_B, we get

$$T_B = \frac{h^2}{2\pi mk}\left(\frac{N}{2.612V}\right)^{2/3}. \tag{18.38}$$

For $T > T_B$, all the bosons are in excited states. As T falls below T_B, an increasing number of bosons occupy the ground state until at $T = 0$, all the bosons are in this state and $N_0 = N$. The fractional number of bosons in the ground state is

$$\frac{N_0}{N} = 1 - \frac{N_{ex}}{N}. \tag{18.39}$$

Dividing Equation (18.36) by Equation (18.37), we have

$$\frac{N_0}{N} = 1 - \left(\frac{T}{T_B}\right)^{3/2}. \tag{18.40}$$

What is a typical value of the Bose temperature? Consider a boson gas made up of 6.02×10^{23} ^{4}He atoms confined to a volume of 22.4×10^{-3} m^3. The mass of a ^{4}He atom is 6.65×10^{-27} kg. Using Equation (18.38), we find that

$$T_B = \frac{(6.63 \times 10^{-34})^2}{2\pi(6.65 \times 10^{-27})(1.38 \times 10^{-23})} \left[\frac{6.02 \times 10^{23}}{(2.612)(22.4 \times 10^{-3})}\right]^{2/3} = 0.036 \text{ K}.$$

In the case of helium, the gas liquifies at 4.21 K at atmospheric pressure, long before the temperature is reduced to 0.036 K. In fact, all real gases liquefy before their Bose temperature is reached.

Plots of N_0/N and N_{ex}/N versus T/T_B are shown in Figure 18.4. A corresponding graph of μ/kT_B, numerically calculated, is shown in Figure 18.5.* The sudden "collapse" into the ground state at very low temperature is called the Bose-Einstein condensation. It is not a condensation in geometrical space but rather in what might be called momentum space.

For many years physicists believed that a Bose-Einstein condensate did not exist in nature. In 1995 researchers created a nearly pure condensate by

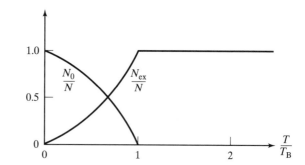

Figure 18.4 Variation with temperature of N_0/N and N_{ex}/N for a boson gas.

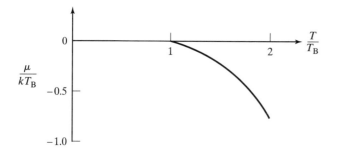

Figure 18.5 Variation with temperature of μ/kT_B for a boson gas.

*See Problem 18-10 for the method of calculating μ/kT_B versus T/T_B.

cooling a vapor of rubidium atoms to a temperature of 1.3×10^{-7} K. The achievement was hailed as "the most interesting development in atomic physics in a decade," opening up an entirely new field of investigation.*

18.4 PROPERTIES OF A BOSON GAS

The bosons in the ground state do not contribute to the internal energy nor to the heat capacity. The first term in the sum $\Sigma N_j \varepsilon_j$ is $N_0 \varepsilon_0$. For $T < T_B$, N_0 may be large but $\varepsilon_0 = 0$, and for $T > T_B$, $N_0 = 0$ in any case.

For temperatures above the Bose temperature, all the bosons are in excited states and we may expect the internal energy to approach $(3/2) NkT$ as the temperature is increased. Below the Bose temperature the number of bosons in the excited states is $N_{ex} = N(T/T_B)^{3/2}$. As a first approximation, we assume that each of these bosons will have a thermal energy of the order of kT. Thus

$$U \approx NkT \left(\frac{T}{T_B}\right)^{3/2} \quad (T < T_B),$$

indicating that U varies as $T^{5/2}$ below the Bose temperature. A more nearly exact result is obtained by noting that

$$U = \int_0^\infty \varepsilon N(\varepsilon) d\varepsilon. \tag{18.41}$$

Using Equations (18.24) and (18.31), we find that

$$U = 2\pi V \left(\frac{2m}{h^2}\right)^{3/2} \int_0^\infty \frac{\varepsilon^{3/2} d\varepsilon}{e^{(\varepsilon - \mu)/kT} - 1}. \tag{18.42}$$

As we have seen, if we choose the energy of the ground state to be zero, the chemical potential μ of the boson gas is very close to zero for temperatures below the Bose temperature. Setting $\mu = 0$ and making the substitution $x = \varepsilon/kT$, we obtain

$$U = \frac{2}{\sqrt{\pi}} kT \left(\frac{2\pi mkT}{h^2}\right)^{3/2} V \int_0^\infty \frac{x^{3/2} dx}{e^x - 1}. \tag{18.43}$$

*See the Richtmyer Memorial Lecture, "Bose-Einstein Condensation in an Ultracold Gas," by Carl E. Wieman, *American Journal of Physics* **64,** pp. 847–855 (1996).

The definite integral is equal to the product of the gamma function $\Gamma(5/2) = 3\pi^{1/2}/4$ and the Riemann zeta function $\zeta(5/2) = 1.34$. Using the expression for T_B, Equation (18.38), we get

$$U = 0.770 \, NkT \left(\frac{T}{T_B}\right)^{3/2} \qquad (T < T_B), \qquad (18.44)$$

not greatly different from our approximation.

It is a simple matter now to obtain the heat capacity:

$$C_V = \frac{dU}{dT} = 1.92 \, Nk \left(\frac{T}{T_B}\right)^{3/2} \qquad (T < T_B). \qquad (18.45)$$

Note that C_V is proportional to $T^{3/2}$.

A graph of the heat capacity as a function of temperature is shown in Figure 18.6. The curve has a change in slope at $T = T_B$, and the heat capacity has its maximum there, equal to 1.92 Nk. At higher temperatures, C_V approaches the classical value of $(3/2) \, Nk$.

The absolute entropy at temperatures below the Bose temperature can be calculated from the heat capacity, since Equation (18.45) holds at absolute zero:

$$S = \int_0^T \frac{C_V dT'}{T'} = 1.28 \, Nk \left(\frac{T}{T_B}\right)^{3/2} \qquad (T < T_B). \qquad (18.46)$$

The entropy goes to zero at $T = 0$ as it should, according to the third law of thermodynamics. The Helmholtz function $F = U - TS$ is

$$F = -0.51 \, NkT \left(\frac{T}{T_B}\right)^{3/2} \qquad (T < T_B), \qquad (18.47)$$

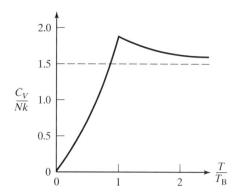

Figure 18.6 Variation with temperature of the heat capacity of a boson gas.

from which we can determine the pressure, using the reciprocity relation $P = -(\partial F/\partial V)_{T,N}$. Expressing F in terms of the volume V using Equation (18.38) for T_B, we have

$$F = -1.33\,kT\left(\frac{2\pi mkT}{h^2}\right)^{3/2}V,\tag{18.48}$$

from which

$$P = 1.33\,kT\left(\frac{2\pi mkT}{h^2}\right)^{3/2}\quad(T < T_B).\tag{18.49}$$

The pressure of a boson gas at low temperatures is proportional to $T^{5/2}$ and is independent of the volume. This is because the bosons in the ground state $\varepsilon = 0$ have no momentum and therefore make no contribution to the pressure. It is easy to show that $P = (2/3)U/V$, the same relation obtained for a classical ideal gas of noninteracting particles.

18.5 APPLICATION TO LIQUID HELIUM

The phase diagram of ordinary helium (^{4}He) is shown in Figure 18.7. The substance has a critical point at a temperature of 5.25 K and exhibits unique behavior in the vicinity of 2 K. Above the critical temperature ^{4}He can't exist as a liquid. When helium gas is compressed isothermally at a temperature below the critical temperature but above 2.18 K, it condenses to a liquid phase

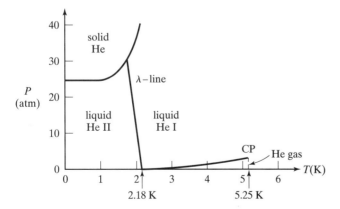

Figure 18.7 Phase diagram of ^{4}He. The two liquid phases are labeled I and II.

called helium I. When the vapor is compressed at temperatures below 2.18 K, a liquid phase called helium II results. Helium II is a superfluid. He I and He II can coexist in equilibrium over a range of pressures and temperatures defined by the so-called lambda line. He II remains a liquid down to absolute zero.

Solid helium can't exist at pressures below 25 atm, and it can't exist in equilibrium with its vapor at any pressure or temperature. Helium has two triple points. At one of them liquid He I and liquid He II are in equilibrium with solid helium. At the other, called the λ-point, the two forms of liquid helium are in equilibrium with the vapor.

The transition between ordinary liquid helium He I and superfluid He II can take place at any point along the λ-line. A graph of the heat capacity versus temperature for the two phases has the general shape of the Greek letter λ (Figure 18.8). As the figure shows, the heat capacity does not change continuously. Its variation with temperature is vastly different in the two phases.

The properties of liquid He II are unique. Its viscosity is virtually zero; hence it is called a superfluid. In the two-fluid theory of He II it is assumed that, at temperatures below the λ-point, liquid He II is a mixture of a normal fluid having viscosity and a superfluid that has no viscosity. The proportion of the superfluid is zero at the λ-point and increases to unity as the temperature approaches zero.

Since the forces of interaction between atoms are weak in liquid helium, we might assume that liquid He II is an ideal boson gas, to a first approximation.* If we identify the λ-point with the Bose temperature, the two-fluid model suggests that the superfluid component consists of the N_0 atoms in the ground

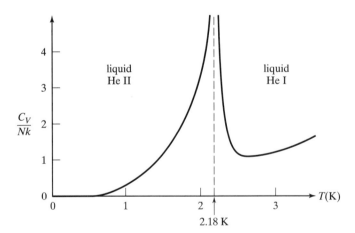

Figure 18.8   The heat capacity anomaly of ^{4}He.

* It might seem strange to apply a theory for an ideal gas to a liquid. However, in some ways, liquid helium behaves more like a gas than a liquid.

state while the normal component is identified with the N_{ex} atoms in the excited states. As the temperature falls below T_B, an increasing number of atoms populate the ground state, and the superfluid component becomes predominant.

Are T_B and the λ-point the same? The volume of a kilomole of liquid ^{4}He is 27×10^{-3} m^3 so that the concentration is

$$\frac{N}{V} = \frac{6.02 \times 10^{26}}{27 \times 10^{-3}} = 2.2 \times 10^{28} \text{ m}^{-3}.$$

Using the value of 6.65×10^{-27} kg for the mass of the ^{4}He atom, we can compute T_B from Equation (18.38). The result is

$$T_B = 3.1 \text{ K},$$

as compared with 2.18 K for the λ-point. We conclude that the superfluid properties of liquid He II could be attributed, in part at least, to a Bose-Einstein-like condensation at the λ-point. The ideal boson gas model, however, is only an approximation to actual liquid helium that neglects interatomic interactions. The complete picture is considerably more complicated.

PROBLEMS

18-1 (a) Calculate the total electromagnetic energy inside an oven of volume 1 m^3 heated to a temperature of 400°F.
   (b) Show that the thermal energy of the air in the oven is a factor of approximately 10^{10} larger than the electromagnetic energy.

18-2 (a) Calculate λ_{max} for the Earth, assuming the Earth to be a blackbody.
   (b) Calculate the temperature at which the human eye is most sensitive to a wavelength of 500 nm.

18-3 Assume that the radiation from the Sun can be regarded as blackbody radiation. The radiant energy per wavelength interval has a maximum at 480 nm.
   (a) Estimate the temperature of the Sun.
   (b) Calculate the total radiant power emitted by the Sun. (The radius of the Sun is approximately 7×10^8 m.)

18-4 (a) Find the frequency at which the radiant energy per unit frequency interval of a blackbody is a maximum. How does this compare with the frequency at which the radiant energy per unit wavelength interval is a maximum?
   (b) Find the frequency at which the cosmic background radiation is a maximum and compare your result with that shown in Figure 18.3. Assume that $T = 2.7$ K.

18-5 (a) Calculate the number of photons in equilibrium in a cavity of volume 1 m^3 held at a temperature $T = 273$ K.

(b) Compare this number with the number of molecules the same volume of an ideal gas contains at STP.

18-6 Assume that the universe is a spherical cavity with radius 10^{26} m and temperature 2.7 K. How many thermally excited photons are there in the universe?

18-7 For photons, where there is no restriction on the total number of particles, the partition function is independent of N, as are the physical properties of a photon gas. The partition function z for a *single oscillator* is given by (see Section 15.2)

$$\ln z = -\ln(1 - e^{-h\nu/kT}),$$

ignoring the zero-point energy. The number of single particle (photon) states in a volume V in the frequency range ν to $\nu + d\nu$ is

$$g(\nu)d\nu = \frac{8\pi V}{c^3}\nu^2 d\nu.$$

Therefore, the partition function Z of the photon *gas* is the sum over states given by

$$\ln Z = -\frac{8\pi V}{c^3}\int_0^\infty \nu^2 \ln(1 - e^{-h\nu/kT})d\nu.$$

(a) Show that integration by parts leads to the equation

$$\ln Z = \frac{8\pi V}{3}\left(\frac{kT}{hc}\right)^3 \int_0^\infty \frac{x^3 dx}{e^x - 1},$$

where $x \equiv h\nu/kT$.

(b) Referring to Appendix D, show that

$$\ln Z = \frac{8\pi^5}{45}\left(\frac{kT}{hc}\right)^3 V.$$

18-8 Use the result of Problem 18.7 to show that:

(a) $$U = kT^2\left(\frac{\partial \ln Z}{\partial T}\right)_{V,N} = \frac{8\pi^5 k}{15}\left(\frac{k}{hc}\right)^3 T^4 V,$$

consistent with Equation (18.8);

(b) $$\mu = -kT\left(\frac{\partial \ln Z}{\partial N}\right)_{V,T} = 0,$$

confirming the statement in Section 18.1;

(c) $$S = \frac{U}{T} + k\ln Z = \left(\frac{32\pi^5 k}{45}\right)\left(\frac{k}{hc}\right)^3 T^3 V,$$

the same result as Equation (18.18); and

(d)
$$P = kT\left(\frac{\partial \ln Z}{\partial V}\right)_{T,N} = \frac{1}{3}\left(\frac{U}{V}\right),$$

which differs from the result $P = (2/3)\,(U/V)$ obtained in Chapter 11 for an ideal gas of molecules.

18-9 In classical thermodynamics, it is found that

$$\left(\frac{\partial U}{\partial V}\right)_T = T\left(\frac{\partial P}{\partial T}\right)_V - P.$$

For blackbody radiation, the energy density $u \equiv U/V$ depends on T only, as does the pressure $P = (1/3)\,u$. Use the above classical result to obtain a differential equation for u whose solution is the Stefan-Boltzmann law.

18-10 Figure 18.5 is a plot of μ/kT_B versus T/T_B. In the region $0 < T/T_B < 1$, μ/kT_B is essentially zero. For $T/T_B > 1$, Equation (18.34) applies, with N_{ex} set equal to N (all N of the bosons are in excited states). Using the definitions $x \equiv \varepsilon/kT_B$, $\xi \equiv \mu/kT_B$, $\eta \equiv T/T_B$, show that Equation (18.34) reduces to

$$2.315 = \int_0^\infty \frac{x^{1/2}dx}{e^{(x-\xi)/\eta} - 1},$$

which can be evaluated numerically to give ξ as a function of η (Figure 18.5).

18-11 The chemical potential of a boson gas at a temperature $T = 2T_B$ is approximately $-0.8\,kT_B$. Determine the mean number of bosons $f(\varepsilon)$ in single particle states having energies of (a) 0, (b) $0.5\,kT_B$, (c) $2.0\,kT_B$, (d) $2.0\,kT_B$, (e) $3.0\,kT_B$.

18-12 (a) Find the chemical potential of a kilomole of ^{4}He gas at STP. Express your answer in joules and in electron-volts.
(b) Use Equation (18.24) to show that the mean occupancy of a single particle state having energy $(3/2)\,kT$ is 8.8×10^{-7} at STP.

18-13 (a) An ideal boson gas consists of ^{4}He atoms whose Bose temperature is 0.087 K. Find the boson concentration, the number of bosons per cubic meter.
(b) What percentage of the bosons are in the ground state at a temperature of 10^{-2} K?

18-14 In a Bose-Einstein condensation experiment, 10^7 rubidium-87 atoms were cooled down to a temperature of 200 nK. The atoms were confined to a volume of approximately 10^{-15} m^3.
(a) Calculate the Bose temperature T_B.
(b) Determine how many atoms were in the ground state at 200 nK.
(c) Calculate the ratio kT/ε_0, where $T = 200$ nK and where the ground state energy ε_0 is given by

$$\varepsilon_0 = \frac{3h^2}{8mV^{2/3}}.$$

18-15 Hydrogen freezes at 14 K and boils at 20 K at atmospheric pressure. The density of liquid hydrogen is 70 kg m^{-3}. Hydrogen molecules are bosons. No evidence has been found for Bose-Einstein condensation in hydrogen. How do you account for this?

18-16 Show that for temperatures below the Bose temperature of a boson gas, (a) $C_V = (5/2) \, U/T$, (b) $S = (5/3) \, U/T$, (c) $F = -(2/3) \, U$, (d) $P = (2/3) \, U/V$.

18-17 A system of N bosons of mass m and zero spin is in a container of volume V at a temperature $T > 0$. The number of particles is

$$N = 2\pi V \left(\frac{2m}{h^2}\right)^{3/2} \int_0^\infty \frac{\varepsilon^{1/2} d\varepsilon}{e^{(\varepsilon - \mu)/kT} - 1},$$

(Equation (18.32)). In the dilute gas approximation, $\exp(-\mu/kT) \gg 1$, and the Bose-Einstein distribution becomes the Maxwell-Boltzmann distribution. Evaluate the integral in this approximation, referring to Appendix D, and show that

$$\exp(-\mu/kT) = \left(\frac{d}{\lambda}\right)^3,$$

where $\lambda = h/(2\pi mkT)^{1/2}$ is the de Broglie wavelength of the particles' thermal motion and $d = (V/N)^{1/3}$. It follows that in the classical limit the average distance d between the particles is very large compared to λ.

Chapter 19

Fermi-Dirac Gases

19.1 THE FERMI ENERGY

Fermi-Dirac statistics governs the behavior of indistinguishable particles of half-integer spin called *fermions*. Fermions obey the Pauli exclusion principle, which prohibits the occupancy of an available quantum state by more than one particle. We consider an ideal gas comprising N noninteracting fermions, each of mass m, in a container of volume V held at temperature T.

The Fermi-Dirac distribution is

$$f_j = \frac{N_j}{g_j} = \frac{1}{e^{(\varepsilon_j - \mu)/kT} + 1}, \tag{19.1}$$

or

$$f(\varepsilon) = \frac{N(\varepsilon)}{g(\varepsilon)} = \frac{1}{e^{(\varepsilon - \mu)/kT} + 1}, \tag{19.2}$$

in the continuum approximation. The right-hand side of Equation (19.2) is referred to as the "Fermi function"; it gives the probability that a single particle state ε will be occupied by a fermion. Clearly, $0 \le f(\varepsilon) \le 1$.

For $\varepsilon = \mu$, $f(\varepsilon)$ has the value $1/2$ at any temperature. Here μ is the chemical potential, which is a function of temperature. Its value at $T = 0$, that is, $\mu(0)$, is called the *Fermi energy*. The Fermi energy is also written as ε_F.

Consider the Fermi function at a temperature of absolute zero. Evidently, at $T = 0$,

$$(\varepsilon - \mu(0))/kT = \begin{cases} -\infty & \text{if } \varepsilon < \mu(0) \\ \infty & \text{if } \varepsilon > \mu(0) \end{cases}.$$

Correspondingly,

$$f(\varepsilon) = \begin{cases} 1 \text{ if } \varepsilon < \mu(0) \\ 0 \text{ if } \varepsilon > \mu(0) \end{cases}. \tag{19.3}$$

This tells us that at $T = 0$ all states with energy $\varepsilon < \mu(0)$ are occupied and all states with $\varepsilon > \mu(0)$ are unoccupied. At absolute zero fermions will occupy the lowest energy states available. The exclusion principle says that only one fermion is allowed per state. So all N particles will be crowded into the N lowest energy levels.

It follows that only one configuration (microstate) is possible for the whole assembly at $T = 0$. Thus the thermodynamic probability w is 1 and $S = k \ln w = 0$. The vanishing of the entropy at absolute zero is consistent with the third law of thermodynamics. The Fermi function at $T = 0$ is shown in Figure 19.1. We ask: how does the Fermi energy $\mu(0)$ depend on m, N, and V? It obviously doesn't depend on the temperature since $T = 0$. We need $g(\varepsilon)$, the density of states of Chapter 12 once again. For particles of spin 1/2, such as electrons, the spin factor is 2, and

$$g(\varepsilon)d\varepsilon = \frac{8\sqrt{2}\pi V}{h^3} m^{3/2}\varepsilon^{1/2}d\varepsilon = 4\pi V \left(\frac{2m}{h^2}\right)^{3/2} \varepsilon^{1/2}d\varepsilon. \tag{19.4}$$

For conservation of particles, $\sum_j N_j = N$, or

$$\int_0^\infty N(\varepsilon)d\varepsilon = \int_0^\infty f(\varepsilon)g(\varepsilon)d\varepsilon = N. \tag{19.5}$$

Substituting Equations (19.3) and (19.4) in Equation (19.5), we get

$$N = \int_0^{\mu(0)} g(\varepsilon)d\varepsilon = 4\pi V \left(\frac{2m}{h^2}\right)^{3/2} \int_0^{\mu(0)} \varepsilon^{1/2}d\varepsilon = \frac{8\pi V}{3}\left(\frac{2m}{h^2}\right)^{3/2} \mu(0)^{3/2}. \tag{19.6}$$

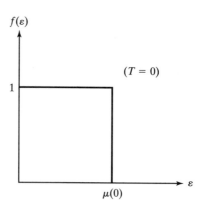

Figure 19.1 The Fermi function at $T = 0$.

Solving for $\mu(0)$, we obtain the result

$$\mu(0) = \frac{h^2}{2m}\left(\frac{3N}{8\pi V}\right)^{2/3}. \tag{19.7}$$

For convenience, we introduce a Fermi temperature T_F such that $\mu(0) = \varepsilon_F = kT_F$. This can be written as

$$T_F = \frac{h^2}{2\pi mk}\left(\frac{N}{1.504V}\right)^{2/3}, \tag{19.8}$$

analogous to the Bose temperature given in Equation (18.38).

19.2 THE CALCULATION OF $\mu(T)$

Figures 19.2 and 19.3 show how the Fermi function changes with temperature. To obtain these curves, we must determine $\mu(T)$. The calculation is considerably more complicated than it was for $T = 0$. We have

$$N = \int_0^\infty f(\varepsilon)g(\varepsilon)d\varepsilon = 4\pi V\left(\frac{2m}{h^2}\right)^{3/2}\int_0^\infty \frac{\varepsilon^{1/2}d\varepsilon}{e^{(\varepsilon-\mu)/kT} + 1}. \tag{19.9}$$

Let the integral be I in this expression and equate Equation (19.6) with Equation (19.9). The result is

$$\frac{2}{3}\mu(0)^{3/2} = I = \int_0^\infty \frac{\varepsilon^{1/2}d\varepsilon}{e^{(\varepsilon-\mu)/kT} + 1}. \tag{19.10}$$

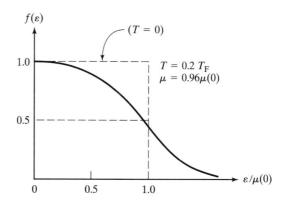

Figure 19.2 The Fermi function at $T = 0.2T_F$.

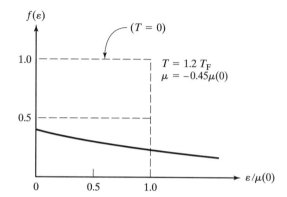

Figure 19.3 The Fermi function at $T = 1.2\, T_F$.

The plus sign in the denominator of the integrand makes all the difference in our ability to evaluate the integral. It can be evaluated numerically. But we can find a very good infinite series approximation for $\mu(T)$ in the following way. Let

$$I = \int_0^\infty \frac{dF(\varepsilon)}{d\varepsilon} f(\varepsilon)\, d\varepsilon, \tag{19.11}$$

where

$$F(\varepsilon) = \frac{2}{3}\varepsilon^{3/2},$$

and

$$f(\varepsilon) = \frac{1}{e^{(\varepsilon-\mu)/kT} + 1}.$$

Integrating Equation (19.11) by parts, we obtain

$$I = f(\varepsilon)F(\varepsilon)\Big|_0^\infty - \int_0^\infty F(\varepsilon)\frac{df(\varepsilon)}{d\varepsilon}\, d\varepsilon.$$

Now $F(\infty)$ is infinite but $f(\infty) = 0$ and its decay is exponential, so $f(\varepsilon)F(\varepsilon) = 0$ at the upper limit. At the lower limit, $f(0)$ is finite but $F(0) = 0$, so the product is zero. Hence

$$I = - \int_0^\infty F(\varepsilon)\frac{df(\varepsilon)}{d\varepsilon}\, d\varepsilon.$$

A quick look at Figure 19.1 shows that at $T = 0$, the function $f(\varepsilon)$ has a zero slope everywhere except at $\varepsilon = \mu(0)$, where the slope is infinite. Thus the derivative $df(\varepsilon)/d\varepsilon$ is a Dirac delta function. At temperatures less than T_F but greater than zero, we might expect the derivative to be a kind of "fuzzy" delta function peaked at $\varepsilon = \mu$. This suggests that we expand $F(\varepsilon)$ in a Taylor series about μ since the only significant contributions to the integral I will be in the vicinity of $\varepsilon = \mu$.* The expansion is

$$F(\varepsilon) = F(\mu) + F'(\mu)(\varepsilon - \mu) + \frac{1}{2!}F''(\mu)(\varepsilon - \mu)^2 + \ldots$$

$$= \frac{2}{3}\mu^{3/2} + \mu^{1/2}(\varepsilon - \mu) + \frac{1}{4}\mu^{-1/2}(\varepsilon - \mu)^2 + \ldots. \qquad (19.12)$$

The contribution to the integral of the linear term is zero since $df(\varepsilon)/d\varepsilon$ is symmetric about μ and the areas under the curve to the left and right of μ will cancel. Thus

$$F(\varepsilon) \approx \frac{2}{3}\mu^{3/2} + \frac{1}{4}\mu^{-1/2}(\varepsilon - \mu)^2. \qquad (19.13)$$

Evaluating the derivative $df(\varepsilon)/d\varepsilon$ in the integrand, we obtain

$$I = \frac{1}{kT}\int_0^\infty \frac{F(\varepsilon)e^{(\varepsilon-\mu)/kT}}{[e^{(\varepsilon-\mu)/kT} + 1]^2}d\varepsilon. \qquad (19.14)$$

We set $y = (\varepsilon - \mu)/kT$ in the integrand, so that

$$I = \int_{-\mu/kT}^\infty \frac{F(y)e^y dy}{(e^y + 1)^2}.$$

The factor e^y is already negligible at $\varepsilon = 0$ where $y = -\mu/kT$, and it's small for $\varepsilon < \varepsilon_F$, at least at low temperatures, which is the region of concern. Hence we can safely replace the lower limit by $-\infty$. Then, using Equation (19.12), we get

$$I = \frac{2}{3}\mu^{3/2}\int_{-\infty}^\infty \frac{e^y dy}{(e^y + 1)^2} + \frac{(kT)^2}{4\mu^{1/2}}\int_{-\infty}^\infty \frac{y^2 e^y dy}{(e^y + 1)^2}.$$

*This method of evaluating the chemical potential and the internal energy for low temperatures is known as the Sommerfeld expansion.

The first integral has the value 1 and the second $\pi^2/3$. Thus, reminding ourselves of Equation (19.10), we have

$$I = \frac{2}{3}\mu^{3/2} + \frac{\pi^2}{12}\frac{(kT)^2}{\mu^{1/2}} = \frac{2}{3}\mu(0)^{3/2}.$$

A little algebra gives

$$\mu = \mu(0)\left[1 + \frac{\pi^2}{8}\left(\frac{kT}{\mu}\right)^2\right]^{-2/3}$$

$$\approx \mu(0)\left[1 - \frac{\pi^2}{12}\left(\frac{kT}{\mu}\right)^2\right].$$

Finally, we replace μ in the "correction term" by $\mu(0) = kT_F$, to obtain

$$\mu \approx \mu(0)\left[1 - \frac{\pi^2}{12}\left(\frac{T}{T_F}\right)^2\right] \qquad (T \ll T_F). \qquad (19.15)$$

A comparison of this approximate expression with the "exact" numerical evaluation is shown in Figure 19.4. The approximation gives $\mu = 0$ at $T/T_F = 1.10$ instead of 0.999, and $\mu = -0.85\,\mu(0)$ at $T/T_F = 1.5$ instead of $-1.1\mu(0)$.

We especially note that μ is positive for temperatures below the Fermi temperature and negative for higher temperatures. As the temperature increases above T_F, more and more of the fermions are in the excited states and the mean occupancy of the ground state falls below 1/2. In this region,

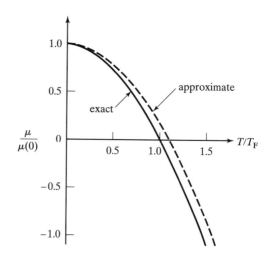

Figure 19.4 Exact and approximate calculations of $\mu/\mu(0)$ versus T/T_F.

$$f(0) = \frac{1}{e^{-\mu/kT} + 1} < \frac{1}{2},$$

which implies that

$$\frac{\mu}{kT} < 0,$$

or $\mu < 0$. We also note that the situation is different for a boson gas, where μ is negative at all temperatures and is zero at absolute zero.

At high temperatures the fermion gas approximates the classical ideal gas. In the classical limit,

$$\mu = -kT \ln \left(\frac{Z}{N} \right), \tag{19.16}$$

with

$$\frac{Z}{N} = 2 \left(\frac{2\pi mkT}{h^2} \right)^{3/2} \frac{V}{N}. \tag{19.17}$$

(The spin degeneracy factor is 2 for fermions.) For $T \gg T_F$, μ/kT takes on a large negative value and $\exp(-\mu/kT) \gg 1$. As an example, consider a kilomole of ^{3}He gas atoms (which are fermions) at standard temperature and pressure (Problem 19-2). The Fermi temperature is 0.069 K, so that $T/T_F = 3900$. Using Equations (19.16) and (19.17), we find that $\mu/kT = -12.7$ and $\exp(-\mu/kT) = 3.3 \times 10^5$. The average occupancy of single particle states is very small, as in the case of an ideal dilute gas obeying the Maxwell-Boltzmann distribution.

19.3 FREE ELECTRONS IN A METAL

Statistical thermodynamics provides profound insights into the behavior of conduction electrons in metals at moderate temperatures. Electrons are spin 1/2 fermions. Each atom in the crystal lattice of the metal is assumed to part with some number of its outer valence electrons, which can then move freely about in the metal. There is an electric field due to the positive ions that varies widely from point to point. However, the effect of the field is canceled out except at the surface of the metal where there is a strong potential barrier, called the work function, that draws back into the metal any electron that happens to make a small excursion outside. The free electrons are therefore confined to the interior of the metal as gas molecules are confined to the interior of a container. We speak of the electrons as an *electron gas.*

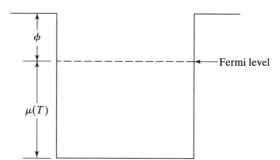

Figure 19.5 Potential well for free electrons in a metal.

In this model, the free electrons move in a potential box or well whose walls coincide with the boundaries of the specimen. They occupy energy states up to the so-called *Fermi level,* which is the chemical potential $\mu(T)$. The work function ϕ is the energy required to remove an electron at the Fermi level from the metal surface. The depth of the potential well is equal to $\mu(T) + \phi$ (Figure 19.5).

The Fermi level of the free electrons in most metals at room temperature is only fractionally less than the Fermi energy $\varepsilon_F \equiv \mu(0)$. It is often assumed that the two are equal, and this leads to confusion. The Fermi level, strictly speaking, is $\mu(T)$, which is an *approximation* to the Fermi energy valid for $T \ll T_F$.

A more realistic picture of the potential well is given in Figure 19.6, which shows how the potential varies in the vicinity of the positive ions in the crystal lattice. The periodicity leads to a band structure in the density of quantum states, which is the foundation of semiconductor physics.

To get an idea of the magnitude of the quantities in the electron gas model, we consider the free electrons in silver, which is monovalent (one free

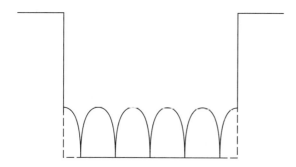

Figure 19.6 Sketch of the potential well showing periodicities associated with the positive ions of the crystal.

electron per atom). The density of silver is 10.5×10^3 kg m^{-3} and its atomic weight is 107. The concentration is therefore

$$\frac{N}{V} = 10.5 \times 10^3 \frac{\text{kg}}{\text{m}^3} \times \frac{1 \text{ kilomole}}{107 \text{ kg}} \times \frac{6.02 \times 10^{26} \text{ atoms}}{\text{kilomole}} = 5.90 \times 10^{28} \text{ m}^{-3}.$$

Since silver is monovalent, this is also the electron concentration. The Fermi energy is given by Equation (19.7):

$$\varepsilon_F = \mu(0) = \frac{h^2}{2m} \left(\frac{3N}{8\pi V} \right)^{2/3}.$$

Here m is the electron mass, equal to 9.11×10^{-31} kg. So

$$\varepsilon_F = \frac{(6.63 \times 10^{-34})^2}{2 \times 9.11 \times 10^{-31}} \left(\frac{3 \times 5.90 \times 10^{28}}{8\pi} \right)^{2/3}$$

$$= 8.85 \times 10^{-19} \text{ J} \times \frac{1 \text{ eV}}{1.6 \times 10^{-19} \text{ J}} = 5.6 \text{ eV}.$$

The Fermi temperature is

$$T_F = \frac{\varepsilon_F}{k} = \frac{5.6 \text{ eV}}{8.62 \times 10^{-5} \text{ eV K}^{-1}} = 65,000 \text{ K}.$$

The ratio T/T_F at room temperature is

$$\frac{T}{T_F} = \frac{300}{6.5 \times 10^4} = 0.00462.$$

At room temperature, therefore, the electron gas is in the so-called *degenerate region* $T \ll T_F$. The chemical potential (the Fermi *level*) can be found from Equation (19.15). The computation gives

$$\mu(T) = 0.999\varepsilon_F.$$

This shows why $\mu(T)$ is often identified with ε_F.

The work function ϕ depends on the metal and the condition of its surface and is typically of the order of 3–4 eV. At very high temperatures, some of the free electrons may have sufficient energies to leave the metal. This results in *thermionic emission*. The condition for their escape is

$$\frac{p_x^2}{2m} > \phi + \mu(T),$$

where p_x is the component of the electron's momentum normal to the surface of the metal.

19.4 PROPERTIES OF A FERMION GAS

The function $N(\varepsilon)d\varepsilon$ is the number of fermions in the single particle energy range ε to $\varepsilon+d\varepsilon$. We know that

$$N(\varepsilon)d\varepsilon = f(\varepsilon)g(\varepsilon)d\varepsilon,$$

where the Fermi function $f(\varepsilon)$ and the degeneracy function $g(\varepsilon)$ behave like the curves of Figure 19.7 for a temperature in the range $0 < T \ll T_F$.

The product of the two curves gives $N(\varepsilon)$ versus ε, as shown in Figure 19.8. The electrons crowd around the Fermi energy because the degeneracy increases with energy; there are therefore more available quantum states to be occupied with 0 or 1 electron(s) per state. (It is the electrons in the tail of this distribution that have the best chance of escaping from the metal in the free electron model.)

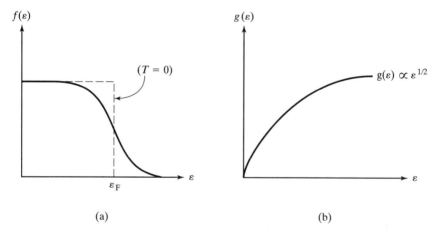

(a)   (b)

Figure 19.7   The variation with single particle energy of (a) the Fermi function, and (b) the degeneracy function. The curves are sketched for $0 < T \ll T_F$.

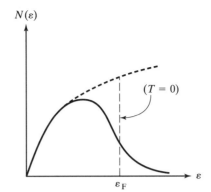

Figure 19.8 The energy distribution of fermions for $0 < T \ll T_F$.

The internal energy of a gas of N fermions is

$$U = \int_0^\infty \varepsilon N(\varepsilon)\,d\varepsilon = \int_0^\infty \varepsilon f(\varepsilon)g(\varepsilon)\,d\varepsilon = 4\pi V\left(\frac{2m}{h^2}\right)^{3/2}\int_0^\infty \frac{\varepsilon^{3/2}d\varepsilon}{e^{(\varepsilon-\mu)/kT}+1}.$$

An approximate evaluation of the integral can be carried out in the same manner in which we evaluated the similar integral in Section 19.2. If the term of order T^4 is included, the result is

$$U \approx \frac{3}{5}N\varepsilon_F\left[1 + \frac{5\pi^2}{12}\left(\frac{T}{T_F}\right)^2 - \frac{\pi^4}{16}\left(\frac{T}{T_F}\right)^4 + \cdots\right]. \qquad (19.18)$$

At $T = 0$, $U = (3/5)N\varepsilon_F$; this energy is large because all the electrons must occupy the lowest energy states up to the Fermi level. The average energy of a free electron in silver at $T = 0$ is

$$\bar{\varepsilon}(0) = \frac{U(0)}{N} = \frac{3}{5}\varepsilon_F = \frac{3}{5}(5.6\text{ eV}) = 3.4\text{ eV}.$$

Note that the mean kinetic energy of an electron, even at absolute zero, is two orders of magnitude greater than the mean kinetic energy of an ordinary gas molecule at room temperature.

The electronic heat capacity C_e can be found by taking the derivative of Equation (19.18):

$$C_e = \frac{dU}{dT} = \frac{\pi^2}{2}Nk\left[\left(\frac{T}{T_F}\right) - \frac{3\pi^2}{10}\left(\frac{T}{T_F}\right)^3 + \cdots\right]. \qquad (19.19)$$

For temperatures that are small compared with the Fermi temperature, we can neglect the second term in the expansion compared with the first and obtain

$$C_e \approx \frac{\pi^2}{2} Nk \left(\frac{T}{T_F} \right) = \frac{\pi^2}{2} Nk \left(\frac{kT}{\varepsilon_F} \right). \tag{19.20}$$

For silver at room temperature,

$$C_e = \frac{\pi^2}{2} Nk \left(\frac{0.025 \text{ eV}}{5.6 \text{ eV}} \right) = 2.2 \times 10^{-2} \, Nk.$$

Thus the electronic specific heat capacity is $2.2 \times 10^{-2} R$. This small value explains a puzzle. Metals have a specific heat capacity of about $3R$, the same as for other solids. It was originally believed that their free electrons should contribute an additional $(3/2)R$ associated with their three translational degrees of freedom. Our last calculation shows that the contribution is negligible.

Why is it so small? While the kinetic energy of the electrons is much greater than the thermal energy of electrons in a gas, the energy of the electrons *changes* only slightly with temperature (dU/dT is small). Only those electrons near the Fermi level can increase their energies as the temperature is raised, and there are precious few of them.

At very low temperatures the picture is different. From the Debye theory, $C_V \propto T^3$ and so the heat capacity of a metal takes the form

$$C_V = AT + BT^3,$$

where the first term is the electronic contribution and the second is associated with the crystal lattice. At sufficiently low temperatures, the former can dominate, as the sketch of Figure 19.9 indicates.

Whereas we have emphasized free electrons in metals, most of our results apply to any ideal gas of fermions. Using the fact that the reversible

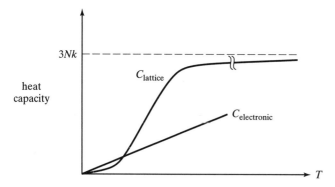

Figure 19.9   Sketch of the heat capacity of a metal as a function of temperature showing the electronic and lattice contributions.

heat flow into a gas at constant volume is given by $T dS = C_V dT$, we can calculate the entropy from Equation (19.19):

$$S = \int_0^T \frac{C_e dT'}{T'} = \frac{\pi^2}{2} Nk \left[\left(\frac{T}{T_F} \right) - \frac{\pi^2}{10} \left(\frac{T}{T_F} \right)^3 + \dots \right]. \qquad (19.21)$$

Therefore $S = 0$ at $T = 0$, as it must be. The Helmholtz function $F \equiv U - TS$ is

$$F = NkT_F \left[\frac{3}{5} - \frac{\pi^2}{4} \left(\frac{T}{T_F} \right)^2 + \frac{\pi^4}{80} \left(\frac{T}{T_F} \right)^4 + \dots \right]. \qquad (19.22)$$

The fermion gas pressure is found from

$$P = -\left(\frac{\partial F}{\partial V} \right)_{T,N}. \qquad (19.23)$$

It is left as Problem 19-9 to prove that

$$P = \frac{2}{5} \frac{NkT_F}{V} \left[1 + \frac{5\pi^2}{12} \left(\frac{T}{T_F} \right)^2 + \dots \right]. \qquad (19.24)$$

Comparison with Equation (19.18) shows that $P = (2/3)U/V$.

For silver we found that $N/V = 5.9 \times 10^{28}$ m^{-3} and $T_F = 65{,}000$ K. Thus

$$P \approx \frac{2}{5} (5.90 \times 10^{28})(1.38 \times 10^{-23})(6.5 \times 10^4) = 2.1 \times 10^{10} \text{ Pa}.$$

Given this tremendous pressure, we can appreciate the role of the surface potential barrier in keeping the electrons from evaporating from the metal.

19.5 APPLICATION TO WHITE DWARF STARS

Very high pressures, of the order of 10^{17} atm, exist in the degenerate electron gas of a white dwarf star. It is this pressure that prevents its gravitational collapse.

A typical star has a hot core with a temperature of the order of 10^7 K. The heat energy is supplied by thermonuclear reactions. The atoms are completely ionized (kT is 900 eV at this temperature), creating a huge electron gas. In young stars the electron pressure is sufficient to withstand the weight of the material pressing on the center, thereby preventing collapse. In old stars, the hydrogen at the core has run out, fusion has stopped, and the core cools. However, the loss of gravitational energy results in an increase in the kinetic energy of the electrons and ions and the cooling process is partially offset. In white dwarfs the collapse beyond a certain point is prevented by the electron gas pressure. White dwarfs typically have the mass of the Sun and the radius of the Earth.

The pressure of the electron gas in Sirius B, a white dwarf, can be estimated using the formula

$$P = \frac{2}{5}\left(\frac{N}{V}\right)\varepsilon_F, \qquad T \ll T_F. \qquad (19.25)$$

We need the following characteristics of Sirius B:

$$\text{Mass } M = 2.09 \times 10^{30} \text{ kg}$$
$$\text{Radius } R = 5.57 \times 10^{6} \text{ m}$$
$$\text{Volume } V = 7.23 \times 10^{20} \text{ m}^3.$$

We assume that nuclear fusion has ceased—that all the core hydrogen has been converted to helium. We further assume that the helium is completely ionized—that each He atom has lost two electrons. In addition, we suppose that the number of nucleons in the star is equal to its mass divided by the mass of a nucleon. Thus

$$\text{No. of nucleons} = \frac{2.09 \times 10^{30}}{1.66 \times 10^{-27}} = 1.26 \times 10^{57}.$$

Since there are four nucleons and two electrons in a helium atom, the number of electrons N in the fermion gas of Sirius B is 0.63×10^{57}. Then, with $m_e = 9.11 \times 10^{-31}$ kg,

$$\varepsilon_F = \frac{h^2}{2m_e}\left(\frac{3N}{8\pi V}\right)^{2/3} = 5.33 \times 10^{-14} \text{ J} = 0.33 \text{ MeV}.$$

Then $T_F = \varepsilon_F/k = 3.9 \times 10^9$ K. If $T = 10^7$ K, then the condition $T \ll T_F$ is satisfied.

To a fair approximation, the electron gas pressure is given by Equation (19.25). Thus

$$P = \frac{2}{5}\left(\frac{0.63 \times 10^{57}}{7.23 \times 10^{20}}\right)(5.33 \times 10^{-14}) = 1.8 \times 10^{22} \text{ Pa} = 1.8 \times 10^{17} \text{ atm.}$$

We'd like to show that this pressure is sufficient to prevent further collapse.

A white dwarf is stable when its total energy is a minimum. The energy is

$$U = U_e + U_{grav},\tag{19.26}$$

where U_e is the energy of the electron gas and U_{grav} is the gravitational energy. We need to express each in terms of the star's radius R, then take the derivative of U with respect to R and set it equal to zero.
    For $T \ll T_F$,

$$U_e \approx U(0) = \frac{3}{5}N\varepsilon_F = \frac{3}{5}\frac{h^2}{2m_e}\left(\frac{3}{8\pi V}\right)^{2/3}N^{5/3}.$$

But $V = (4/3)\pi R^3$, so

$$U_e = \frac{a}{R^2},\tag{19.27}$$

where

$$a = \frac{3h^2}{10m_e}\left(\frac{9}{32\pi^2}\right)^{2/3}N^{5/3}.\tag{19.28}$$

The gravitational energy of a solid sphere of mass M and radius R is known from classical mechanics to be

$$U_{grav} = -\frac{b}{R},\tag{19.29}$$

where

$$b = \frac{3}{5}GM^2.\tag{19.30}$$

Thus

$$U = \frac{a}{R^2} - \frac{b}{R}.\tag{19.31}$$

Minimizing this function, we find that

$$R_{min} = \frac{2a}{b}.\tag{19.32}$$

Substituting the values of a and b from Equations (19.28) and (19.30), respectively, we find that (see Problem 19-5)

$$R_{\text{min}} \approx 7 \times 10^6 \text{m}. \tag{19.33}$$

Since the observed value of the radius of Sirius B is roughly equal to this, we conclude that the star has contracted down to its stable minimum size and is truly a white dwarf.

White dwarfs are at present viewed as comprising a core of degenerate electron gas surrounded by a nondegenerate $(T > T_F)$ outer layer. Nuclear reactions have stopped. The star is cooling slowly. The time before it becomes invisible is estimated to be 10^{10} years. Since the age of the universe is believed to be 1.5×10^{10} years, it follows that very few white dwarfs have had time to become invisible.

The stability of white dwarfs is but one example of the very extensive use of quantum statistics in the study of stellar evolution.*

PROBLEMS

19-1 Assume that for $T = 3T_F$ the value of the chemical potential is $-5.6\, \varepsilon_F$. Calculate the value of the Fermi function $f(\varepsilon)$ at the temperature T for values of $\varepsilon/\varepsilon_F$ of (a) 0, (b) 0.5, (c) 1.0, and (d) 2.0.

19-2 Consider a kilomole of ^{3}He gas atoms under STP conditions.
  (a) What is the Fermi temperature of the gas?
  (b) Calculate μ/kT and $\exp(-\mu/kT)$.
  (c) Find the average occupancy $f(\varepsilon)$ of a single particle state that has an energy of $(3/2)\, kT$.

19-3 For a system of noninteracting electrons, show that the probability $f(\varepsilon)$ of finding an electron in a state with energy Δ *above* the chemical potential μ is the same as the probability of finding an electron absent from a state with energy Δ *below* μ at any given temperature T.

19-4 **(a)** Verify that the average energy per fermion is $(3/5)\varepsilon_F$ at absolute zero by making a direct calculation of $U(0)/N$.
  (b) Similarly, prove that the average speed of a fermion gas particle at $T = 0$ is $(3/4)v_F$, where the Fermi velocity v_F is defined by $\varepsilon_F = (1/2)mv_F^2$.

---

*The discussion of this section was motivated by Chapter 12 of *Statistical Physics* by W.C.V. Rosser, Ellis Horwood, 1982. Rosser gives an extensive treatment of white dwarfs and neutron stars.

19-5 Calculate the first two terms of the series expansion of U in Equation (19.18). Use the Sommerfeld expansion method applied to the calculation of $\mu(T)$ in Section 19.2. Write $F(\varepsilon) = (2/5)\varepsilon^{5/2}$ and expand $F(\varepsilon)$ in a Taylor series about μ, retaining the first three terms only. You will get an expression for U in terms of μ. As a final step, substitute the expression for μ of Equation (19.15).

19-6 At very low temperatures, the electronic specific heat capacity of a metal is approximately

$$c_e = AT,$$

where A can be determined by experiment. For gold, the constant A is found to be $A = 0.73$ J kilomole^{-1} K^{-2}. Compare this result to the value obtained from Equation (19.20). (The atomic weight of gold is 197 and its density is 18.9×10^3 kg m^{-3}.)

19-7 **(a)** Calculate ε_F for aluminum assuming three electrons per aluminum atom.
   (b) Show that for aluminum at $T = 1000$ K, μ differs from ε_F by less than 0.01%. (The density of aluminum is 2.69×10^3 kg m^{-3} and its atomic weight is 27.)
   (c) Calculate the electronic contribution to the specific heat capacity of aluminum at room temperature and compare it to $3R$.

19-8 In sodium there are approximately 2.6×10^{28} conduction electrons per cubic meter, which behave as a free electron gas. Give an approximate value for the electronic specific heat of sodium at room temperature. (The atomic weight of sodium is 23.) In the series expansion for c_e, show that at this temperature the cubic term is negligible compared with the linear term.

19-9 Derive the expression for the fermion gas pressure, Equation (19.24), using Equations (19.22) and (19.23).

19-10 Calculate the isothermal compressibility of the fermion gas consisting of the free electrons in silver. Compare your answer with the experimental value for silver of 0.99×10^{-11} Pa^{-1}.

19-11 Show that the gravitational energy of a uniformly solid sphere of mass M and radius R is $U_{grav} = -(3/5)\, GM^2/R$. (Consider the gravitational interaction between a solid sphere (core) of radius r and a surrounding shell of thickness dr.)

19-12 For the white dwarf star Sirius B, do the computation leading to $R_{min} \approx 7 \times 10^6$ m. Use Equations (19.28) and (19.30) with the appropriate parameter values.

19-13 Consider the collapse of the Sun into a white dwarf. For the Sun, $M = 2 \times 10^{30}$ kg, $R = 7 \times 10^8$ m, $V = 1.4 \times 10^{27}$ m^3.
   (a) Calculate the Fermi energy of the Sun's electrons.
   (b) What is the Fermi temperature?
   (c) What is the average speed of the electrons in the fermion gas (see Problem 19-4). Compare your answer with the speed of light.
   (d) What is the density of the electron gas? Compare it with the density of water. (Note: For a star with mass greater than a critical value of 1.44 times the mass of the Sun, the collapse continues beyond the white dwarf stage as gravitational attraction overcomes the electron pressure. The result is a very dense neutron star that eventually ends in a supernova explosion.)

Chapter **20**

Information Theory

20.1 INTRODUCTION

No reasonably up-to-date introduction to statistical thermodynamics would be complete without some mention of information theory. We have seen that the entropy is a measure of the degree of randomness or disorder of a system. A system that is in its single lowest quantum state is one in perfect order. On the other hand, a system that is likely to be in any of a number of different states is a disorderly system. The greater the number of states it might occupy, the greater the disorder. Disorder, in this sense, implies a lack of information regarding the exact state of the system. A disordered system is one about which we lack complete information. The reason for briefly discussing information theory is that it casts further light on the meaning of entropy.

20.2 UNCERTAINTY AND INFORMATION

In his classic 1948 paper, *A Mathematical Theory of Communication,* Claude Shannon laid down the foundations of information theory.[*] It was further developed by Leon Brillouin[†] and applied to statistical thermodynamics by E.T. Jaynes.[‡] Shannon's single greatest accomplishment was to provide a mathematical measure of information. The cornerstone of the theory is the observation that information is a combination of the certain and the uncertain, of the expected and the unexpected. The first part carries no real information; the measurable information is contained in the second part, the unexpected. The degree of surprise generated by a *certain* event—say, one that has already occurred—is zero. If a less probable event is reported, the information conveyed is greater. The information should therefore *increase* as the probability *decreases.*

Uncertainty is reduced by relevant information. If I throw a die many times, I can expect that the average score will be 3 1/2. If I throw it once there

---

[*] C.E. Shannon, *Bell System Tech. J.* 27, 379 and 623 (1948).

[†] L. Brillouin, *Science and Information Theory,* Academic Press, New York, 1956.

[‡] E.T. Jaynes, "Information Theory and statistical mechanics" *Phys. Rev.* 106, 620 (1957).

is an equal probability of getting any number between 1 and 6. But if someone tells me before I throw it that the same number 3 appears on all six sides, my uncertainty about the outcome of the experiment is totally removed. I can ask the question, before I throw the die, how much uncertainty is there about the outcome? Or, correspondingly, how much information is transmitted by the actual result?

For a given experiment, consider a set of possible outcomes whose probabilities are $p_1, p_2, \ldots p_n$. Shannon discovered that it is possible to find a quantity $H(p_1 \ldots p_n)$ that measures in a unique way the amount of uncertainty represented by the given set of probabilities. Only three conditions are needed to specify the function $H(p_1 \ldots p_n)$ to within a constant factor. They are:

1. H is a continuous function of the p_i.
2. If all the p_i's are equal, $p_i = 1/n$; then $H(1/n, \ldots, 1/n)$ is a monotonic increasing function of n. (With equally likely outcomes, there is more uncertainty when there are more possible events.)
3. If the possible outcomes of a particular experiment depend on the possible outcomes of n subsidiary experiments, then H is the sum of the uncertainties of the subsidiary experiments.

The last condition is known as the composition law and needs some elaboration. Consider three possible outcomes of an experiment with probabilities $p_1 = 1/2$, $p_2 = 1/3$, and $p_3 = 1/6$. We note that the sum of the probabilities is 1, as must be the case. Now consider a second experiment broken down into two successive experiments. Let the first subsidiary experiment have two equally probable outcomes, with $p_1 = 1/2$ and $p_2 = 1/2$. Let the second subsidiary experiment have two possible outcomes with probabilities $p_1' = 2/3$ and $p_2' = 1/3$ (Figure 20.1). We note that the probability of two independent outcomes occurring together is the *product* of the probabilities of the separate outcomes. We also observe that in the two cases the final outcomes have the same probabilities. Thus, condition 3 requires that

$$H\left(\frac{1}{2}, \frac{1}{3}, \frac{1}{6}\right) = H\left(\frac{1}{2}, \frac{1}{2}\right) + \frac{1}{2}H\left(\frac{2}{3}, \frac{1}{3}\right). \qquad (20.1)$$

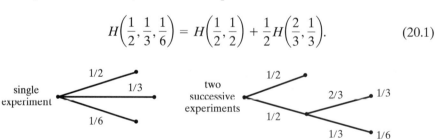

Figure 20.1   Decomposition of an experiment. For the first experiment, the probabilities of the outcomes are $p_1 = 1/2$, $p_2 = 1/3$, and $p_3 = 1/6$. In the second experiment, the first subsidiary experiment has probabilities $p_1 = 1/2$ and $p_2 = 1/2$, while the second has probabilities $p_1' = 2/3$ and $p_2' = 1/3$.

The coefficient of the second term on the right-hand side is 1/2 because the second outcome occurs only half the time (see Problem 20-1).

Shannon showed that the simplest choice for H consistent with the three conditions is

$$H(p_1 \ldots p_n) = \sum_{i=1}^{n} f(p_i), \tag{20.2}$$

where the function f is unknown. Since $f(p_i)$ is a continuous function, it suffices to determine f for the special case of equal probabilities. For $p_i = 1/n$ for all i, Equation (20.2) gives

$$H\left(\frac{1}{n}, \ldots, \frac{1}{n}\right) = nf\left(\frac{1}{n}\right). \tag{20.3}$$

We note that condition 2 requires that

$$\frac{d}{dn}\left[nf\left(\frac{1}{n}\right)\right] \geq 0. \tag{20.4}$$

Invoking the composition law, condition 3, we consider the case where the total number of equally possible outcomes of the first subsidiary experiment is r and that of the second subsidiary experiment is s. Thus $n = rs$ and

$$H\left(\frac{1}{r}, \ldots, \frac{1}{r}\right) + H\left(\frac{1}{s}, \ldots, \frac{1}{s}\right) = H\left(\frac{1}{n}, \ldots, \frac{1}{n}\right) = H\left(\frac{1}{rs}, \ldots, \frac{1}{rs}\right). \tag{20.5}$$

Using Equation (20.3) we obtain

$$rf\left(\frac{1}{r}\right) + sf\left(\frac{1}{s}\right) = rsf\left(\frac{1}{rs}\right). \tag{20.6}$$

Let $R \equiv 1/r$ and $S \equiv 1/s$. Equation (20.6) then becomes

$$\frac{1}{R}f(R) + \frac{1}{S}f(S) = \frac{1}{RS}f(RS). \tag{20.7}$$

Setting $g(R) \equiv (1/R)f(R)$, $g(S) \equiv (1/S)f(S)$, etc., we have

$$g(R) + g(S) = g(RS).$$

If we differentiate this equation first with respect to R and then with respect to S, we obtain

$$g'(R) = Sg'(RS)$$

and

$$g'(S) = Rg'(RS)$$

From these relations, it follows that

$$Rg'(R) = Sg'(S). \qquad (20.8)$$

Since R and S are independent variables, the only way in which Equation (20.8) can be satisfied is for both sides to be equal to the same constant:

$$Rg'(R) = A \qquad (A = \text{constant}),$$

or

$$g(R) = A \ln R + C \qquad (C = \text{constant}).$$

Recalling the definitions of R and $g(R)$, we see that this gives

$$f\left(\frac{1}{r}\right) = -\frac{A}{r} \ln r + \frac{C}{r}.$$

If the probability is 1, the uncertainty must be zero. That is $f(1) = 0$, so $C = 0$, and for equal probabilities,

$$nf\left(\frac{1}{n}\right) = -A \ln n. \qquad (20.9)$$

The remaining issue is the sign of A. Applying the condition of Equation (20.4), we find that $-A/n \geq 0$, so A must be negative. Setting $A = -K$ with K positive, and writing $n = 1/p$, we obtain

$$f(p) = -Kp \ln p. \qquad (20.10)$$

Using this result in Equation (20.2), we obtain the expression for the uncertainty we have been seeking:

$$H = -K \sum_{i=1}^{n} p_i \ln p_i, \qquad (20.11)$$

where K is a positive constant.

Let's check our example of the composite experiment. We have

$$H\left(\frac{1}{2},\frac{1}{3},\frac{1}{6}\right) = -K(-0.346 - 0.366 - 0.299) = 1.01K,$$

$$H\left(\frac{1}{2},\frac{1}{2}\right) + \frac{1}{2}H\left(\frac{2}{3},\frac{1}{3}\right) = -K(-0.346 - 0.346) - \frac{K}{2}(-0.270 - 0.366)$$

$$= 1.01K.$$

Also, for equal probabilities, $p_i = 1/n$ for all i, and $H = K \ln n$. It follows that Equation (20.11) satisfies all the conditions imposed on the measure of uncertainty.

It is instructive to consider the *binary* case of two possible outcomes of an experiment with probabilities p_1 and p_2 such that $p_1 + p_2 = 1$. Then

$$H/K = -p_1 \ln p_1 - p_2 \ln p_2$$
$$= -p_1 \ln p_1 - (1 - p_1)\ln(1 - p_1). \qquad (20.12)$$

What is H when p_1 is either 0 or 1? We need L'Hôpital's rule in the form

$$\lim_{x\to 0}\frac{u(x)}{v(x)} = \lim_{x\to 0}\frac{u'(x)}{v'(x)}.$$

Here

$$\lim_{x\to 0}(x \ln x) = \lim_{x\to 0}\left(\frac{\ln x}{\frac{1}{x}}\right) = \lim_{x\to 0}\left(\frac{\frac{1}{x}}{-\frac{1}{x^2}}\right) = 0.$$

Thus the uncertainty is zero when either p_1 or p_2 is zero and the other is consequently unity. Also, if we differentiate Equation (20.12) with respect to p_1 and set the derivative equal to zero, we find that H/K has a maximum at $p_1 = 1/2$ $(=p_2)$. These results are illustrated in Figure 20.2.

These inferences can easily be extended to the case of n possible outcomes. In particular, we can use the method of Lagrange multipliers to maximize H subject to the condition $\sum_i p_i = 1$ (Problem 20.2). The result is $H_{max} = K \ln n$.

20.3 UNIT OF INFORMATION

We need to choose a value for K so that H has some convenient unit. Consider the simplest possible experiment with equally likely outcomes—heads or tails, or the binary digits 0 or 1. Then

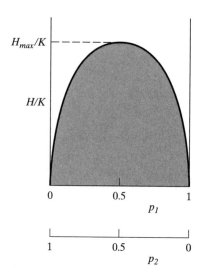

Figure 20.2 Uncertainty
measure for the case of two
possibilities with probabilities
p_1 and p_2.

$$\frac{H}{K} = -\frac{1}{2}\ln\frac{1}{2} - \frac{1}{2}\ln\frac{1}{2} = \ln 2.$$

If we use 2 as the base of the logarithm and take $K = 1$, we obtain

$$H = \log_2 2 = 1.$$

We call the unit of information a *bit* for *B*inary dig *IT*.*

Consider the following examples:

1. *Decimal digit*
Here

$$H = \sum_{i=1}^{10}\frac{1}{10}\log_2(10) = \log_2(10) = 3.32\log_{10}(10) = 3.32.$$

Thus a decimal digit contains about 3 1/3 bits of information.

2. *Foregone conclusion*
The message "the sun will rise between midnight and noon tomorrow"
has a probability of one so that $H = 0$ bits.

3. *Coin flipping*
Flip a coin three times. The information will be three times that associated with one flip of the coin. Thus $H = 3\log_2 2 = 3$ bits. This can be

---

*The "bit" was coined by the statistician John Tukey during a lunchtime discussion at
AT&T Bell Laboratories.

TABLE 20.1
Known probabilities
of occurrence for
symbols in the
English alphabet.

Symbol	p_i
blank	0.200
E	0.105
T	0.072
O	0.065
A	0.063
⋮	⋮
J, Q, Z	0.001

written $H = \log_2 2^3$. By extension, flipping a coin N times yields N bits of information.

4. *English alphabet*
   Suppose that a message comprises words using the 26 letters of the alphabet and a blank, totaling 27 characters. If the probabilities were equal, then $H = \log_2 27 = 4.76$ bits per character. If, on the other hand, we use known probabilities of occurrence, H is reduced to 4.03 bits per character (Table 20.1). Because of the redundancy built into the structure of the language, the actual information content is in the range of 1 to 2 bits per character. The calculation involves the use of conditional probabilities.

20.4 MAXIMUM ENTROPY

We have seen that the amount of uncertainty represented by a discrete probability distribution is

$$H(p_1 \ldots p_n) = -K \sum_{i=1}^{n} p_i \ln p_i, \qquad (20.13)$$

where $K > 0$. In Section 20.5 it will be shown that, apart from the constant, this is just the expression for the entropy in statistical thermodynamics. Therefore, we can consider the terms "entropy" and "uncertainty" to be synonymous. Indeed, entropy is a term used in information theory.

The connection between entropy and uncertainty is illustrated by the following example. Imagine that an insulated container is divided into two chambers by a partition. Initially there is gas on one side of the partition and none on the other. Suppose the partition is suddenly removed so that the gas expands to fill the whole container. We know from our study of the free expansion of a gas that the internal energy is unchanged but the entropy increases.

When the molecules of the gas are all in one of the two chambers, the entropy is less than when they are distributed throughout the container. Our knowledge of the possible positions of the molecules is greater when they are all on one side of the partition than when they are somewhere in the entire container. The more detailed our knowledge is concerning a physical system, the less uncertainty we have about it, and the less the entropy is. Conversely, more uncertainty means greater entropy.

To make inferences based on only partial information, it is necessary to determine the probability distribution that has maximum entropy subject to what is already known. Suppose that we know the mean value of some particular variable x that can assume the discrete values $x_i, i = 1, 2, \ldots, n$. The mean value is

$$\bar{x} = \sum_{i=1}^{n} p_i x_i, \tag{20.14}$$

where the unknown probabilities satisfy the condition

$$\sum_{i=1}^{n} p_i = 1. \tag{20.15}$$

In general, there will be a large number of probability distributions $\{p_1, p_2, \ldots p_n\}$ consistent with the information given. Our problem is to determine that distribution which maximizes the uncertainty (i.e., the entropy). We introduce the Lagrange multipliers α and β and follow the procedure we used in Chapter 13. We write

$$\frac{\partial H}{\partial p_j} + \alpha \frac{\partial \phi}{\partial p_j} + \beta \frac{\partial \psi}{\partial p_j} = 0, \tag{20.16}$$

where

$$\phi = \sum_i p_i = 1 \tag{20.17}$$

and

$$\psi = \sum_i p_i x_i = \bar{x}. \tag{20.18}$$

Then

$$-K\frac{\partial}{\partial p_j}\left(\sum_i p_i \ln p_i\right) + \alpha\frac{\partial}{\partial p_j}\left(\sum_i p_i\right) + \beta\frac{\partial}{\partial p_j}\left(\sum_i p_i x_i\right) = 0,$$

or

$$-K \ln p_j - K\frac{p_j}{p_j} + \alpha + \beta x_j = 0.$$

Solving for $\ln p_j$ and simplifying the constants, we obtain

$$\ln p_j = -\lambda - \mu x_j,$$

or

$$p_j = e^{-\lambda - \mu x_j}.$$

We can determine the new Lagrange multipliers λ and μ from the constraints. Equation (20.17) gives

$$e^{-\lambda} = \frac{1}{\sum_j e^{-\mu x_j}},$$

so that

$$p_j = \frac{e^{-\mu x_j}}{\sum_j e^{-\mu x_j}}. \qquad (20.19)$$

We define the partition function

$$Z \equiv \sum_j e^{-\mu x_j}. \qquad (20.20)$$

Then

$$p_j = \frac{e^{-\mu x_j}}{Z}, \qquad (20.21)$$

and

$$\lambda = \ln Z. \qquad (20.22)$$

Invoking the remaining constraint, Equation (20.18), we can derive an expression from which we can determine μ. Noting that

$$\frac{\partial Z}{\partial \mu} = -\sum_j x_j e^{-\mu x_j},$$

we have

$$\bar{x} = -\frac{1}{Z}\frac{\partial Z}{\partial \mu} = -\frac{\partial}{\partial \mu}\ln Z. \tag{20.24}$$

Equation (20.21) gives the probabilities that maximize the uncertainty. Thus

$$H_{\max} = -K\sum_j p_j \ln p_j$$

$$= \frac{K\mu}{Z}\sum_j x_j e^{-\mu x_j} + K\ln Z$$

$$= K(\mu\bar{x} + \ln Z). \tag{20.25}$$

This result can easily be generalized to include additional information associated with the knowledge of the mean value of other variables.

20.5 THE CONNECTION TO STATISTICAL THERMODYNAMICS

The randomness or disorder of a system implies an uncertainty regarding its state. Boltzmann described the entropy as a measure of the "missing" information.

The similarity between the development of the last section and the previous chapters is striking. Consider a system of N distinguishable particles obeying Boltzmann statistics. Assume that the quantum states are nondegenerate. Then the thermodynamic probability is

$$w = \frac{N!}{\displaystyle\prod_{j=1}^{n} N_j!}, \tag{20.26}$$

and the entropy is

$$S/k = \ln w = \ln N! - \sum_j \ln N_j!$$

$$\approx N \ln N - N - \sum_j N_j \ln N_j + \sum_j N_j$$

$$= \ln N \sum_j N_j - \sum_j N_j \ln N_j$$

$$= \sum_j N_j [\ln N - \ln N_j]$$

$$= -\sum_j N_j \ln\left(\frac{N_j}{N}\right).$$

Then

$$S = -kN \sum_{j=1}^{n} \left(\frac{N_j}{N}\right) \ln \left(\frac{N_j}{N}\right). \tag{20.27}$$

Imagine that the system is in perfect order, meaning that *all* the particles are in the lowest quantum state. Then

$$\frac{N_1}{N} = 1 \text{ and } \frac{N_2}{N} = \frac{N_3}{N} = \ldots = \frac{N_n}{N} = 0,$$

and the sum in Equation (20.27) reduces to one term equal to zero. Hence the entropy is zero for the perfectly ordered state.

A disordered system would be likely to be in any number of different quantum states; the larger the number of states available to it, the greater the disorder. If $N_j = 1$ for N different states and $N_j = 0$ for all other available states, then

$$S = -kN \sum_{j=1}^{N} \left(\frac{1}{N}\right) \ln \left(\frac{1}{N}\right) = kN \ln N. \tag{20.28}$$

This function is positive and increases with increasing N.

If we associate N_j/N with the probability p_j that N_j particles are in the *j*th quantum state, it follows from Equation (20.27) that

$$S = -kN \sum_j p_j \ln p_j = \left(\frac{kN}{K}\right) H. \tag{20.29}$$

Apart from a constant, the entropy of statistical thermodynamics and the entropy of information theory are the same. If we attempt to determine fully the macrostate of a system but are given only partial information, then the expected amount of information we would gain would be a measure of our lack of knowledge of the state of the system. Because the information is missing, it is sometimes called *negative* entropy or *negentropy* in the literature. However, the constant of proportionality between S and H in Equation (20.29) is positive.

The Boltzmann distribution for nondegenerate energy states is

$$\frac{N_j}{N} = \frac{e^{-\varepsilon_j/kT}}{Z},$$

where

$$Z = \sum_j e^{-\varepsilon_j/kT}.$$

If we substitute the distribution in Equation (20.27), we obtain

$$S_{\max} = \frac{N}{T}\left(\frac{1}{Z}\right)\sum_j \varepsilon_j e^{-\varepsilon_j/kT} + Nk \ln Z.$$

The internal energy is

$$U = \sum_j N_j \varepsilon_j = N\sum_j \left(\frac{N_j}{N}\right)\varepsilon_j = \frac{N}{Z}\sum_j \varepsilon_j e^{-\varepsilon_j/kT}.$$

Thus

$$S_{\max} = \frac{U}{T} + Nk \ln Z.$$

This expression has exactly the same form as Equation (20.25). Hence the internal energy plays the role of a variable whose mean value is known. "Partial information" in this case is knowledge of the energy of the system.

20.6 INFORMATION THEORY AND THE LAWS OF THERMODYNAMICS

We have seen that the thermodynamic concept of entropy as defined by Clausius is physically related to the information theory concept of entropy as defined by Shannon. This characterization is clearly more than a mere analogy.

Accordingly, some investigators have attempted to show that the fundamental equations of equilibrium thermodynamics can be derived from information theory. In classical thermodynamics heat, temperature, and work are taken as primary concepts, while energy and entropy are derivative ideas. The information theory treatment inverts the procedure. Heat is introduced without the need to define it in terms of reservoirs, adiabatic walls, etc. Information theory is free from all artificial constructs such as heat baths or Carnot engines. The zeroth, first, second, and third laws become consequences rather than premises. The third law, for example, is true by definition since, for a perfectly ordered state at absolute zero, there is no missing information ($S = 0$).

Information theory is thus an alternative way of presenting the postulates of thermodynamics. Its elucidation as such is beyond the scope of this text. Suffice it to say that its far-reaching consequences unify our understanding of thermal phenomena.

20.7 MAXWELL'S DEMON EXORCISED

How seriously Maxwell took his demon is hard to say. In any case, he neither conducted nor promoted experiments to test his hypothesis. What is certain is that the paradox posed by his demon bothered generations of physicists.

In the early 1900s it was noted that Brownian agitation of the trap door (recall that it is massless and frictionless) would result in a random opening and closing of the door, thereby rendering ineffective its long-term operation. After the development of quantum mechanics in the 1920s, it was suggested that the uncertainty principle might play a role in the problem; later it was shown that this would not be the case for heavy molecules at reasonably low pressures.

In his seminal 1929 paper on Maxwell's demon, Szilard observed that an amount of energy at least as great as whatever amount of energy can be gained, must necessarily have been expended by the demon in his detection, inspection, and routing activities. Demers and Brillouin (1944–51) called attention to the fact that in an isolated enclosure it would be impossible for the demon to *see* the individual molecules. To make the molecules visible against the background blackbody radiation, the demon would have to use a flashlight. The entropy produced in the irreversible operation of the flashlight would always exceed the entropy destroyed by the demon's sorting procedure. Thus a real demon could not produce a violation of the second law of thermodynamics.

Richard Feynman, looking at the demon as a mechanical system, wrote "If we assume that the specific heat of the demon is not infinite, it must heat up. . . . Soon it is shaking from Brownian motion so much that it can't tell whether it is coming or going, much less whether the molecules are coming or going, so it does not work."

Information theory provides a new framework in which to view the paradox. A knowledge of the momentum and position of the molecules constitutes the information needed to achieve a given entropy decrease. But this is counterbalanced by the entropy increase necessary to acquire this information. Lehringer observes that about 10^{24} bits of information are required to reduce the entropy of 1 kilomole of gas by 1 calorie/K.

Recently, Landauer[*] and Bennett[†] have interpreted Maxwell's demon as a computing automaton, subject to limitations imposed by the rules of data processing and computational logic. Their view is that the *gathering* of information by the demon does not necessarily increase the entropy. But after the demon has used the information to allow the passage of fast molecules, he must forget it if he is to continue doing his job. It can be shown that *erasing* the information raises the entropy more than any previous lowering that arises from its use. The association of computation with thermodynamics suggests that computational procedures might fruitfully be expressed in terms of the behavior of some thermodynamic system.

In *The Nature of Thermodynamics*, Bridgman states that "If the Maxwell demon had been invented yesterday instead of in the last century I believe he would not have caused as much consternation."[‡] Perhaps not, but the language of statistical thermodynamics is beginning to permeate fields as diverse as computer science, complexity theory, biology, and animal behavior. Maxwell's demon may be exorcised, but the ideas that conjured him up live on.

PROBLEMS

20-1 Referring to the compound experiment of Section 20.2, show explicitly that

$$H\left(\frac{1}{2},\frac{1}{3},\frac{1}{6}\right) = H\left(\frac{1}{2},\frac{1}{2}\right) + \frac{1}{2}H\left(\frac{2}{3},\frac{1}{3}\right).$$

20-2 Use the method of Lagrange multipliers to maximize H subject to the condition $\sum_{i=1}^{n} p_i = 1$. Show that $H_{\max} = K \ln n$.

20-3 Make a table of the logarithms to the base 2 of the numbers from 1 to 10. In adjoining columns list the logarithms to the base e and to the base 10.

*R. Landauer, "Computation: A Fundamental Physical View," *Maxwell's Demon: Entropy, Information, Computing,* edited by H.S. Leff and A.F. Rex, Princeton University Press, Princeton, New Jersey, 1990, pp. 260 *ff.*

†C.H. Bennett, "The Thermodynamics of Computation—A Review," *Ibid.*, pp. 213 *ff.*

‡P. W. Bridgman, *The Nature of Thermodynamics,* Peter Smith, Gloucester, Mass. 1969.

Note that

$$\log_a x = \log_a b \cdot \log_b x$$

20-4 Consider a loaded die. Let the probabilities of throwing the numbers $1, 2, 3, 4, 5$, and 6 be $0.1, 0.1, 0.1, 0.1, 0.1$, and 0.5, respectively. Calculate the uncertainty H in bits for a throw of the die. What would H be if the die were "honest"?

20-5 A perfect shuffle of a deck of 52 playing cards is one in which all of the possible orderings of the cards are equally probable.
(a) How much information in bits is there in a perfect shuffle?
(b) How many times do you have to cut and interleave a deck in order to achieve something approximating a perfect shuffle?

20-6 Suppose that a measurement of temperature gives $T = 246.3$ K. Assume the "alphabet" of possible values as being multidimensional: the hundreds of degrees being one dimension, the tens value a second dimension, etc. Associated with each dimension is a probability. Reasonable values might be

$$p_{100}(2) = 1; \quad p_{10}(4) = 0.2; \quad p_1(6) = 0.1; \quad p_{0.1}(3) = 0.1.$$

With these values, calculate the amount of information in the measurement.

20-7 Consider a substitution cryptogram in which, for each letter of the alphabet, some other letter is substituted. The number of possible keys is $26!$. The cryptogram can be viewed as a compound experiment consisting of two parts: x, the communication of the clear text, and y, the choice of a key from one of the $26!$ possibilities. Thus the total information associated with the compound experiment is $H(x) + \log_2 26!$ bits. A substitution cryptogram of 40 letters can usually be solved. What does this imply about the maximum information in bits per letter of an English message?

20-8 Imagine that system 1 has probability $p_j^{(1)}$ of being forced into a state j and system 2 has probability $p_k^{(2)}$ of being forced into state k. Then

$$H^{(1)} = -K \sum_j p_j^{(1)} \ln p_j^{(1)}, \quad H^{(2)} = -K \sum_k p_k^{(2)} \ln p_k^{(2)}.$$

Each state of the composite system consisting of systems 1 and 2 can then be labeled by the pair of numbers j, k. Let the probability that the composite system is found in this state be p_{jk}. Then

$$H = -K \sum_j \sum_k p_{jk} \ln p_{jk}.$$

If the two systems are only weakly interacting, so that they are statistically independent, then $p_{jk} = p_j^{(1)} p_k^{(2)}$. In this case show that

$$H = H^{(1)} + H^{(2)}.$$

20-9 P.T. Landsberg (*Thermodynamics and Statistical Mechanics*, Dover, New York, 1990) draws a distinction between entropy and disorder. For a system of n distinguishable states, he defines the disorder $D(n)$ as

$$D(n) = \frac{S(n)}{c \ln n}.$$

Here $S(n)$ is the entropy and c is a constant. The rate of change of disorder can be found by differentiating this equation with respect to the time. Show that

$$\dot{D}(n) = \left[\frac{\dot{S}(n)}{S(n)} - \frac{\dot{n}}{n \ln n} \right] D(n).$$

Here the dots denote time derivatives. Since the second term in the brackets can conceivably be greater than the first term, the result suggests that it is possible to have decreasing disorder even though the entropy increases. The expression has a possible application to the growth of biological systems.

Appendix A

Review of Partial Differentiation

A.1 PARTIAL DERIVATIVES

Consider a function of three variables

$$f(x, y, z) = 0. \tag{A.1}$$

Since only two variables are independent, we can write $x = x(y, z)$ and $y = y(x, z)$. Then

$$dx = \left(\frac{\partial x}{\partial y}\right)_z dy + \left(\frac{\partial x}{\partial z}\right)_y dz, \tag{A.2}$$

and

$$dy = \left(\frac{\partial y}{\partial x}\right)_z dx + \left(\frac{\partial y}{\partial z}\right)_x dz. \tag{A.3}$$

The subscripts denote the variables that are held constant in the differentiation and are often omitted. Substituting Equation (A.3) in Equation (A.2), we obtain

$$dx = \left(\frac{\partial x}{\partial y}\right)_z \left(\frac{\partial y}{\partial x}\right)_z dx + \left[\left(\frac{\partial x}{\partial y}\right)_z \left(\frac{\partial y}{\partial z}\right)_x + \left(\frac{\partial x}{\partial z}\right)_y\right] dz. \qquad \text{(A.4)}$$

If we choose x and z as the independent variables, then Equation (A.4) holds for all values of dx and dz. Thus, if $dz = 0$ and $dx \neq 0$, we have

$$\left(\frac{\partial x}{\partial y}\right)_z = \frac{1}{\left(\dfrac{\partial y}{\partial x}\right)_z}. \qquad \text{(A.5)}$$

This expression is known as the *reciprocal relation*.

If $dx = 0$ and $dz \neq 0$, we get

$$\left(\frac{\partial x}{\partial y}\right)_z \left(\frac{\partial y}{\partial z}\right)_x = -\left(\frac{\partial x}{\partial z}\right)_y. \qquad \text{(A.6)}$$

The reciprocal relation gives

$$\left(\frac{\partial x}{\partial z}\right)_y = \frac{1}{\left(\dfrac{\partial z}{\partial x}\right)_y}.$$

Substituting this in Equation (A.6) yields

$$\left(\frac{\partial x}{\partial y}\right)_z \left(\frac{\partial y}{\partial z}\right)_x \left(\frac{\partial z}{\partial x}\right)_y = -1. \qquad \text{(A.7)}$$

Equation (A.7) is the *cyclical rule*, or *cyclical relation*.

A function u of three variables x, y, and z can be written as a function of only two variables when they are related through Equation (A.1), that is, when only two variables are independent. Thus

$$u = u(x, y). \qquad \text{(A.8)}$$

Alternatively, we can write

$$x = x(u, y). \qquad \text{(A.9)}$$

Then

$$dx = \left(\frac{\partial x}{\partial u}\right)_y du + \left(\frac{\partial x}{\partial y}\right)_u dy. \qquad \text{(A.10)}$$

If we divide Equation (A.10) by dz while holding u constant, we obtain

$$\left(\frac{\partial x}{\partial z}\right)_u = \left(\frac{\partial x}{\partial y}\right)_u \left(\frac{\partial y}{\partial z}\right)_u. \tag{A.11}$$

This is the *chain rule* of differentiation.

A.2 EXACT AND INEXACT DIFFERENTIALS

Consider the function $z = z(x)$. The fundamental theorem of integral calculus is

$$\int_a^b \frac{d}{dx} z(x) \, dx = \int_a^b dz = z(b) - z(a). \tag{A.12}$$

If the integration is taken around the closed path C (a to b and b to a), we have

$$\oint_C dz = \int_a^b dz + \int_b^a dz = \int_a^b dz - \int_a^b dz = 0. \tag{A.13}$$

This can be generalized to two independent variables. Let $z = z(x, y)$. Then

$$dz = M(x, y)dx + N(x, y)dy. \tag{A.14}$$

We wish to determine the condition under which the integral of dz around a closed path is zero in this case.

Consider the integral of $\partial N/\partial x$ over a rectangle A in the x-y plane. From Figure A.1, the integral is

$$\int\int_A \frac{\partial N}{\partial x} \, dxdy = \int_c^d \int_a^b \frac{\partial N}{\partial x} \, dxdy = \int_c^d [N(b, y) - N(a, y)]dy. \tag{A.15}$$

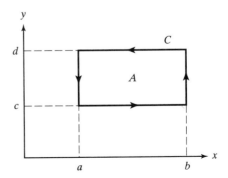

Figure A.1   Contour C enclosing the rectangular area A.

We now evaluate the integral $\oint_C N(x, y)dy$ around the path C in a counter-clockwise direction:

$$\oint_C N(x, y)dy = \int_c^d N(b, y)dy + \int_d^c N(a, y)dy$$

$$= \int_c^d [N(b, y) - N(a, y)]dy. \qquad (A.16)$$

The contribution along the horizontal segments of the path are zero since $dy = 0$ there. Combining Equations (A.15) and (A.16) gives

$$\int\int_A \frac{\partial N}{\partial x} dx dy = \oint_C N dy. \qquad (A.17)$$

Similarly, it can be shown that

$$-\int\int_A \frac{\partial M}{\partial y} dx dy = \oint_C M dx. \qquad (A.18)$$

Adding these two equations, we obtain the result

$$\oint_C (M dx + N dy) = \oint_C dz = \int\int_A \left(\frac{\partial N}{\partial x} - \frac{\partial M}{\partial y}\right) dx dy. \qquad (A.19)$$

It follows that the condition under which the integral of dz around a closed path is zero is

$$\frac{\partial M}{\partial y} = \frac{\partial N}{\partial x}. \qquad (A.20)$$

When this condition is satisfied, the differential dz is said to be *exact*.
   Note also that

$$dz = \left(\frac{\partial z}{\partial x}\right)dx + \left(\frac{\partial z}{\partial y}\right)dy, \qquad (A.21)$$

and

$$\frac{\partial}{\partial y}\left(\frac{\partial z}{\partial x}\right) = \frac{\partial^2 z}{\partial y \partial x} = \frac{\partial^2 z}{\partial x \partial y} = \frac{\partial}{\partial x}\left(\frac{\partial z}{\partial y}\right).$$

Identifying the partial derivatives in Equation (A.21) with M and N, we again obtain Equation (A.20).

If the differential dz is exact, we wish to be able to integrate it to obtain z. If $M = M(x)$ and $N = N(y)$, the problem is trivial:

$$z = \int M(x)dx + \int N(y)dy. \tag{A.22}$$

However, this will not be the case in general. We define

$$u \equiv \int_y M\,dx,$$

the partial integral of M with y fixed. That the function $N - \partial u/\partial y$ depends on y only can be seen as follows:

$$\frac{\partial}{\partial x}\left(N - \frac{\partial u}{\partial y}\right) = \frac{\partial N}{\partial x} - \frac{\partial}{\partial x}\left(\frac{\partial u}{\partial y}\right) = \frac{\partial N}{\partial x} - \frac{\partial}{\partial y}\left[\frac{\partial}{\partial x}\int_y M\,dx\right]$$

$$= \frac{\partial N}{\partial x} - \frac{\partial M}{\partial y} = 0.$$

The last step is a consequence of our assumption that dz is exact. As a trial solution, we take

$$z = u + \int\left(N - \frac{\partial u}{\partial y}\right)dy$$

$$= \int_y M\,dx + \int\left(N - \frac{\partial u}{\partial y}\right)dy$$

$$= \int_y M\,dx + g(y). \tag{A.23}$$

This gives us a recipe for finding z: carry out the integration indicated in Equation (A.23), take the differential dz including the term $g'(y)dy$, and compare the result with the original differential to find g'. A final integration gives g.

As an example, consider the exact differential

$$dz = (x + y)dx + (x + 2y)dy. \tag{A.24}$$

Then

$$z = \int (x + y)dx + g(y)$$
$$= \frac{x^2}{2} + xy + g(y).$$

So

$$dz = xdx + ydx + xdy + g'(y)dy$$
$$= (x + y)dx + [x + g'(y)]dy.$$

Comparing this with the original differential, we see that $g'(y) = 2y$, so that $g(y) = y^2 + C$. Therefore

$$z = \frac{x^2}{2} + xy + y^2 + C. \tag{A.25}$$

Alternatively, we could have written

$$z = \int_x Ndy + h(x), \tag{A.26}$$

which leads to the same result.

What if dz is inexact? Consider the example

$$dz = ydx - xdy,$$

which doesn't satisfy Equation (A.20), the criterion for exactness. To indicate that dz is not exact, we write

$$đz = ydx - xdy.$$

This cannot be integrated. However, we notice that if we multiply the equation by $1/y^2$, we obtain an exact differential:

$$\frac{đz}{y^2} = \frac{ydx - xdy}{y^2} = d\left(\frac{x}{y}\right).$$

The term $1/y^2$ is known as an *integrating factor*. It can be proved that it is always possible to find an integrating factor for an inexact differential that is a function of two independent variables, and hence to obtain a differential that is integrable. This fact is of great importance in thermodynamics.

An integrating factor is not unique; for a given inexact differential, many integrating factors can often be found. The following are all integrating factors for the differential $ydx - xdy$:

$$\frac{1}{y^2}, \frac{1}{x^2}, \frac{1}{xy}, \frac{1}{x^2 + y^2}, \frac{1}{x^2 - y^2}, \frac{1}{(x + y)^2}, \frac{1}{(x - y)^2}.$$

A formal method exists for finding an integrating factor μ. Let

$$\mu dz \equiv dw = \mu M dx + \mu N dy. \tag{A.27}$$

If dw is exact, then

$$\frac{\partial}{\partial y}(\mu M) = \frac{\partial}{\partial x}(\mu N),$$

or

$$\frac{\partial \mu}{\partial y}M + \mu\frac{\partial M}{\partial y} = \frac{\partial \mu}{\partial x}N + \mu\frac{\partial N}{\partial x}.$$

In general, this cannot be solved for μ. However, suppose that μ is a function of x only. Then $\partial\mu/\partial y = 0$, and

$$\mu\frac{\partial M}{\partial y} = \frac{d\mu}{dx}N + \mu\frac{\partial N}{\partial x},$$

or

$$\frac{d\mu}{\mu} = \frac{1}{N}\left(\frac{\partial M}{\partial y} - \frac{\partial N}{\partial x}\right)dx.$$

Integrating, and setting the constant of integration equal to zero, we obtain

$$\mu = \exp\left[\int\frac{1}{N}\left(\frac{\partial M}{\partial y} - \frac{\partial N}{\partial x}\right)dx\right]. \tag{A.28}$$

Note that if the differential is exact to begin with, the integral is zero and $\mu = 1$.

In practice, the formal integration of Equation (A.28) is unnecessary, as we can see from the following example. Let

$$dz = \sin y dx + \cos y dy.$$

Then

$$\mu \, dz = \mu \sin y \, dx + \mu \cos y \, dy \equiv dw.$$

For exactness,

$$\frac{\partial}{\partial y}(\mu \sin y) = \frac{\partial}{\partial x}(\mu \cos y),$$

or

$$\frac{\partial \mu}{\partial y} \sin y + \mu \cos y = \frac{\partial \mu}{\partial x} \cos y.$$

If $\mu = \mu(x)$, $\partial \mu / \partial y = 0$, and $d\mu / \mu = dx$, or $\mu = e^x$. Thus

$$dw = e^x \sin y \, dx + e^x \cos y \, dy.$$

which is exact. If we choose to select $\mu = \mu(y)$, then $\mu = 1/\sin y$ and

$$dw = dx + \cot y \, dy.$$

Thus we have obtained two integrating factors; usually one is easier to calculate than the other.

From this analysis it is evident that the differential $dz = M dx + N dy$ is exact if the equivalent statements hold:

1. $\dfrac{\partial M}{\partial y} = \dfrac{\partial N}{\partial x}$

2. $\displaystyle\oint_C dz = 0$ (A.29)

3. $\displaystyle\int_a^b dz$ is independent of the path.

This means that if a problem can be formulated in terms of exact differentials, or differentials made exact through the use of an integrating factor, then the path of integration between the end points is arbitrary, and we are free to select one that leads to a simple evaluation.

In thermodynamics all state variables are, by definition, exact differentials whose integrals are path-independent. However, there exist certain thermodynamic quantities such as the differentials of *work* and *heat* whose integrals depend on the path of integration. Thus the search for appropriate integrating factors that allow us to express fundamental relationships in terms of variables of state is an important goal of the mathematical theory.

PROBLEMS

A-1 Test the following differentials for exactness. For the cases in which the differential is exact find the function $z(x, y)$.

(a) $dz = 2x \ln y \, dx + (x^2/y) dy$.

(b) $dz = (y - 1)dx + (x - 3)dy$.
(c) $dz = (2y^3 - 3x)dx - 4xy \, dy$.

A-2 (a) Show that $dz = y \, dx + (x + 2y)dy$ is exact and integrate it to find $z(x, y)$.
(b) Show that $\bar{d}z = x \, dx + (x + 2y)dy$ is inexact.
(c) Integrate the two previous differentials counterclockwise around the triangle whose vertices are the three points $(0,0), (1,1)$, and $(0,1)$.

A-3 Consider the differential $dz = (2xy - 3)dx + x^2 dy$.
(a) Show that dz is an exact differential.
(b) Find $z = z(x, y)$.

(c) Referring to Figure A.2, evaluate the line integral $\int dz$ from $(1,0)$ to $(0,1)$ along

    Path A: lines parallel to the axes;
    Path B: circular arc;
    Path C: straight line diagonal.

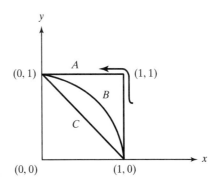

Figure A.2 Paths of integration for Problem A-3.

A-4 In the following cases show that $đz$ is not an exact differential. Find an integrating factor such that $dw \equiv \mu đz$ is an exact differential and check that this is so.
   (a) $đz = 2y \, dx + 3x \, dy.$
   (b) $đz = y \, dx + (2x - y^2) dy.$
   (c) $đz = (x^4 + y^4) dx - xy^3 dy.$

A-5 Find an integrating factor μ and integrate $dw = \mu đz$, where

$$đz = (y \cos^3 x - 1) dx + \sin x \, \cos^2 x \, dy.$$

Appendix B

Stirling's Approximation

The definition of $n!$, the factorial of a positive integer n, is

$$n! = n(n-1)(n-2)\cdots 1. \tag{B.1}$$

From this and the fact that

$$(n+1)! = (n+1)n!, \tag{B.2}$$

it follows that $1! = 0! = 1$.

We are interested in the natural logarithm of $n!$ when n is a very large number. From Equation (B.1) we see that

$$\ln n! = \underbrace{\ln 1}_{0} + \ln 2 + \ln 3 + \cdots + \ln n = \sum_{k=1}^{n} \ln k. \tag{B.3}$$

Equation (B.3) is clearly the area under the step curve shown by the dashed lines between $n = 1$ and $n = n$ in Figure B.1. The rectangles all have unit width; the height of the first is $\ln 2$, that of the second is $\ln 3$, etc. This area is approximately equal to the area under the smooth curve $y = \ln n$ between the limits 1 and n if n is large. For small values of n the area under the step curve is appreciably different from and exceeds that of the smooth curve. As n

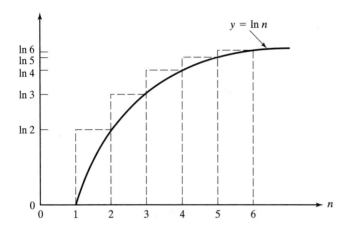

Figure B.1   Graph of ln n versus n.

increases, the smooth curve flattens out and the difference becomes negligibly small. Thus we can approximate Equation (B.3) by

$$\ln n! \approx \int_1^n \ln n \, dn = (n \ln n - n)\Big|_1^n = n \ln n - n + 1.$$

Since n is assumed large, we can neglect unity and get

$$\ln n! \approx n \ln n - n. \tag{B.4}$$

This is Stirling's approximation, valid for large n. For $n = 10$, Equation (B.4) gives a value low by 13.8 percent and for $n = 30$, a value low by 3.5 percent.

As an example, we can use Stirling's formula to estimate 52!, the number of possible rearrangements of cards in a standard deck of playing cards. Here

$$\ln 52! \approx 52 \ln 52 - 52 = 153.$$

Then

$$\log_{10} 52! = (0.4343)\ln 52! = (0.4343)(153) = 66,$$

so

$$52! \approx 10^{66}.$$

$$n! = 10^{\left(n \ln(n) - n \log(n)\right)}$$

Appendix C

Alternative Approach To Finding the Boltzmann Distribution

The expression defining Boltzmann statistics is

$$w = N! \prod_{j=1}^{n} \frac{g_j^{N_j}}{N_j!}. \tag{C.1}$$

We wish to show that

$$\frac{N_j}{g_j} \propto e^{\beta \epsilon_j}. \tag{C.2}$$

To simplify the argument we assume no degeneracy for any energy level, so that $g_j = 1$ for all j. Then Equations (C.1) and (C.2) become

$$w = \frac{N!}{N_1! \cdots N_j! \cdots N_n!} \tag{C.3}$$

and

$$N_j \propto e^{\beta \epsilon_j}. \tag{C.4}$$

We also assume, initially at least, that the energy levels are equally spaced. We wish to make a *minimal* change in the given configuration while holding N and U constant.* Thus we transfer one particle from level j to level k (gaining energy) and one particle from level l to level k (losing the same amount of energy) (Figure C.1). Then the new (primed) configuration is:

$$N_j' = N_j - 1, \quad N_k' = N_k + 2, \quad N_l' = N_l - 1.$$

Note that we still have $N = \sum N_j$ and $U = \sum N_j \varepsilon_j$.
The new value of w is then

$$w' = \frac{N!}{N_1! \cdots (N_j - 1)!(N_k + 2)!(N_l - 1)! \cdots N_n!}. \tag{C.5}$$

If w, as given by Equation (C.1), is the predominant (equilibrium) configuration for which ln w reaches a maximum value, then the relatively infinitesimal change in configuration described previously produces essentially no change in w. Thus we write $w = w'$. Equating Equations (C.3) and (C.5), we obtain

$$\frac{N!}{N_1! \cdots N_j! N_k! N_l! \cdots N_n!} = \frac{N!}{N_1! \cdots (N_j - 1)!(N_k + 2)!(N_l - l)! \cdots N_n!},$$

so

$$N_j! N_k! N_l! = (N_j - 1)!(N_k + 2)!(N_l - 1)!,$$

or

$$\frac{N_j!}{(N_j - 1)!} \frac{N_l!}{(N_l - 1)!} = \frac{(N_k + 2)!}{N_k!},$$

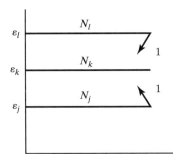

Figure C.1 Adjacent energy levels.

*Here $N = \sum N_j$ is the total number of particles and $U = \sum N_j \varepsilon_j$ is the total energy of the system.

or

$$N_j N_l = (N_k + 2)(N_k + 1).$$

But $N_k \gg 2$ or 1, so

$$N_j N_l = N_k^2$$

to a *very* good approximation. This can be written

$$\frac{N_j}{N_k} = \frac{N_k}{N_l}. \tag{C.6}$$

If we had chosen i, j, k instead of j, k, l, we would have obtained

$$\frac{N_i}{N_j} = \frac{N_j}{N_k},$$

so we evidently have a continuing geometric series

$$\cdots = \frac{N_i}{N_j} = \frac{N_j}{N_k} = \frac{N_k}{N_l} \cdots. \tag{C.7}$$

Now, if the levels are *not* equally spaced, but are given by

$$\frac{(\varepsilon_l - \varepsilon_k)}{(\varepsilon_k - \varepsilon_j)} = \frac{p}{q},$$

then it can easily be shown that Equation (C.6) is replaced by

$$\left(\frac{N_j}{N_k}\right)^p = \left(\frac{N_k}{N_l}\right)^q,$$

so that

$$\frac{p}{q}\ln\left(\frac{N_j}{N_k}\right) = \ln\left(\frac{N_k}{N_l}\right),$$

or

$$\frac{1}{(\varepsilon_k - \varepsilon_j)}\ln\left(\frac{N_j}{N_k}\right) = \frac{1}{(\varepsilon_l - \varepsilon_k)}\ln\left(\frac{N_k}{N_l}\right).$$

Again, we would have found equalities like this if we had used i, j, k instead of j, k, l, etc. Thus the function must be a constant—call it $-\beta$. Finally, we have

$$\frac{1}{(\varepsilon_k - \varepsilon_j)} \ln\left(\frac{N_j}{N_k}\right) = -\beta,$$

or

$$\frac{N_j}{N_k} = e^{-\beta(\varepsilon_k - \varepsilon_j)}$$

or

$$N_j \propto e^{\beta \varepsilon_j}, \tag{C.8}$$

which is the result we have been seeking. The Boltzmann distribution and the Maxwell-Boltzmann distribution are identical.

Appendix D

Various Integrals

$$\int_{-\infty}^{+\infty} e^{-ax^2}dx = 2\int_0^{\infty} e^{-ax^2}dx = \left(\frac{\pi}{a}\right)^{1/2}.$$ (D.1)

$$\int_{-\infty}^{+\infty} x^{2n}e^{-ax^2}dx = 2\int_0^{\infty} x^{2n}e^{-ax^2}dx = \frac{1\cdot3\cdot5\cdots(2n-1)}{2^n a^n}\left(\frac{\pi}{a}\right)^{1/2}.$$ (D.2)

$$\int_{-\infty}^{+\infty} x^{2n+1}e^{-ax^2}dx = 0.$$ (D.3)

$$\int_0^{\infty} x^n e^{-ax}dx = \frac{n!}{a^{n+1}}.$$ (D.4)

(For Equations (D.2), (D.3), and (D.4), $n = 0, 1, 2, 3 \ldots$)

$$\int_0^{\infty} \frac{x^{s-1}dx}{e^x - 1} = \Gamma(s)\zeta(s), \quad s > 0.$$ (D.5)

Here $\Gamma(n)$ is the gamma function: $\Gamma(n) \equiv \int_0^\infty x^{n-1} e^{-x} dx$.

$$\Gamma(n+1) = n\Gamma(n) = n!, \qquad \text{where } n \text{ is a positive integer.}$$
$$\Gamma(1) = \Gamma(2) = 1.$$
$$\Gamma\left(\frac{1}{2}\right) = \sqrt{\pi}.$$
$$\Gamma\left(\frac{3}{2}\right) = \sqrt{\pi}/2.$$

$\zeta(a)$ is the zeta function: $\zeta(a) \equiv \sum_{n=1}^{\infty} n^{-a}$.

$\zeta(1) = \infty.$	$\zeta(3/2) = 2.612.$
$\zeta(2) = \pi^2/6 = 1.645.$	$\zeta(5/2) = 1.341.$
$\zeta(3) = 1.202.$	$\zeta(7/2) = 1.127.$
$\zeta(4) = \pi^4/90 = 1.082.$	$\zeta(9/2) = 1.055.$

$$\int_0^\infty \frac{x^3 dx}{e^x - 1} = \frac{\pi^4}{15}. \tag{D.6}$$

$$\int_0^\infty \frac{x^4 e^x dx}{(e^x - 1)^2} = \frac{4\pi^4}{15}. \tag{D.7}$$

$$\int_0^\infty \frac{x^2 e^x dx}{(e^x + 1)^2} = \frac{\pi^2}{6}. \tag{D.8}$$

$$\int_{-\infty}^\infty \frac{e^x dx}{(e^x + 1)^2} = 1. \tag{D.9}$$

$$\int_{-\infty}^\infty \frac{x^2 e^x dx}{(e^x + 1)^2} = \frac{\pi^2}{3}. \tag{D.10}$$

Note that
$$\int_{-\infty}^{\infty} \frac{x^{2n}e^x dx}{(e^x + 1)^2} = 2 \int_0^{\infty} \frac{x^{2n}e^x dx}{(e^x + 1)^2}, \quad n \text{ an integer,}$$

since
$$\frac{x^{2n}e^x}{(e^x + 1)^2} = \frac{x^{2n}}{(e^x + 1)(e^{-x} + 1)},$$

and therefore the integrand is seen to be an even function of x.

In evaluating integrals associated with the Maxwell-Boltzmann speed distribution, consider the integral

$$I \equiv \int_0^{\infty} e^{-ax^2} dx, \quad a > 0. \tag{D.11}$$

The equation could equally well be written

$$I = \int_0^{\infty} e^{-ay^2} dy,$$

so that

$$I^2 = \int_0^{\infty} e^{-ax^2} dx \int_0^{\infty} e^{-ay^2} dy$$

$$= \int_0^{\infty} \int_0^{\infty} e^{-a(x^2+y^2)} dx dy.$$

Changing the integration variables to polar coordinates, we have

$$r^2 = x^2 + y^2,$$
$$dx dy = r dr d\theta.$$

The region of integration is the first quadrant. Thus

$$I^2 = \int_0^{\pi/2} \int_0^{\infty} e^{-ar^2} r dr d\theta$$

$$= \frac{\pi}{2} \int_0^{\infty} e^{-ar^2} r dr$$

$$= \frac{\pi}{2} \int_0^{\infty} \left(-\frac{1}{2a} \right) d(e^{-ar^2})$$

$$= -\frac{\pi}{4a}e^{-ar^2}\bigg|_0^\infty$$

$$= \frac{\pi}{4a}.$$

Taking the square root, we obtain

$$I = \int_0^\infty e^{-ax^2}dx = \frac{\sqrt{\pi}}{2}a^{-1/2}. \tag{D.12}$$

Integrals of the form

$$\int_0^\infty x^{2n}e^{-ax^2}dx, \quad n = 1, 2, 3, \ldots,$$

can be found by differentiating Equation (D.12) with respect to a:

$$-\frac{dI}{da} = \int_0^\infty x^2 e^{-ax^2}dx = \frac{\sqrt{\pi}}{4}a^{-3/2},$$

and

$$-\frac{d^2I}{da^2} = \int_0^\infty x^4 e^{-ax^2}dx = \frac{3\sqrt{\pi}}{8}a^{-5/2}.$$

Successive differentiation leads to the general formula Equation (D.2).

For odd powers of x in the algebraic term of the integrand, we can set $y = x^2$. For example,

$$J = \int_0^\infty xe^{-ax^2}dx$$

$$= \frac{1}{2}\int_0^\infty e^{-ax^2}d(x^2)$$

$$= \frac{1}{2}\int_0^\infty e^{-ay}dy$$

$$= \frac{1}{2a} e^{-ay} \Big|_0^\infty$$

$$= \frac{1}{2a}.$$

Then

$$-\frac{dJ}{da} = \int_0^\infty x^3 e^{-ax^2} dx = \frac{1}{2a^2}.$$

With successive differentiations, we can generate the general formula

$$\int_0^\infty x^{2n+1} e^{-ax^2} dx = \frac{n!}{2a^{n+1}}, \quad n = 0, 1, 2, \ldots. \tag{D.13}$$

Bibliography

CLASSICAL THERMODYNAMICS

C.J. Adkins, *Equilibrium Thermodynamics,* 3rd edition, Cambridge University Press, Cambridge, 1983.

W.P. Allis and M.A. Herlin, *Thermodynamics and Statistical Mechanics,* McGraw-Hill, New York, 1952.

M. Bailyn, *A Survey of Thermodynamics,* American Institute of Physics, New York, 1994.

R.P. Bauman, *Modern Thermodynamics with Statistical Mechanics,* Macmillan, New York, 1992.

P.W. Bridgman, *The Nature of Thermodynamics,* Harper, New York, 1961.

H.B. Callen, *Thermodynamics,* Wiley, New York, 1960.

G. Carrington, *Basic Thermodynamics,* Oxford University Press, Oxford, 1994.

E. Fermi, *Notes on Thermodynamics and Statistics,* University of Chicago Press, Chicago, 1966.

E. Fermi, *Thermodynamics,* Dover, New York, 1956.

C.B.P. Finn, *Thermal Physics,* 2nd edition, Chapman and Hall, London, 1993.

W. Greiner, L. Neise, H. Stöcker, *Thermodynamics and Statistical Mechanics,* Springer-Verlag, New York, 1995.

W. Kauzmann, *Thermodynamics and Statistics: With Applications to Gases,* Benjamin, New York, 1967.

C. Kittel and H. Kroemer, *Thermal Physics,* 2nd edition, Freeman, San Francisco, 1980.

R. Kubo, *Thermodynamics,* John Wiley and Sons, New York, 1960.

P.T. Landsberg, *Thermodynamics and Statistical Mechanics,* Dover, New York, 1990.

D.F. Lawden, *Principles of Thermodynamics,* Wiley, New York, 1987.

M.C. Martin, *Elements of Thermodynamics,* Prentice-Hall, Englewood Cliffs, New Jersey, 1986.

P.M. Morse, *Thermal Physics,* 2nd edition, Benjamin, New York, 1969.

A.P. Pippard, *The Elements of Classical Thermodynamics,* Cambridge University Press, Cambridge, 1987.

M. Planck, *Treatise on Thermodynamics,* 3rd edition, Dover, New York, 1945.

H. Reiss, *Methods of Thermodynamics,* Dover, New York, 1996.

B.N. Roy, *Principles of Modern Thermodynamics,* Institute of Physics, London, 1995.

F.W. Sears and G.L. Salinger, *Thermodynamics, Kinetic Theory, and Statistical Thermodynamics,* 3rd edition, Addison-Wesley, Reading, Massachusetts, 1975.

A. Sommerfeld, *Thermodynamics and Statistical Mechanics,* Academic Press, New York, 1964.

M. Sprackling, *Heat and Thermodynamics,* MacMillan, London, 1993.

M. Sprackling, *Thermal Physics,* American Institute of Physics, New York, 1991.

M.V. Sussman, *Elementary General Thermodynamics,* Addison-Wesley, Reading, Massachusetts, 1972.

H.C. Van Ness, *Understanding Thermodynamics,* Dover, New York, 1969.

M.W. Zemansky and R.H. Dittman, *Heat and Thermodynamics,* 7th edition, McGraw-Hill, New York, 1997.

KINETIC THEORY OF GASES

L. Boltzmann, *Lectures on Gas Theory,* Dover, New York, 1995.

C.H. Collie, *Kinetic Theory and Entropy,* Longman, London, 1982.

W. Kauzmann, *Kinetic Theory of Gases,* W.A. Benjamin, New York, 1966.

STATISTICAL MECHANICS

R. Bowley and M. Sanchez, *Introductory Statistical Mechanics,* Clarendon Press, Oxford, 1996.

R.H. Fowler and E.A. Guggenheim, *Statistical Thermodynamics,* Cambridge University Press, Cambridge, 1952.

C. Garrod, *Statistical Mechanics and Thermodynamics,* Oxford University Press, Oxford, 1995.

T. Grenault, *Statistical Physics,* 2nd edition, Chapman and Hall, London, 1995.

T.L. Hill, *An Introduction to Statistical Thermodynamics,* Dover, New York, 1986.

K. Huang, *Statistical Mechanics,* 2nd edition, Wiley, New York, 1987.

E.L. Knuth, *Introduction to Statistical Thermodynamics,* McGraw-Hill, New York, 1966.

R. Kubo, *Statistical Mechanics,* North-Holland, Amsterdam, 1965.

L.D. Landau and E.M. Lifschitz, *Statistical Physics,* Addison-Wesley, Reading, Massachusetts, 1958.

R.B. Lindsay, *Introduction to Physical Statistics,* Wiley, New York, 1941.

L.K. Nash, *Elements of Statistical Thermodynamics,* 2nd edition, Addison-Wesley, Reading, Massachusetts, 1972.

R.K. Pathria, *Statistical Mechanics,* Pergamon Press, Oxford, 1972.

M. Plischke and B. Bergerson, *Equilibrium Statistical Physics,* 2nd edition, World Scientific, Singapore, 1994.

F. Reif, *Fundamentals of Statistical and Thermal Physics,* McGraw-Hill, New York, 1965.

W.G.V. Rosser, *An Introduction to Statistical Physics,* 2nd edition, Horwood, Chichester, UK, 1985.

K. Stowe, *Introduction to Statistical Mechanics and Thermodynamics,* Wiley, New York, 1984.

R.C. Tolman, *Principles of Statistical Mechanics,* Oxford University Press, Oxford, 1938.

G.H. Wannier, *Statistical Physics,* Dover, New York, 1966.

SPECIAL TOPICS

P. Coveney and R. Highfield, *The Arrow of Time,* W.H. Allen, London, 1990.

J.S. Dugdale, *Entropy and Its Physical Meaning,* Taylor and Francis, London, 1996.

J.D. Fast, *Entropy,* McGraw-Hill, New York, 1962.

M. Goldstein and I.F. Goldstein, *The Refrigerator and the Universe,* Harvard University Press, Cambridge, Massachusetts, 1993.

D.L. Goodstein, *States of Matter,* Dover, New York, 1985.

H.S. Leff and A.F. Rex, editors, *Maxwell's Demon: Entropy, Information, Computing,* Princeton University Press, Princeton, New Jersey, 1990.

M. Mott-Smith, *The Concept of Heat and Its Workings Simply Explained,* Dover, New York, 1962.

J.R. Pierce, *An Introduction to Information Theory: Symbols, Signals and Noise,* 2nd edition, Dover, New York, 1980.

G. Raisbeck, *Information Theory,* M.I.T. Press, Cambridge, Massachusetts, 1964.

C.E. Shannon and W. Weaver, *The Mathematical Theory of Communication,* University of Illinois Press, Urbana, Illinois, 1963.

M.W. Zemansky, *Temperatures Very Low and Very High,* Dover, New York, 1964.

Answers to Selected Problems

Chapter 1

1-1   (a) closed;   (b) isolated;   (c) open.

1-2   (a) quasistatic, reversible, isobaric.
(b) quasistatic, irreversible, isothermal.
(c) irreversible, adiabatic (if fast enough so no heat is exchanged).
(d) irreversible, isochoric.
(e) reversible, isobaric, isothermal, isochoric.
(f) irreversible, adiabatic.

1-5   260°.

1-6   348.65 K; 75.50°C.

1-7   4.00 K.

1-8 $a = 0.5\,\text{mV}/°\text{C}; b = -10^{-3}\,\text{mV}/(°\text{C})^2; T = 430.3°\text{C}.$

1-10   574.

1-11 $-195.80°\text{C}; -320.44°\text{F}; 139.23\,\text{R}.$

Chapter 2

2-1 $1.5 \times 10^{-4}\,\text{kg}.$

2-2   (a) 0.308 kilomole;   (b) 9.86 kg;   (c) $3.96 \times 10^6\,\text{Pa}$;   (d) 0.277 kilomole.

2-3   **(c)** 800 K.

2-6   2.67.

2-7   3.69.

Chapter 3

3-1 -5.74×10^7 J.

3-2 1.91×10^5 J.

3-3   **(b)** $T_1/4$;   **(c)** $-(3/8)\, nRT_1$.

3-4 -92 J.

3-5   **(b)** 4.26×10^6 J;   **(c)** 4.30×10^6 J.

3-6   **(b)** 3.12×10^5 J.

3-7   **(a)** 3.45×10^6 J;   **(b)** 3.45×10^6 J.

3-8   **(a)** 60 J;   **(b)** -70 J;   **(c)** 50 J, 10 J.

3-9   **(a)** -3.12×10^5 J;   **(b)** -4.32×10^5 J;   **(c)** 150 K;   **(d)** 1.25×10^5 Pa.

3-10 9.71×10^{-2} J.

3-11   **(a)** $A = -1.6 \times 10^{-5}$ m³Pa^{-1}, $B = 2.6$ m³;   **(b)** 270 K;   **(c)** 6×10^4 J;
    (d) 3.6×10^4 J.

Chapter 4

4-1   **(a)** 59.2 J kilomole^{-1} K^{-1}, 7400 J kilomole^{-1} K^{-1};   **(b)** 1.85×10^5 J;
    (c) 3730 J kilomole^{-1}.

4-5 10^{-12}, approximately.

4-6   **(a)** 8.31×10^6 J;   **(b)** 11.6×10^6 J.

4-7 $Q = n[aT + (3/2)bT^2 - c/2T]$.

4-11   **(a)** 1 m³, $\Delta U = 0$, -5.62×10^5 J;   **(b)** 1.22 m³, 357 K, -4.48×10^5 J,
    4.48×10^5 J.

4-13 3.66×10^6 J.

4-14 4.28×10^3 J.

4-15   696 m.

Chapter 5

5-3   **(a)** $\eta = -a/c_v v^2$;   **(b)** $h = u_0 + c_v T - 2a/v + RTv/(v - b)$;
    (d) $\kappa = [RTv/(v - b)^2 - 2a/v^2]^{-1}$

5-6   **(a)** -900 kcal;   **(b)** -1600 kcal;   **(c)** 300 kcal, -400 kcal;   **(d)** 25%, 3.

5-7   340 watts.

5-8   It is more efficient to lower the temperature of the cold reservoir.

5-9   71 K, 227 K.

5-10   **(a)** 2.49×10^6 J, 10.70×10^6 J, -2.49×10^6 J, -8.02×10^6 J;
         (b) 2.68×10^6 J;   **(c)** 20.3%.

5-11   200 J, -1200 J.

5-13   **(a)** $T_2/(T_2 - T_1)$;   **(b)** 14.6.

5-14   **(b)** 55.3%.

Chapter 6

6-1 294 JK^{-1}.

6-2   **(a)** 1074 JK^{-1};   **(b)** 1220 JK^{-1};   **(c)** -6060 JK^{-1}.

6-3 127 JK^{-1}.

6-4 $0, 0.893$ JK^{-1}.

6-5   **(a)** 19 K;   **(b)** 0.27 JK^{-1}, 0.27 JK^{-1}.

6-9   **(a)** 5.76×10^3 JK^{-1}, 0;   **(b)** 5.76×10^3 JK^{-1}, 5.76×10^3 JK^{-1}.

6-10   **(b)** $aT + bT^3/3$.

Chapter 7

7-1   **(a)** -165 J;   **(b)** 6.03×10^6 J;   **(c)** 2.21×10^4 JK^{-1}.

7-2 660 JK^{-1}.

7-3   **(a)** 6.3 JK^{-1};   **(b)** 1.38 JK^{-1};   **(c)** ≈ 0;   **(d)** 5.73×10^3 JK^{-1};   **(e)** 0.

7-4   **(a)** 6.93 JK^{-1}, -5.0 JK^{-1};   **(b)** -6.93 JK^{-1}; 10.0 JK^{-1}
         (c) 1.93 JK^{-1}, 3.07 JK^{-1}.

7-5   **(a)** 9.13×10^3 JK^{-1};   **(b)** 9.13×10^3 JK^{-1}.

7-7   **(a)** -4.9×10^3 JK^{-1};   **(b)** 4.9×10^3 JK^{-1};   **(c)** 16.4 JK^{-1}.

7-9 1.7 JK^{-1}.

7-11 $c_v \ln T + R \ln(v - b) + s_0$.

7-12 664 J kilomole^{-1}K^{-1}.

7-13   **(b)** $0.4/P$.

7-16 331 ms^{-1}.

7-17   **(b)** 12.5 J kg^{-1}K^{-1}.

Chapter 8

8-1   The change is greater for an ideal gas.

8-3 -1.62×10^6 J.

8-6 $nRT + BP + CP^2 + DP^3$.

8-8   (a) $-RT \ln\left(\dfrac{v_2 - b}{v_1 - b}\right) + a\left(\dfrac{1}{v_1} - \dfrac{1}{v_2}\right)$;   (b) $a\left(\dfrac{1}{v_1} - \dfrac{1}{v_2}\right)$.

8-13 (a) 200 K;   (c) 2R.

Chapter 9

9-3   (b) -1.50×10^7 J, 5×10^4 JK^{-1}.

9-5   0.

9-6   2.

9-8   (a) 300 K, 2.5 atm;   (b) -1.56×10^6 J;   (c) 5.19×10^3 JK^{-1}.

Chapter 10

10-4 2.38×10^{-8} J kilomole^{-1} K^{-1}, 23.8 J kilomole^{-1} K^{-1}.

10-5   (b) $a(T - T_0) + (b/3)(T^3 - T_0{}^3)$.

Chapter 11

11-1   (a) $N/v_0{}^2$;   (b) 2 v_0, 2.45 v_0;   (c) v_0;   (d) $1.41v$.

11-2 1.14×10^{28} m^{-2}s^{-1}.

11-3 1.25×10^{13} m^{-3}.

11-4   112 ms^{-1}, 517 ms^{-1}, 215 ms^{-1}.

11-5   0.129 eV, 2.13×10^5 ms^{-1}, about 0.2%.

11-6   0.682 $(kT/m)^{1/2}$, 2.31 $(kT/m)^{1/2}$.

11-7   (c) 0.427.

11-8   394 ms^{-1}, 445 ms^{-1}, 483 ms^{-1}; 2278 ms^{-1}, 2571 ms^{-1}, 2790 ms^{-1}.

11-9   (b) 1.50.

11-10 $\sqrt{2}\ell$, zero.

11-11 3.86×10^{-8} m, 83.9, 6.16×10^9 s^{-1}.

11-12 2.00×10^{13} m, one collision every 20 centuries.

11-13 4.0×10^{-5} s.

11-14 (a) 14 Pa;   (b) 975 s.

11-15 0.485 P.

11-16 $(c_2/c_1)(m_1/m_2)^{1/2}$.

11-17 About 10 days.

11-18 2.80×10^{-10} m.

11-19 (a) 1.10×10^{-5} Pa s, 7.39×10^{-3} Jm^{-1}s^{-1}K^{-1}, 8.53×10^{-6} m^{2}s^{-1}.

Chapter 12

12-1 (a) 1.13×10^{15}; (b) 1.26×10^{14}; (c) 0.112.

12-2 (a) 1.27×10^{30}; (b) 1.01×10^{29}; (c) 0.0796.

12-3 49.

12-6 (b) 84; (c) $1.33, 1.00, 0.71, 0.48, 0.29, 0.14, 0.05$.

12-7 (c) 336,798; (d) $2.73, 1.64, 0.908, 0.454, 0.195, 0.065, 0.013$.

12-8 (a) $k \ln 3$.

12-11 8.66×10^{7}.

12-12 $5.6 \times 10^{52}, 10^{26}$.

Chapter 13

13-1 (a) $4, 4$; (b) $2ab, 64\%$.

13-2 (a) 2002; (b) 252.

13-3 (a) $1.33, 1.00, 0.78, 0.44, 0.22, 0.11, 0.11$; (b) $1, 1, 1, 1, 0, 0, 0$.

13-9 (a) $25.3, 0.0211$; (b) $0.980, 0.0202$.

13-10 (b) 10^{5}.

13-11 (a) $2.78, 1 : 1.10 : 0.675, 265\ k$; (b) 587 K.

Chapter 14

14-1 (a) $U/T + Nk \ln Z, -NkT \ln Z$.

14-3 (b) 2.66×10^{8}.

14-4 $1.44 \times 10^{5}\ \text{JK}^{-1}$.

14-5 -0.29 eV.

14-6 (a) $3.74 \times 10^{6}\ \text{J}, 0.0388\ \text{eV}$; (b) 6.06×10^{33}; (c) -0.417 eV;
(d) 2.22×10^{-8}.

14-7 $4\,NkT, NkT/V, Nk[5 + \ln(aVT^{4}/N)]$.

Chapter 15

15-1 (a) $0.982, 0.0180, 0.00033, 0.632, 0.232, 0.085$.

15-3 (a) $865, 117, 16$; (b) $656\ k\theta$.

15-4 (b) $6Nk(\theta_{\text{rot}}/T)\exp(-2\theta_{\text{rot}}/T)(1 + T/2\theta_{\text{rot}})$.

15-6 (a) $7.53 \times 10^{-11}\ \text{m}$; (b) $11.2 \times 10^{-11}\ \text{m}$.

15-8 (a) $5.67 \times 10^{6}\ \text{J}$; (b) $-4.34 \times 10^{7}\ \text{J}$; (c) $1.79 \times 10^{5}\ \text{JK}^{-1}$.

15-10 $2.48 \times 10^{4}\ \text{J kilomole}^{-1}\ \text{K}^{-1}$.

Chapter 16

16-6 1.11×10^3 J kilomole^{-1} K^{-1} (Einstein), 2.68×10^3 J kilomole^{-1} K^{-1} (Debye).

16-7   **(b)** 293 K.

Chapter 17

17-2   **(d)** 4.13×10^{16} s^{-1}, 9.24×10^{-24} Nm, 9.24×10^{-24} Am^2.

17-3 2.12×10^{-6}.

17-4 2.27×10^{-4} eVT^{-1}, 2.90×10^{-4} eVT^{-1}.

17-5 1.87×10^4 JKT^{-2}.

17-6   **(a)** 0.212 JT^{-1};   **(b)** 92.7 JT^{-1}.

17-12   **(a)** 7.87×10^5 Am^{-1};   **(b)** 1550 T.

17-13   507 Am^{-1}.

Chapter 18

18-1   **(a)** 3.89×10^{-5} J.

18-2   **(a)** 10.6 microns;   **(b)** 5800 K.

18-3   **(a)** 6040 K;   **(b)** 4.65×10^{20} megawatts.

18-4   **(a)** $2.82\, kT/h$, smaller by a factor of 1.76;   **(b)** 1.58×10^{11} Hz.

18-5   **(a)** 4.1×10^{14};   **(b)** photons/molecules $= 1.7 \times 10^{-11}$.

18-6   Approximately 10^{87}.

18-11   **(a)** 2.032;   **(b)** 1.092;   **(c)** 0.685;   **(d)** 0.327;   **(e)** 0.176.

18-12   **(a)** -4.67×10^{-20} J, -0.292 eV.

18-13   **(a)** 1.01×10^{26} m^{-3};   **(b)** 96%.

18-14   **(a)** 8.59×10^{-6} K;   **(b)** 9.96×10^6;   **(c)** 242.

Chapter 19

19-1   **(a)** 0.134;   **(b)** 0.116;   **(c)** 0.0997;   **(d)** 0.0735.

19-2   **(a)** 0.070 K;   **(b)** $-12.7, 3.28 \times 10^5$;   **(c)** 6.80×10^{-7}.

19-6 0.648 J kilomole^{-1} K^{-1}. Equation (19.20) gives a value 12% lower than the experimental value.

19-7   **(a)** 1.88 eV;   **(c)** 1.89×10^3 J kilomole^{-1} K^{-1}, $c_e/3R = 0.046$.

19-8   331 J kilomole^{-1} K^{-1}.

19-10 2.9×10^{-11} Pa^{-1}.

19-12 7.2×10^6 m.

19-13   **(a)** 0.21 MeV;   **(b)** 2.4×10^9 K;   **(c)** 2×10^8 ms^{-1};   **(d)** 0.39 kg m^{-3}.

Answers to Selected Problems

Chapter 20

20-4 2.161 bits, 2.583 bits.

20-5 **(a)** 226 bits; **(b)** 5.

20-6 1.33 bits.

20-7 2.5 bits/letter.

Appendix A

A-1 **(a)** $z = x^2 \ln y + C$.

(b) $z = y(x - 3) - x + C$.

(c) Inexact.

A-2 **(a)** $z = xy + y^2 + C$.

(c) $0, 1/2$.

A-3 **(b)** $z = x^2 y - 3x + C$.

(c) 3 for all three paths.

A-4 **(a)** $x^{-1/3}$ or $y^{1/2}$.

(b) y.

(c) x^{-5}.

A-5 **(a)** $\sec^2 x, w = y \sin x - \tan x + C$.

Index

CONVERSION FACTORS

QUANTITY	CONVERSION
Length	$1 \text{ m} = 100 \text{ cm}$
	$= 3.281 \text{ ft}$
	$= 39.37 \text{ in}$
Mass	$1 \text{ kg} = 10^3 \text{ g}$
	$= 2.205 \text{ lb}_m$
Volume	$1 \text{ m}^3 = 10^6 \text{ cm}^3$
	$= 10^3 \text{ liter}$
	$= 35.31 \text{ ft}^3$
Force	$1 \text{ N} = 1 \text{ kg} \cdot \text{m} \cdot \text{s}^{-2}$
	$= 10^5 \text{ dyne}$
	$= 0.2248 \text{ lb}_f$
Pressure	$1 \text{ Pa} = 1 \text{ N} \cdot \text{m}^{-2}$
	$= 10 \text{ dyne} \cdot \text{cm}^{-2}$
	$= 9.872 \times 10^{-6} \text{ atm}$
	$= 10^{-5} \text{ bar}$
	$= 7.502 \times 10^{-3} \text{ torr}$
	$= 14.50 \times 10^{-5} \text{ psia}$
Energy	$1 \text{ J} = 1 \text{ N} \cdot \text{m}$
	$= 10^7 \text{ erg}$
	$= 2.390 \times 10^{-4} \text{ kcal}$
	$= 6.242 \times 10^{18} \text{ eV}$
	$= 9.478 \times 10^{-4} \text{ Btu}$